AF825967

ANTIDIABETIC PLANTS FOR DRUG DISCOVERY

Pharmacology, Secondary Metabolite Profiling, and Ingredients with Insulin Mimetic Activity

Cover photo credits:

Top left: *Azadirachta indica.* Photo by J.M.Garg. https://en.wikipedia.org/wiki/Azadirachta_indica#/media/File:Neem_(*Azadirachta_indica*)_in_Hyderabad_W_IMG_6976.jpg. https://creativecommons.org/licenses/by/3.0/

Bottom left: *Gymnema sylvestre*. Photo by Lalithamba from India. https://en.wikipedia.org/wiki/Gymnema_sylvestre#/media/File:*Gymnema_sylvestre*_R.Br_-_Flickr_-_lalithamba.jpg. https://creativecommons.org/licenses/by/2.0/

Top right: Mulberry fruit in Libya. Photo by Tarek Siala from Tripoli, Libya. https://en.wikipedia.org/wiki/Morus_(plant)#/media/File:Mulberry_in_Libya.jpg. https://creativecommons.org/licenses/by/2.0/

Bottom right: Raw cinnamon bark. Photo by Bertrand Thiry. https://en.wikipedia.org/wiki/Cinnamon#/media/File:Baton_de_cannelle.jpg. https://creativecommons.org/licenses/by-sa/3.0/

ANTIDIABETIC PLANTS FOR DRUG DISCOVERY

Pharmacology, Secondary Metabolite Profiling, and Ingredients with Insulin Mimetic Activity

Edited by

Deepu Pandita, MPhil

Anu Pandita, MSc

Chander Bhanu, MBBS, PGDDCE

AAP | APPLE ACADEMIC PRESS

First edition published 2023

Apple Academic Press Inc.
1265 Goldenrod Circle, NE,
Palm Bay, FL 32905 USA

760 Laurentian Drive, Unit 19,
Burlington, ON L7N 0A4, CANADA

CRC Press
6000 Broken Sound Parkway NW,
Suite 300, Boca Raton, FL 33487-2742 USA

4 Park Square, Milton Park,
Abingdon, Oxon, OX14 4RN UK

© 2023 by Apple Academic Press, Inc.

Apple Academic Press exclusively co-publishes with CRC Press, an imprint of Taylor & Francis Group, LLC

Reasonable efforts have been made to publish reliable data and information, but the authors, editors, and publisher cannot assume responsibility for the validity of all materials or the consequences of their use. The authors, editors, and publishers have attempted to trace the copyright holders of all material reproduced in this publication and apologize to copyright holders if permission to publish in this form has not been obtained. If any copyright material has not been acknowledged, please write and let us know so we may rectify in any future reprint.

Except as permitted under U.S. Copyright Law, no part of this book may be reprinted, reproduced, transmitted, or utilized in any form by any electronic, mechanical, or other means, now known or hereafter invented, including photocopying, microfilming, and recording, or in any information storage or retrieval system, without written permission from the publishers.

For permission to photocopy or use material electronically from this work, access www.copyright.com or contact the Copyright Clearance Center, Inc. (CCC), 222 Rosewood Drive, Danvers, MA 01923, 978-750-8400. For works that are not available on CCC please contact mpkbookspermissions@tandf.co.uk

Trademark notice: Product or corporate names may be trademarks or registered trademarks and are used only for identification and explanation without intent to infringe.

Library and Archives Canada Cataloguing in Publication

Title: Antidiabetic plants for drug discovery : pharmacology, secondary metabolite profiling, and ingredients with insulin mimetic activity / edited by Deepu Pandita, PhD, Anu Pandita, PhD, Chander Bhanu, PhD.
Names: Pandita, Deepu, 1977- editor. | Pandita, Anu, 1977- editor. | Bhanu, Chander, 1973- editor.
Description: First edition. | Includes bibliographical references and index.
Identifiers: Canadiana (print) 20220282307 | Canadiana (ebook) 20220282374 | ISBN 9781774910061 (hardcover) | ISBN 9781774910078 (softcover) | ISBN 9781003282938 (ebook)
Subjects: LCSH: Medicinal plants. | LCSH: Botany, Medical. | LCSH: Diabetes—Alternative treatment. | LCSH: Hypoglycemic agents.
Classification: LCC SB293 .A58 2023 | DDC 633.8/8—dc23

Library of Congress Cataloging-in-Publication Data

CIP data on file with US Library of Congress

ISBN: 978-1-77491-006-1 (hbk)
ISBN: 978-1-77491-007-8 (pbk)
ISBN: 978-1-00328-293-8 (ebk)

About the Editors

Deepu Pandita, MPhil
Senior Lecturer, Government Department of School Education, Jammu, Union Territory of Jammu and Kashmir, India

Deepu Pandita is working as a Senior Lecturer in the Government Department of School Education, Jammu, Union Territory of Jammu and Kashmir, India. Deepu Pandita has more than 18 years of teaching experience and earned her Master of Botany (MSc) at the University of Kashmir, Jammu and Kashmir, India, and Master of Philosophy (MPhil) in Biotechnology from the University of Jammu, Jammu and Kashmir, India. Deepu Pandita has a number of international and national courses to her credit. Deepu Pandita has qualified fellowships including JRF NET and SRF from the Council of Scientific & Industrial Research (CSIR), New Delhi, India, Biotechnology Fellowship, Government Department of Science and Technology, Jammu & Kashmir, India; and IASC-INSA-NASI Summer Research Teacher Fellowship, India. Deepu Pandita has presented her research at work at both the national and international conferences and was awarded a Best Oral Presentation Award in the International Conference on Biotechnology for Better Tomorrow, a Women Researcher Award, and Research Excellence Award by two professional associations in India. Deepu Pandita is a life member of various scientific societies. Deepu Pandita acts as a reviewer (18 journals) and associate editor and editor of a number of international journals. Deepu Pandita has published a number of editorials, 25 book chapters (Springer, Elsevier, CRC, etc.), reviews, and research articles in various journals of national and international repute like Cells (MDPI) (IF=7.666), Frontiers in Plant Sciences (IF=6.627), and Frontiers in Physiology (IF=4.566) and has currently six books in the process of production.

Anu Pandita, MSc
Dietician, Vatsalya Clinic, New Delhi, India

Anu Pandita is working as a Senior Dietician at Vatsalya Clinic, E-5/11, Krishna Nagar, New Delhi, India. Previously she worked as a Lecturer at the Bee Enn College of Nursing, Talab Tillo, Jammu, India, and Dietician at the Ahinsa Dham Bhagwan Mahavir Charitable Health Centre, New Delhi, India. Anu Pandita has done her MSc internship and a course in the Dietetics Department of PGI, Chandigarh, India. Anu Pandita has taken a case study at the Pediatric Gastroenterology Ward at "Nehru Hospital" PGI, Chandigarh, India, on a patient suffering from chronic liver disease. She earned a certificate for a course in food and nutrition as well. She has a number of training sessions, refresher courses, and workshops to her credit. Anu Pandita is a life-time member of the Indian Dietetic Association and Indian Science Congress Association, Kolkata, India. Anu Pandita has presented her research at work at both the national and international conferences. Anu Pandita has published various book chapters in Springer, CRC, etc. publications and has a number of review and research articles in various journals of national and international repute, including Cells (MDPI) (IF=7.666) and Frontiers in Physiology (IF=4.566). She currently has six books in the process of production.

Chander Bhanu, MBBS, PGDDCE
Senior Consultant Physician, Vatsalya Clinic; Dr. Kailash Chander Jain Memorial Clinic; Ahinsa Dham Bhagwan Mahavir Medical Center, New Delhi, India

Chander Bhanu, MBBS, PGDDCE, is working as a Senior Consultant Physician at Vatsalya Clinic, Krishna Nagar, New Delhi, Dr. Kailash Chander Jain Memorial Clinic, and Ahinsa Dham Bhagwan Mahavir Medical Center, Madhu Vihar, New Delhi, India. Dr. Bhanu earned his MBBS (Bachelor of Medicine and Bachelor of Surgery) from Govt. Medical College, Calicut, Kerala, and PGDDCE (Post Graduate Diploma in Diabetes and Clinical Endocrinology) from the Royal College of Physicians, United Kingdom. Dr. Bhanu has a certification in diabetes mellitus accredited by the Royal Liverpool Academy, United Kingdom. Previously he has worked as a neuro associate in the Institute of Brain and Spine; resident medical officer at the VIMHANS (Vidya Sagar Institute of Mental Health and Neurosciences) Hospital, New Delhi; and a chief medical officer in Primus Hospital; and Medical Officer in Neurology and Neurosurgery Departments in Fortis Hospital. He has expertise and experience in various clinical procedures such as pleural tapping, ascitic tapping, lumbar puncture, suturing, and other procedures both in pediatrics and adult groups and also deals with stroke and accident cases as well.

Contents

Contributors

Iffat Zareen Ahmad
Natural Products Laboratory, Department of Bioengineering, Integral University, Dasauli, Kursi Road, Lucknow – 226026, Uttar Pradesh, India, E-mail: iffat@iul.ac.in

Khyatie Anjaria
Department of Botany, Ramniranjan Jhunjhunwala College, Ghatkopar (W), Mumbai – 400 086, Maharashtra, India

Ruchi Bansal
Division of Germplasm Evaluation, ICAR-National Bureau of Plant Genetic Resources, New Delhi – 110012, India, E-mail: ruchibansal06@gmail.com

Masarat Bashir
College of Temperate Sericulture, Sheri-Kashmir University of Agricultural Sciences and Technology, Srinagar, Jammu and Kashmir, India, E-mail: masarat.bashir4@gmail.com

Lepakshi Md. Bhakshu
Department of Botany, PVKN Government College (Autonomous), Chittoor – 517 002, Andhra Pradesh, India, E-mails: lmbakshu@gmail.com; bakshu@pvkngcchittoor.info

Ifagbemi Bienvenue Chabi
Department of Nutrition and Food Sciences, Faculty of Agricultural Science, University of Abomey-Calavi, 01 BP 526 Cotonou, Benin

Ashraf Dar
Department of Biochemistry, University of Kashmir, Srinagar, Jammu and Kashmir – 190006, India, E-mail: ashrafdar@kashmiruniversity.ac.in

Himanshu Dawda
Department of Botany, Ramniranjan Jhunjhunwala College, Ghatkopar (W), Mumbai – 400 086, Maharashtra, India

Jayant Dewangan
Genotoxicity Lab, Division of Toxicology and Experimental Medicine, CSIR-Central Drug Research Institute, Lucknow, Uttar Pradesh, India

Vincent Ezin
Department of Crop Production, Faculty of Agricultural Science, University of Abomey-Calavi, 01 BP 526 Cotonou, Benin, E-mails: ishola.vincent@yahoo.com; awevincent@gmail.com

Farrukh Jamal
Department of Biochemistry, Dr. Rammanohar Lohia Avadh University, Ayodhya, Uttar Pradesh, India

Diksha Kapoor
Department of Pharmaceutical Chemistry and Quality Assurance, SVKM's Dr. Bhanuben Nanavati College of Pharmacy, Mumbai, Maharashtra, India

Shuchi Kaushik
M. P. Forensic Science Laboratory, Madhya Pradesh, India, E-mail: shuchi.kaushik2@gmail.com

Tabassum Khan
Department of Pharmaceutical Chemistry and Quality Assurance,
SVKM's Dr. Bhanuben Nanavati College of Pharmacy, Mumbai, Maharashtra, India,
E-mail: tabassum.khan@bncp.ac.in

D. Vijaya Lakshmi
Department of Microbiology, Yogi Vemana University, Vemanapuram, Kadapa – 516005,
Andhra Pradesh, India

Usha Mukundan
Department of Botany, Ramniranjan Jhunjhunwala College, Ghatkopar (W), Mumbai – 400 086,
Maharashtra, India, E-mail: usha.rjc@gmail.com

Madheswaran Muthusamy
Muthayammal Engineering College, Rasipuram, Tamil Nadu, India,
E-mail: prof.madheswaran@gmail.com

Ramachandra Reddy Pamuru
Department of Biochemistry, Yogi Vemana University, Kadapa-516005, Andhra Pradesh, India,
Phone: +91-08562-225425, Fax: +91-08562-225419, E-mail: reddyprbiotech@yvu.edu.in

Prabhash Kumar Pandey
Department of Biochemistry, Faculty of Science, University of Allahabad, Prayagraj,
Uttar Pradesh, India; Genotoxicity Lab, Division of Toxicology and Experimental Medicine,
CSIR-Central Drug Research Institute, Lucknow, Uttar Pradesh, India,
E-mails: prabhashpandey@allduniv.ac.in; pandey.prabhash21@gmail.com

Anu Pandita
Vatsalya Clinic, Krishna Nagar, New Delhi, India

Deepu Pandita
Government Department of School Education, Jammu, Jammu and Kashmir, India,
E-mail: deepupandita@gmail.com

Swati Priya
Department of Botany, Kurukshetra University, Kurukshetra, Haryana – 136119, India,
E-mail: swatipriya9013@gmail.com

Karishma Rajbhar
Department of Botany, Ramniranjan Jhunjhunwala College, Ghatkopar (W), Mumbai – 400 086,
Maharashtra, India

Sasireka Rajendran
Department of Biotechnology, Mepco Schlenk Engineering College, Sivakasi,
Tamil Nadu, India

R. R. Venkata Raju
Department of Botany, Sri Krishnadeveraya University, Ananthapuramu – 515003,
Andhra Pradesh, India

Kavita Rani
Department of Botany, Kurukshetra University, Kurukshetra, Haryana – 136119, India,
E-mail: kavitadahiya05@gmail.com

Srikanta Kumar Rath
Genotoxicity Lab, Division of Toxicology and Experimental Medicine,
CSIR-Central Drug Research Institute, Lucknow, Uttar Pradesh, India

Vinoth Rathinam
Department of Electronics and Communication Engineering, P.S.R. Engineering College,
Sivakasi, Tamil Nadu, India

K. Venkata Ratnam
Department of Botany, Rayalaseema University, Kurnool – 518 007, Andhra Pradesh, India

Neha Sharma
Amity Institute of Biotechnology, Amity University, Gwalior, Madhya Pradesh, India,
E-mail: nsharma2@gwa.amity.edu

Narender Singh
Department of Botany, Kurukshetra University, Kurukshetra, Haryana – 136119, India,
E-mail: nsheorankuk@yahoo.com

Ranjan Singh
Department of Microbiology, Dr. Rammanohar Lohia Avadh University, Ayodhya,
Uttar Pradesh, India

Rajesh Singh Tomar
Amity Institute of Biotechnology, Amity University, Gwalior, Madhya Pradesh, India

Shambhoo Sharan Tripathi
Department of Biochemistry, Faculty of Science, University of Allahabad, Prayagraj,
Uttar Pradesh, India

Sugumari Vallinayagam
Department of Biotechnology, Mepco Schlenk Engineering College, Sivakasi, Tamil Nadu, India

Pulala Raghuveer Yadav
Department of Biotechnology, Indian Institute of Technology Hyderabad, Kandi – 502 285,
Telangana, India

Ambedkar Yarrapalle
Department of Biochemistry, Yogi Vemana University, Vemanapuram, Kadapa – 516005,
Andhra Pradesh, India

Abbreviations

AAL	α-amylase
ABA	abscisic acid
ABG	aged black garlic
AD	Alzheimer's disease
ADA	American Diabetes Association
AGL	α-glucosidase
ALT	alanine transaminase
ARE	antioxidant response element
AST	aminotransferase
Aβ	β-amyloid
BCRP	breast cancer resistance protein
BMI	body mass index
BS	β-sitosterol
BTH	butylated hydroxytoluene
BW	body weight
BWG%	body weight gain percentage
CAT	catalase
COX-2	cyclooxygenase-2
CRP	C-reactive protein
DADS	diallyl disulfide
DATS	diallyl trisulfide
DB	evoked diabetic
DM	diabetes mellitus
DMAPP	dimethylallyl diphosphate
DPPH	diphenyl-1-picrylhydrazyl
DTS	diallyl trisulfide
EJ	*E. jambolana*
ERK	extracellular signal-related kinase
EST	expressed sequence tag
FBG	fasting blood glucose
FBPase	fructose-1, 6-bisphosphatase
FER	food efficiency ratio

FFA	free fatty acid
FOSHU	food for specified health use
G6Pase	glucose-6-phosphatase
GAGs	glycosaminoglycans
GBM CSC	glioblastoma multiforme cancer stem cells
GCLM	glutamate-cysteine ligase modifier
GDM	gestational diabetes mellitus
GLUT	glucose transporters
GLUT4	glucose transporter type 4
GPX	glutathione peroxidase
GR	glutathione reductase
GSH	glutathione
GSH-Px	glutathione peroxidase
GST	glutathione S-transferase
GTT	glucose tolerance test
HAP	hydroxyapatite
Hb	hemoglobin
HCMV	human cytomegalovirus
HDL	high-density lipid
HO-1	heme oxygenase-1
HPLC	high-performance liquid chromatography
IBS	irritable bowel syndrome
IDF	international diabetes federation
IP	intraperitoneal
IPP	isopentyl diphosphate
IR	insulin receptor
LDL	low-density lipoprotein
LPL	lipoprotein lipase
LPO	lipid peroxidation
MAPK	mitogen-activated protein kinase
MD	mild diabetic
MDA	malondialdehyde
MLP	mulberry leaf polysaccharide
MPLC	medium pressure liquid chromatography
mRNA	messenger RNA
MS	mass spectroscopic
MSG	monosodium L-glutamate
MTP	microsomal triglyceride transfer protein

NA	nicotinamide
NADPH	nicotinamide adenine dinucleotide
NGS	next-generation sequencing
Nrf2	nuclear factor erythroid-2
NSE	*N. sativa* ethanol
NSO	*N. sativa* oil
OGTT	oral glucose tolerance test
OLETF	Otsuka long-Evans Tokushima fatty
OPLS	orthogonal partial least square
PAS	periodic acid-Schiff
PCA	principal component analysis
PGE2	prostaglandin E2
PI3K	phosphatidylinositol-3 kinase
PLS-DA	partial least square discriminant analysis
POPs	persistent organic pollutants
PPARs	peroxisome proliferator-activated receptors
PTP1B	protein tyrosine phosphatase 1B
RANKL	nuclear factor-kappa B ligand
RBCs	red blood cells
RNSs	reactive nitrogen species
ROS	reactive oxygen species
SAC	S-allyl cysteine
SACS	S-allyl cysteine sulfoxide
SB	stem-bark
SC	*Syzygium cumini*
SD	severe diabetic
SGLT2	sodium/glucose co-transporter2
SGOT	serum glutamic-oxaloacetic transaminase
SGPT	serum glutamic-pyruvic transaminase
SJ	*S. jambolanum*
SOD	superoxide dismutase
STZ	streptozotocin
T1DM	type 1 diabetes mellitus
T2DM	type 2 diabetes mellitus
TAC	total antioxidant capacity
TBARS	thiobarbituric acid–reactive substances
TC	total cholesterol
TG	total glyceride

TQ	thymoquinone
TTP	tristetraprolin
VLDL	very-low-density lipoprotein
WC	waist circumference
WE	water extract
XBP-1s	X-box binding protein 1

Preface

Plant natural products, or phytoconstituents, are promising leading candidates for antidiabetic pharmacological actions. The phytoconstituents like flavonoids, terpenoids, saponins, carotenoids, alkaloids, and glycosides play vital roles in the new potent antidiabetic drug development programs. *Azadirachta indica*, *Gymnema sylvestre*, *Syzygium cumini* (L.) Skeels, *Cinnamomum zeylanicum*, *Costus pictus*, *Trigonella foenum-graecum*, *Morus* spp., *Nigella sativa* L., *Aegle marmelos* (L.), *Allium sativum,* and *Ficus benghalensis* produce antihyperglycemic effects by improving the performance of pancreatic tissue. These antidiabetic plants possess natural, insulin-mimetic products such as azadirachtin, Fagomine-3,1,4-dideoxy-1,4-imino-D-arabinitol, 1-deoxynojirimycin, trigonelline, allicin, N-feruloytyrosine, N-feruloyltyramine, gymnemic acids, gurmarin, jambolin or antimellin, camphor, eugenol, cinnamaldehyde, corosolic acid, diosgenin, stigmasterol, thymoquinone, thymol, thymohydroquinone, dithymoquinone, nigellone, marmelosin, luvangetin, aurapten, marmelide, aegeline, aegelinine, leucopelargonidin-3-O-α-L-rhamnoside, leucocynidin 3-O-β-D-galactosyl cellobioside, bengalensinone, benganoic acid, etc., and can be used for the treatment of diabetes mellitus and its comorbidities.

This book discusses antihyperglycemic, hypoglycemic, anti-lipidemic, and current genome editing research perspectives of 11 antidiabetic plants.

CHAPTER 1

Azadirachta indica: Its Biological, Pharmacological, Antidiabetic Potential, and Omics Applications

VINCENT EZIN[1] and IFAGBEMI BIENVENUE CHABI[2]

[1]*Department of Crop Production, Faculty of Agricultural Science, University of Abomey-Calavi, 01 BP 526 Cotonou, Benin, E-mails: ishola.vincent@yahoo.com; awevincent@gmail.com*

[2]*Department of Nutrition and Food Sciences, Faculty of Agricultural Science, University of Abomey-Calavi, 01 BP 526 Cotonou, Benin*

ABSTRACT

Azadirachta indica possesses multifunctional activities including anti-diabetic, parasiticidal, anti-ulcer, antimalarial, antimicrobial, anticancer, spermicide, anti-hormonal, and anti-inflammatory, antipyretic, stomachic, pediculicide, antiarrhythmic, carminative and insecticidal activities which are well known worldwide. Different parts from roots to the apical meristems of the tree are used in folk medicine and agriculture. The research on neem began in the 1920s by an Indian researcher who first discovered its natural insect repellent ability. Pharmaceutically and medicinally, neem ingredients are respectively used in drug production and in curing a lot of diseases like the disruption of normal metabolism, diabetes, cancers pathogens, the treatment of many infectious, diabetic, metabolic, and cancer diseases. Drinking the filter leaf decoction is helpful in treating wounds, sore throats, diabetes, chickenpox, skin disorder, and eczema. However, several defies have been observed as regards the biological functions of

Antidiabetic Plants for Drug Discovery: Pharmacology, Secondary Metabolite Profiling, and Ingredients with Insulin Mimetic Activity. Deepu Pandita, MPhil, Anu Pandita, MSc, & Chander Bhanu, MBBS, PGDDCE (Eds.)
© 2023 Apple Academic Press, Inc. Co-published with CRC Press (Taylor & Francis)

each secondary metabolite in neem extracts and the challenges related to genome and proteome associated with biological functions. A better understanding of the metabolic pathways of phytomedicals has been possible thanks to the sequencing of the neem genome since 2015. This chapter discusses the biology and ecology of neem. We highlight its benefits and pharmacological properties. We depict its richness in secondary metabolites. The current report also describes the ingredients of antidiabetic *Azadirachta indica* with Insulin mimetic activity. We finally synthesize the omics strategies involved in metabolic pathways of phytomedicines associated with secondary metabolite production. The sequenced genome of neem has significantly helped for the better understanding of metabolic pathways.

1.1 INTRODUCTION

Neem, called *Azadirachta indica A* in Latin, is a multipurpose plant with persistent leaves originating from Asia, and it has been traditionally used since the ancient days (Kumar and Navaratnam, 2013). At the beginning of the 20th century, the species was introduced in the African continent and other developing countries. In fact, owing to its enormous benefits, neem has several names such as a divine tree, miraculous tree, village pharmacy, natural pharmacy, wonder tree, tree to solve a world problem, miracle of love. In Asia and Africa, and other developing countries, the extract products obtained from neem plants have been extensively used in folk medicine and as bio-pesticides in agriculture for centuries (Bray et al., 1990; Kelsey et al., 2015). It has the potential in controlling farm and household pests and diseases. The oil extracts and chemical compounds of neem are intensively hunted worldwide for industrial purposes. its oils are very efficient against cancer, diabetic, bacterial, fungal, parasitic, malarial and inflammatory activities (Patel et al., 2016; Tapanelli et al., 2016; Umar et al., 2014). In arid and semi-arid regions, it is the most planted tree due to its adaptation to hot and dry climates. In the African Sahel, neem is mostly grown as a shade tree and as a windbreak; however the production of fuelwood by pruning and the use of neem in the local pharmacopeia are important. Moreover, the plantation of neem trees helps reduce the increasing rate of deforestation and soil erosion either by wind or water, and to cool down the high temperature due to climate change and provide relief from the sunshine.

In Africa, the boiled leaves of *Azadirachta indica* is used to fight against malaria by drinking the extracts for some days. The bitterness of *Azadirachta indica* A. Juss., with its cooling properties allows to treat fever (Leti et al., 2010). Neem leaf is traditionally used to effectively control le levels of sugar in the human blood. Neem is a natural product known for its potent low sugar and anti-cholesterol effects useful in different types of diabetes. It is proven efficient against bacteria, virus, fungal, stress and as powerful detoxicants. The therapeutic potentiality of neem is based on the primary metabolites, which include lipid, protein, and carbohydrates, and on specialized metabolites such as isoprenoids or terpenoids, tetranortriterpenoids, alkaloids, flavonoids, polyphenols and many other organic compounds (Bhambhani et al, 2017; Saleem et al., 2018). Many clinical tests have been performed to ascertain the medicinal significance of *Azadirachta indica.*

The world after the second war has witnessed the development and uses of synthetic pesticides with highly significant efficiency in protecting crops against pests and pathogens, but these synthetic pesticide products have also been shown to be harmful to the environment through their persistence in soils and their accumulation in water and especially in the food chain. These synthetic pesticides include organophosphates and organochlorines called persistent organic pollutants are known to be very toxic, bio-accumulative through the food chain and can be transported over a long distance. For years, these toxic agro-chemicals have caused serious health issues to humans, insects, microorganisms, and animals by destroying their nervous, immune and reproductive systems, even causing the death of many. These issues have become a real challenge in the world which has led to the putting in place of many rules and regulations regarding the marketing, the use, and the ban of these synthetic pesticides so as to keep healthy and friendly environments and maintain better human health. Some of them have been banned in developed countries since 1972. Therefore, these restrictions stirred the researchers towards plant extracts endowed with phytosanitary activity (Butterworth and Morgan, 1972). Thus, the traditional uses of neem to control pests and diseases have been revived, and nowadays, neem extracts have attracted the attention of many for its uses in organic farming. Many biologically active molecules have been extracted from neem leaves, roots, flowers, seeds, fruits and barks (Balandrin et al., 1988; Breuer et al., 2003; Chattopadhyay et al., 2004).

Several are the secondary metabolites that contribute significantly to the multifunction of neem plant in the treatment of many diseases. These secondary metabolites possess multifunctional properties which include anti-oxidation, antidiabetic, parasiticidal, anti-ulcer, antimalarial, anti-microbial, anticancer, spermicide, anti-hormonal, and anti-inflammatory, antipyretic, stomachic, pediculicide, antiarrhythmic activities (Mamta et al., 2013). Several defies have been observed as regards the biological functions of each secondary metabolite in neem extracts and the challenges related to genome and proteome associated with biological functions. A better understanding of the metabolic pathways of phytomedicals has been possible thanks to the sequencing of neem genome since 2015 (Kuravadi et al., 2015; Nagesh et al., 2015). However, scanty research have focused on the molecular aspects of the tree.

This review discusses the biology and ecology of neem. We highlight its benefits and pharmacological properties. We depict its richness in secondary metabolites. The current report also describes the ingredients of antidiabetic *Azadirachta indica* with Insulin mimetic activity. We finally synthesize the omics strategies involved in metabolic pathways of phytomedicinals associated with secondary metabolite production. The sequenced genome of neem has significantly helped for the better understanding of metabolic pathways.

1.2 BIOLOGY AND ECOLOGY OF *AZADIRACHTA INDICA*

A. indica is a fast-growing tropical and meliaceous plant with annual average temperatures of 21–32°C. Its leaves are always green except for severe drought during the dry season; it can reach a height of 20 meters, and more rarely 35 to 40 meters. Neem has a rounded spreading crown and a straight cylindrical bole with thick reddish bark and rough, often deeply fissured. The tree is well-known for its fast growth almost anywhere in the lowland tropics and subtropics. According to Radwanski (1977), 66% of total tree growth take place during the first three years to reach a height of 4–7 m, and 5–11 m in the fifth year and can live for two centuries. They thrive in climates with annual rainfall ranging from 400 to 800 mm and an extended dry season, even with poor soils (Schmutterer, 1990; Ajayi, 2002). However, it has been introduced successfully even in areas where the rainfall is as low as 150 to 250 mm (Ogbuewu et al., 2011). The tree grows on altitudes up to 1,500 m (Jattan et al., 1995). *A. indica* has been introduced to various arid ecosystems throughout the world because of its ability to successfully tolerate severe drought and establish in nutrient-deficient soils (Vikas and Tandon, 2011). Neem tree has the ability to withstand frost due to much suberized back layer and can easily survive desert conditions of temperature above 50°C (Panhwar and Panhwar, 2003).

It is cultivated in Southeast Asia, Africa, Europe, Central America, and Australia but mainly in the southern regions of Asia and Africa, and used to cure many diseases (Islas et al., 2020). The species is hermaphrodite and is pollinated by bees and other insects. It does propagate sexually through seed and vegetatively through suckers, seedlings and tissue culture (Anonymous, 1983). The reproduction by seeds is mainly secured by birds and bats when they feed on the fruits, the sweet pulp and then spit the seed out as a mean of grain dispersals. Phenological observations reported by Vikas and Tandon (2011) highlighted the simultaneous flowering of neem plants belonging to the same population, nevertheless, species of different populations have various periods of setting flowers, and some species do produce flowers twice per year. The protandrous flowers and fruits are organized on axillary panicles; each fleshy fruit is made up of only one seed which are small, oblong and yellow (Mordue and Nisbet, 2000). The tree takes about five years to produce the first fruits but can produce a good yield in the third year (Panhwar and Panhwar, 2003). Fruits are usually produced only once in a year and cannot be stored for longer duration (Prakash and Srivastava, 2008). The fruit yield is extremely variable, ranging from 10 to 5 0kg per tree with an average of 20 kg (Radwanski, 1977). Several studies mentioned the geographical origin of the neem tree, but most of these agree that it is indigenous to India. The neem tree is present almost in 80 countries around the world and its global population is estimated at more than 91 million trees (Koul, 2004). The neem population worldwide given by Forster and Moser (2000) put Africa at the top of the census, with about 45 million trees.

1.3 BENEFITS OF *AZADIRACHTA INDICA*

Neem is the tropical and sub-tropical tree highly valuable for mankind. It is a useful tree with multifunctional usages and has health, environmental and socio-economic benefits. Among these many benefits, the one that is most important and widely well-known is the medicinal profit. In health point of view, neem has been extensively beneficial in healing or preventing of some vast array of ailments as well as chronic diseases. Traditional medicine reports beneficial effects on the joints, the immune system, digestion, the nervous system, blood sugar, hypertension, and diabetes. Neem produces many fruits, and it is grown as a shade tree which provides shelter and cooler environments for inhabitants and animals during the hottest periods of dry seasons, especially in the Sahel, where it is well adapted (NRC, 1992). It is planted to afforest many regions depleted with deforestation and to mitigate climate variability and play a vital function as sinks for carbon dioxide and produce ingredients for organic farming. Neem possesses numerous active ingredients used in bio-insecticides and a very cost effective input to control pests and diseases in sustainable agriculture. These different substances in its fruits, leaves and wood, are shown to be powerful biopesticides, effective on more than 400 species of insects, parasites, pests, and fungi of agricultural and forestry crops. Neem leaf is known as a potent repellent for some pests. The *A. indica* extract can also serves as a bio-fertilizer in agricultural system production (Brahmachari, 2004). Applied as biofertilizers, it helps reduce the activity of microorganisms (especially denitrifiers) responsible for the losses of nitrogen in the atmosphere or through leaching (Musalia et al., 2000; Mohanty et al., 2008). It is also very efficient in controlling the concentration of aflatoxin through the reduction of fungal population in various crops such as pepper, maize, groundnuts, and many other food commodities (NRC, 1992). The oil from *A. indica* is employed in wood industries, as bonding agents, greases, repugnancy, in soap processing, and polishes (Martínez-Romero et al., 2016). The wood of neem is used to make cabinets, bookcases and cupboards, as well as packing cases because its insect repellent quality helps protect the contents from insect damage. Its trunk is also widely used for making posts for construction or fencing because the wood is termite resistant (World Agroforestry Center, 2008).

1.4 SECONDARY METABOLITE PROFILING OF *AZADIRACHTA INDICA*

Throughout the world, especially in the middle-income and developing countries of Africa, Asia, and other tropical parts, neem tree is considered to be a medicinally important plant due to its huge and valuable bioactive secondary metabolites. These active molecules are terpenoids, flavonoids, phytosterols, alkaloids, polyphenols which cure microbial and inflammatory infections and prevent damages to organs (Ghatule et al., 2012). Over 300 biologically active molecules of various structures have been identified and are responsible for the biological activities of medicinal plants (Kumar et al., 2010). These compounds are not evenly distributed in plant parts. For instance, fruits accumulated higher levels of the bioactive phytochemicals than other tissues (Narnoliya et al., 2014). AzaA is the main active molecule present in bio-chemicals for the control pests (Isman et al., 1991). Apart from AzaA isolated from leaves, many other molecules of medicinal uses have been isolated and include nimbin, salanin, azadirachtin, Nimbocinone, nimolinone, kulactone, nimocinolides, isonimocinolide, flavonoids, myricetin, meldenindiol, vilasinin, margosinolide, isomargosinolide, desacetyldihydronimbic acid (Basak and Chakroborty, 1969; NRC, 1992). Other active elements from leaves reported are glycosides nimbanene, 6-desacetylnimbinene, nimbandiol, nimbolide, ascorbic acid, n-hexacosanol and amino acid, 7-desacetyl-7-benzoylazadiradione, 7-desacetyl-7-benzoylgedunin, 17-hydroxyazadiradione, and nimbioland (Alzohairy, 2016). The limonoids are also reported to be vital ingredients in *A. indica* (Passos et al., 2019). In the root parts of the tree, novo terpenoids, 17-hydroxy-sandaracopimar-8,15-dien-11-one, limonoid morenolide and diterpene were added to nimbinene, nimbinal, nimbandiol and salannin, nimbidiol, ferruginol, and 6,7-dehydroferruginol (Passos et al., 2019). Polyphenolic compounds such as gallic, chlorogenic, ferulic, tannic acids are major phenolic acids found in various parts of neem (Singh et al., 2005). Seven polyphenolic molecules including the three new ones: (2,3-(S)-hexahydroxydiphenoyl-(α/β)-D-glucopyranose, Avicularin and Castalagin and the four old ones: Gallic acid, Ellagic acid, Quercetin and Quercetin-3-O-glucoside were also found as active ingredients (Abdelhady et al., 2015).

1.5 PHARMACOLOGY OF *AZADIRACHTA INDICA*

The traditional knowledge about the different organs of plant trees is passed on from one generation to another worldwide, especially in developing countries, which has considerably and positively imparted the health of the populations (Kumar and Navaratnam, 2013). The traditional medicinal customs have led to the use of several plants and their by-products, which has efficiently worked for the health of the end-users (Arumugam et al., 2014). Among these natural plant resources, the neem tree is highly reputed for its effectiveness against ailments and diseases and for its preventive effects against chronic diseases. Approximately 4,000 years ago, neem has been used in folk medicine in India (Vijayan et al., 2013). The internal neem leaves are employed in the treatment of gastric ulcers and malaria. A decoction of a maximum of 30 grams of leaves per liter can be made, which will be drunk at the rate of half a liter during the day. The external use of the leaves, fruits, or oil of neem is suggested for the treatment of skin diseases in general (mycoses in particular) and as an insecticide against scabies and lice and other agricultural pests and diseases. The use of the oil in massage is recommended for joint pain. The leaves have anti-inflammatory, analgesic, anti-ulcer, anti-contraceptive, antipyretics, antibacterial, antifungal, antiparasitic, anticovid-19 and insecticidal properties (Atawodi and Atawodi, 2006; Islas et al., 2020). The seed oil is very effective against inflammations and in particular pain problems (Kumar et al., 2015). Neem oils are purgative, anthelmintic, external antiparasitic, antimycotic, and antidiabetic. It contains antiviral molecule, antioxidants, and antimalarials. Its seeds can be processed into powerful insecticide, azadirachtin, completely harmless to humans and warm-blooded animals, but vulnerable to light. It blocks metamorphosis from larval stage to adult stage, and paralyzes its digestive tract. It is also used as a toothpaste; its antiseptic properties make it a formidable agent against tartar. Neem is used in traditional agriculture to control insect pests, and to treat the soil against insect larvae and other parasites; wood, leaves, grounded fruits, oil in aqueous solution with a little detergent, and oil-free fruit cake are also used. The neem tree is almost considered a sacred tree as its virtues are numerous because it cures virtually all diseases. It is also called the village pharmacy because everything seems good in it: leaves, fruits, bark, seeds, roots (Atawodi and Atawodi, 2006; Islas et al., 2020).

It has been observed a certain numbers of positive reactions in vivo as a potent impact when applied the neem extracts: nucleic acid restoration, removal of toxic substances and damage cells, changes in cell cycle, prevention and curing of tumor growth, inhibition of metastasis and cancer, anti-apoptosis, reinforcement of immune system anti-allergic (Patel et al., 2016). The biological properties of neem are due to the numerous active components of different parts of this tree. An aqueous preparation from neem leaf was extensively used for centuries in most countries of sub-Sahara Africa to treat malaria. This preparation has demonstrated to be non-hazardous, blood- and immune-stimulants (Haque et al., 2006). Above and beyond all other considerations, neem leaf glycoprotein is regarded as a strong immunomodulatory agent (Chakraborty et al., 2008; Mallick et al., 2013). The neem extract has also been proven inhibitory to secretory and ulcer effects without negative consequences (Bandyopadhyay et al., 2002). It does virtually heal peptic ulcer when 30–60 mg of NBA is taken daily for approximately 10 weeks (Bandyopadhyay et al., 2004). When the extracts made from leaves have been administered along with vitamin C, there is prevention of tissue injury due to dysfunction of vascular by enhancing the antioxidant production and decline in ROS production in the cells (Omóbòwálé et al., 2016). It also augments the expressions of certain genes and proteins so as to inhibit programmed cell death, apoptosis, tumor multiplication and cell senescence (Arumugam et al., 2014). Nimbolide extracted from neem is a strong anti-proliferative and apoptosis proxy (Kumar et al., 2010). Meanwhile, Azadiramide A has been proven to be very efficient in inhibiting cellular proliferation growth and enhance apoptosis in cells (Zhu et al., 2017).

1.6 INGREDIENTS OF ANTIDIABETIC *AZADIRACHTA INDICA* WITH INSULIN MIMETIC ACTIVITY

Neem extracts are used to cure diabetes and also help to reduce the risks associated with diabetes such as acidosis, eye disorders, infections and wounds that do not heal. The effects of diabetes on the body are manifested in the form of serious complications and many other disorders such as infectious, metabolic, degenerative, cancer, severe respiratory sicknesses, diabetes, ketoacidosis, cardiovascular, and kidney diseases. These incurable diseases are very common among poor and wealthy people and kill more than millions per annum in the world (Horton, 2005). The sugar diabetes characterized by aberrant high blood sugar concentrations with deficiency in insulin secretion (Typ-1) or insulin action (Type-2) gains much attention. In the world, approximately 439 million peoples especially among adults will be diabetes patients of which 300 million would suffer from type-2 diabetes mellitus by 2030 according to projection and the estimation (Manukumar et al., 2016). Curing diabetes diseases is very complex because it is controlled by many genes (Bhat et al., 2011). The international health and therapeutic practices strive for the combination of complementary medicine and modern medicine so as to better comprehend the metabolic pathways and its impact in human (Islas al., 2020) in order to fight effectively against these chronic diseases. The leaves were the most plant parts used in the treatment of this chronic disease (Table 1.1). It was observed a decline in blood sugar levels during an experiment conducted on rats and rabbits where diabetes was induced and aqueous alcohol obtained from neem leaves was orally administered (Subapriya and Nagini, 2005). Similarly, Shukla et al. (1973) showed that the oral administration of 5 g of aqueous extract or a dry leaf equivalent in capsules reduced the insulin dose of patients by more than 30–50%. Luscombe and Taha (1974) used a 10% liquid water extract of young, tender leaves for their hypoglycemic effects on rabbits and a decrease in blood glucose concentration was observed. The principles neem ingredients extracted from root and bark were very efficient in diabetes treatments (Patil et al., 2013). It has been demonstrated that after 24h of administering 250 mg/kg of *A. indica* leaves to diabetic rats there was effectively reduction in triglycerides (32%), creatinine (23%), glucose (18%), cholesterol (15%), lipids (15%) and urea (13%) (Shravan et al., 2011). One of the functional mechanisms of action of neem antidiabetogenic activity involves the glucosidase inhibitory activity against pancreatic and intestinal glucosidases. Therefore, the activity of glucosidases in the man pancreas is in relation with the rise in sugar levels after a meal, the regulation of which is very vital in curing type- 2 diabetes (Eichler et al., 1984). The methanol extracted from *A. indica* inhibited glucosidase by 62.44 and 41.07% in the experimental mice (Bhat al., 2008). Moreover; neem cures blood sugar through their antioxidant properties. As a matter of fact, superoxide dismutase, catalase, and glutathione peroxidase are the biological antioxidant enzymes that directly scavenge free radicals and prevent their conversion to toxic products (Freeman and Crapo, 1982). The G6PD accelerates the first pentose phosphate pathway reaction leading to a decline in nicotinamide adenine dinucleotide (NADPH) (Carette et al., 2011; Shailey and Basir, 2012). Thus, any change in G6PD will alter NADPH levels and negatively impact the entire antioxidant system. The endocrine cells are very sensitive to chemical oxidation and inflammation (Mesgari-Abbasi et al., 2019), which activate the resistance process of insulin, leading to an ultimate diabetic state (Shailey and Basir, 2012; Gautam et al., 2015; Omóbòwálé et al., 2016). *A. indica* leaf and bark extracts at 500 mg/kg body weight and 100 mg/kg body weight, respectively re-established the above altered factors very nearly to the controls (Shailey and Basir, 2012). Moreover, it was observed a decrease of free radicals and lipid peroxides production through antioxidant enzymes restoration which avert the diabetes-associated tissue damage (Shailey and Basir, 2012).

Furthermore, the antidiabetogenic activity of *A. indica* would also be due to their certain glycoproteins content. In this case, the structural and functional understanding is of paramount importance in increasing the knowledge of engineering of glycol for the healing of diabetes through glycoproteins (Naseri et al., 2020). Various examples of in-vivo studies on the organs of *Azadirachta indica* to cure diabetes mellitus are discussed in Table 1.1.

TABLE 1.1 *In vivo* Studies of Organs of *Azadirachta indica* Used in Curing Diabetes Mellitus

Plant Part Extract	Extract or Active Molecules	Dose (mg/kg)	*In vivo*	Potential Effect	References
Leaves	Chloroform extract	100 μg/200 μL	+	Antihyperglycemic activity	Bhat et al. (2011)
	Aqueous extract	500	+	Increased activity of antioxidant enzymes, antihyperglycemic activity, Hb1Ac reduction	Shailey and Basir (2012)
	Aqueous extract	250	+	Hypoglycemic activity	Kar et al. (2003)
	Aqueous extract (flavonoids, tannins, sterols, saponins, anthraquinones and alkaloids)	25–350	+	Hypoglycemic activity	Arika et al. (2016)
	Aqueous extract	200	+	Hypoglycemic activity	Salehi et al. (2019)

TABLE 1.1 *(Continued)*

Plant Part Extract	Extract or Active Molecules	Dose (mg/kg)	*In vivo*	Potential Effect	References
	Hot aqueous extract	–	+	Diabetes wound healing activity, angiogenesis effect	Bhagavathy and Kancharla (2016)
	Aqueous extract	200	+	Increased activity of antioxidant enzymes, Hypoglycemic effect	Halim (2003)
	Aqueous extract	–	+	Increased activity of antioxidant enzymes, enzymes inhibitory effects	Shori and Baba (2011)
	Aqueous extract	250	–	Hypoglycemic effect	Shravan et al. (2011)
	Aqueous extract	250	+	Hypoglycemic effect	Dholi et al. (2011)
		500	+	Hypoglycemic effect	Khosla et al. (2000)
	Alcoholic extracts	1	+	Hypoglycemic activity	Akter et al. (2013)
	Aqueous extract	620	+	beta-cell regeneration, increase serum insulin	McCalla et al. (2016)
	Alcoholic extract	500	+	Angiogenesis effect, inhibition of hyperglycemia, the increased activity of antioxidant enzymes	Gautam et al. (2015)
	Aqueous extract	500	+	Increased activity of antioxidant enzymes, hypoglycemic effect	Upreti et al. (2013)
	Aqueous extract	400	+	Improvement the insulin signaling molecules and glucose utilization in the skeletal muscle	Satyanarayana et al. (2015)
	Aqueous extract	400	+	Hypoglycemic activity	Chattopadhyay (1999)
Flower	Methanolic flower extracts	250	+	Hypoglycemic effect	Babandi et al. (2016)

TABLE 1.1 *(Continued)*

Plant Part Extract	Extract or Active Molecules	Dose (mg/kg)	*In vivo*	Potential Effect	References
Bark	Aqueous extract	100	+	Increased activity of antioxidant enzymes, Antihyperglycemic activity, Hb1Ac reduction	Shailey and Basir (2012)
Stem	Neem sticks (toothbrush)	–	+	Reduction in serum MCP-1	Anbalagan et al. (2017)
Fruit/seed	Seed oil	5 ml/kg	+	Hypoglycemic effect	Khosla et al. (2000)
	Petroleum ether extracts of kernel	2	+	Oxidative stress prevention	Gupta et al. (2004)
	Petroleum ether extracts of husk	0.9	+		
Roots	Alcoholic extract of bark (Nimbidin, nimbin)	800	+	Antihyperglycemic and hypoglycemic activity Alloxan reduction	Patil et al. (2013)

1.7 OMICS APPLICATION IN *AZADIRACHTA INDICA*

1.7.1 GENOME

The advanced technologies such as molecular tools were not employed earlier enough to unveil and depict the potential of neem as other areas of biological sciences. This was not done until recently, for instance, the molecular mechanism explaining the insulin signal transduction and glucose homeostasis was uncovered in 2015 by Satyanarayana et al. (2015). Omics science such as genomics, transcriptomics, and proteomics coupled with bioinformatics have recently focused on neem to better our understanding of the tree and therefore supporting a paradigm shift in the manner of considering the compound involved in the neem potential. Thus, there is much more understanding of the structure and functions of the neem plant genome through the progress in genetic and genomic dissection. However, more than half of the genes sequenced are still "orphan" of function even in *Azadirachta indica* tree.

The determination of the nucleotides as a whole through next generation sequencing has contributed greatly in detecting the presence of all genes in neem plants and to better understand the synthetic pathways of the compounds produced by the neem tree. The study of the *Azadirachta indica* genome is also worthwhile in comprehending the mechanism of tolerance to abiotic and abiotic stress (Kuravadi and Gowda, 2019). An attempt was made by Krishnan et al. (2012) to sequence *Azadirachta indica* genome as earlier as 2012. They (Krishnan et al., 2012) reported that neem genome is of 20,000 genes with small amount of repetitive DNA but profuse in A-T nucleotides which annotations are not available in the public domain. It was in 2015 that the whole sequencing, total ARN expression, and the metabolite profiling studies were completed and published by Kuravadi et al. From their studies, 32,278 genes were differentially expressed out of 44,495 genes predicted and 87 Mb of repetitive DNA were recorded from *Azadirachta indica* genome. They further stated that from the Rutaceae family, neem and *citrus sinensis* are classified together on the basis of the molecular phylogenetic analyzes. Kuravadi et al. (2015) also reported 267 Mb as the size of neem genome at 70%. Bhambhani et al. (2017) in their molecular docking studies demonstrated that there was a perfect bond of AzaA and AzaB with AiCYP16671 and AiCYP16365 indicating the participation of the two CYPs in AzaA biosynthesis. Genetic variability was also evaluated and established using AFLP marker techniques (Singh et al., 2002).

1.7.2 TRANSCRIPTOMICS

Transcriptomic tool analyzes mRNA which produces genes expression in the cells or tissues or organs of an organism and it is of paramount importance for better comprehension of the underlying forces in the metabolism of cell and tissue and organs and eventually evaluates the transcriptome profile for health and disease issue (Manzoni et al., 2018). Earlier on in 2014, some works on gene expressions were carried out by Rajakani et al. (2014); and Narnoliya et al. (2014) where a small amount of Expressed sequence tags were recorded. The differentially expressed genes through up and down-regulation are illustrated in all the studies of transcriptomic analysis of neem as an evidence of neem properties. Many classes of differentially expressed genes are recorded including glucosamine metabolism and carbohydrate metabolism (Song, X et al., 2017). The acaricidal properties of neem through the action of octadecanoic acid-3,4-tetrahydrofuran diester to control *Sarcoptes scabiei var. cuniculi* has been attributed to the disruption of the biosynthetic pathway of energy metabolism mainly oxidative phosphorylation and citrate cycling (Song, X et al., 2017). The work of Dawkar et al. (2019) offered us new insight into proteomic profiling, metabolomics and gene expression of neem regarding its insecticidal properties on insects especially *Helicoverpa armigera*. Death, delay in growth, decreased in larval weight, difficulties in ecdysis, darkening in skin, and the exploding of larvae of *H. armigera* which ate AzaA-based diet. This is an indication of the potentiality of neem in controlling agricultural pests for high productivity. The effect of the ingested AzaA in *H. armigera* was revealed through LCM spectrometry which demonstrated disruption and incomplete metabolism. The results of transcriptomic evaluation revealed proteins that were differentially regulated from which some few (15) candidate genes were evaluated for their expressions. Out of the 15 candidate genes, 6 encoded for aminopeptidase, trypsin isoforms, GAPDH, ACBP and HSP were augmented in *H. armigera* larvae treated with AzaA and other genes were downregulated. Similarly, Wang et al. (2020) reported that 22 candidate genes quantitatively expressed were involved in AzaA biosynthesis and include acyltransferase, cytochrome P450, 2,3-oxidosqualene cyclase, alcohol dehydrogenase, and esterase. Different candidate genes differentially expressed implied in AzaA biosynthesis were observed by the two authors, an indication of the complexity in the molecular architecture of AzaA. Bhambhani et al. (2017) found three AiCYPs in AzaA biosynthesis due to the fact highly significant correlation between the accumulation of Aza in the fruit tissue and the expression of *AiCYP16671*, *AiCYP16365* and *AiCYP18835*. They further explained that differentially expressed genes were observed in fruits and leaves of neem contributing to triterpenoid biosynthesis. Meanwhile, the works of Lai et al. (2014) on the effect of AzaA on Drosophila larvae using the analysis of transcriptome showed 28 genes differentially expressed demonstrating up and down-regulation. These genes encoded defense response, signal transduction, starch and sucrose metabolism affecting all developmental stages of the insect.

1.7.3 PROTEOMICS AND METABOLOMICS

The most efficient tool for protein characterization today is mass spectrometry compared to the method of two-dimensional electrophoresis. MALDI-TOF type spectrometers measure the mass of the peptides derived from protein digestion using protein trypsin during the process. Top throughput devices allow the rapid identification of numerous proteins when rich genomic information is available for the species studied. In fact, the data collected ("peptide mass fingerprints") are compared with those from Swissprot or TrEMBL type databases. Mass spectrometers of the ESI-MS/MS type make it possible, by successive fragmentations, to go to the amino acid sequence of the peptides obtained. As with Edman sequencing, they then allow comparison with the sequences of other species and are therefore more suitable when the genome of the species studied is poorly characterized (Thiellement et al., 1999).

The results of song et al. (2017) on transcriptomics and proteomic analysis showed a positive correlation between the mRNA and the targeted proteins such as UCCR, Cytochrome c oxidase, nicotinamide adenine dinucleotide dehydrogenase, adenosine triphosphate synthase, enolase and superoxide dismutase. This was revealed by the modification of these proteins associated with the oxidative phosphorylation route in energy metabolism. In the azadirachtin proteomics study of Wang et al. (2014) 21 proteins were up and down-regulated and are hormonal regulation, cytoskeleton, transcription and translation, and energy metabolic process. Azadirachtin significantly decreased protein/enzyme, glycogen, and lipid contents in *Plodia interpunctella* (Rharrabe et al., 2008). AzaA acts by preventing cell division and protein synthesis (Luntz and Nisbet, 2000; Zhou et al., 2020).

Metabolomics are invented to respond to the challenge facing functional genomics related to complex metabolites in plants including neem (Hall et al., 2002). Metabolomics-type approaches are more particularly interested in chemical compounds (complex metabolites) present within the biological systems of neem, after the involvement of complex phenomena of transcription and translation. Overall, these are molecules of low molecular weight, most often less than 1,000 Da, which may correspond to endogenous (primary or secondary metabolites) or exogenous substances that could arise out of gene function, disease processes and drug toxicity (Klassen et al., 2017). Moreover, metabolomics is an advanced technology tool used in bioanalysis to unveil the mode of action of bio-pesticides against agricultural insects (Aliferis and Jabaji, 2011). The metabolomics study of azadirachtin extracted from neem, the most assessed metabolites in neem tissues, showed alterations of metabolism of enzymes, carbohydrates, vitamins and amino acids in *B. dorsalis* larvae (Zhou et al., 2020).

1.8 CONCLUDING REMARKS

Neem is a tree of numerous biological and pharmacological properties. Many secondary metabolites were isolated from neem and play key roles in treating malaria, diabetes, cancer, chickenpox, skin disorder and eczema, inflammation, etc. Nowadays, advanced technologies in genomics, transcriptomics, and proteomics have been employed therefore to answer a large number of fundamental and applied questions of neem about its biological and pharmacological activities and also the structure and functions of genes or gene expression or proteins in neem tree. Attempts of gene sequencing was made in 2012, but it was in 2015 the comprehensive gene sequencing was generated. Among the tetranortriterpenoids isolated thus far from neem, more omics research was carried out on azadirachtin when compared to salanin and nimbin and others. There is a need for further investigations to better apprehend the biochemical route of the main metabolites such as azadirachtin, limonoids, Azadiradione, salanin, and nimbin in neem tissues due to their medicinal, insecticidal, and pharmaceutical activities. Moreover, plantations of neem should be encouraged to avoid genetic erosion of the species under global warming.

KEYWORDS

- ***Azadirachta indica***
- **genome**
- **medicinal and pharmaceutical properties**
- **neem benefits**
- **nicotinamide adenine dinucleotide**
- **transcriptome**

REFERENCES

Abdelhady, M. I. S., Bader, A., Shaheen, U., El-Malah, Y., Abourehab, M. A. S., Barghash, M. F. (2015). *Azadirachta indica* as a source for antioxidant and cytotoxic polyphenolic compounds. *Biosci., Biotech. Res. Asia, 12*(2), 1209–1222.

Ajayi, F. A., (2002). Formulation and promotion of neem-seed derived biopesticides as alternative to persistent organic pollutants (POPs) for the control of vegetable crop pests. In: Ukwe, C. N., Folorunso, A. O., Ibe, A. C. Z. Z. N. E. S., & Sieghart, L., (eds.), Sustainable Industrial Utilization of Neem Tree (*Azadirachta indica*) in Nigeria (pp. 83–90). UNIDO Regional Development Centre, Lagos.

Aliferis, K. A., & Jabaji, S., (2011). Metabolomics – a robust bioanalytical approach for the discovery of the modes-of-action of pesticides: A review. *Pestic. Biochem. Phys., 100*, 105–117.

Alzohairy, M. A., (2016). Therapeutics role of *Azadirachta indica* (neem) and their active constituents in diseases prevention and treatment. *Evidence-Based Complementary and Alternative Medicine*. https://doi.org/10.1155/2016/7382506.

Arumugam, A., Agullo, P., Boopalan, T., Nandy, S., Lopez, R., Gutierrez, C., & Rajkumar, L., (2014). Neem leaf extract inhibits mammary carcinogenesis by altering cell proliferation, apoptosis, and angiogenesis. *Cancer Biology and Therapy, 15*(1), 26–34. https://doi.org/10.4161/cbt.26604.

Atawodi, S. E., & Atawodi, J. C., (2006). *Azadirachta indica* (neem): A plant of multiple biological and pharmacological activities. *Phytochem Rev., 8*, 601.

Balandrin, M. F., Lee, S. M., & Klocke, J. A., (1988). Biologically active volatile organosulfur compounds from seeds of the neem tree, *Azadirachta indica* (Meliaceae). *Journal Agric. Food Chem., 36*, 1048–1054.

Bandyopadhyay, U., Biswas, K., Chatterjee, R., Bandyopadhyay, D., Chattopadhyay, I., Ganguly, C. K., Chakraborty, T., et al., (2002). Gastroprotective effect of neem (*Azadirachta indica*) bark extract: Possible involvement of H+-K+-ATPase inhibition and scavenging of hydroxyl radical. *Life Sciences, 71*, 2845–2865.

Bandyopadhyay, U., Biswas, K., Sengupta, A., Moitra, P., Dutta, P., Sarkar, D., Debnath, P., et al., (2004). Clinical studies on the effect of neem (*Azadirachta indica*) bark extract on gastric secretion and gastroduodenal ulcer. *Life Sciences, 75*, 2867–2878.

Basak, S. P., & Chakroborty, D. P., (1969). "Chemical investigation of *Azadirachta indica* leaf (M. *azadirachta*)." *Journal of the Indian Chemical Society, 45*, 466–467.

Bhambhani, S., Lakhwani, D., Gupta, P., Pandey, A., Dhar, Y. V., Bag, S. K., Asif, M. H., & Trivedi, P. K., (2017). Transcriptome and metabolite analyses in *Azadirachta indica*: Identification of genes involved in biosynthesis of bioactive triterpenoids. *Scientific Reports, 7*, 5043. doi: 10.1038/s41598-017-05291-3.

Bhat, M., Zinjarde, S. S., Bhargava, S. Y., Ravikumar, A., & Joshi, B. N., (2011). "Antidiabetic Indian plants: A good source of potent amylase inhibitors." *Evidence-Based Complementary and Alternative Medicine, 6*. doi: 10.1093/ecam/nen040.

Bray, D. H., Warhurst, D. C., Connolly, J. D., O'Neill, M. J., & Phillipson, J. D., (1990). Plants as sources of antimalarial drugs. Part 7. Activity of some species of *Meliaceae* plants and their constituent limonoids. *Phytotherapy Research, 4*(1), 29–35. doi: 10.1002/ptr.2650040108.

Butterworth, J. H., & Morgan, E. D., (1972). Investigation of the locust feeding inhibition of the seeds of neem tree, *Azadirachta Indica. J. Insect. Physiol., 17*, 969–977.
Campos, E. V. R., De Oliveira, J. L., Pascoli, M., De Lima, R., & Fraceto, L. F., (2016). Neem oil and crop protection: From now to the future. *Frontiers in Plant Science, 7*. doi: 10.3389/fpls.2016.01494.
Chakraborty, K., Bose, A., Pal, S., Sarkar, K., Goswami, S., Chattopadhyay, U., et al., (2008). Neem leaf glycoprotein restores the impaired chemotactic activity of peripheral blood mononuclear cells from head and neck squamous cell carcinoma patients by maintaining CXCR3/CXCL10 balance. *Int. Immunopharmacol., 8*, 330–340.
Chen, J., Fan, X., Zhu, J., Song, L., Li, Z., Lin, F., et al., (2018). Limonoids from seeds of *Azadirachta indica* A. Juss. and their cytotoxic activity. *Acta Pharmaceutica Sinica B., 8*(4), 639–644. https://doi.org/10.1016/j.apsb.2017.12.009.
Dawkar, V. V., Barage, S. H., Barbole, R. S., Fatangare, A., Grimalt, S., Haldar, S., Heckel, D. G., et al., (2019). AzadirachtinA from *Azadirachta indica* impacts multiple biological targets in cotton bollworm *Helicoverpa armigera. ACS Omega, 4*, 9531–9541. doi: 10.1021/acsomega.8b03479.
Devasagayam, T. P. A., & Sainis, K. B., (2002). Immune system and antioxidants, especially those derived from Indian medicinal plants. *Ind. J. Exptl. Biol., 40*, 639–655.
Eichler, H. G., Korn, A., Gasic, S., Prison, W., & Businger, J., (1984). The effect of new specific α-amylase inhibitor on post-prandial glucose and insulin excursions in normal subjects and Type 2 (non-insulin-dependent) diabetic patients. *Diabetologia, 26*(4), 278–281.
Elumalai, P., Gunadharini, D. N., Senthilkumar, K., Banudevi, S., Arunkumar, R., Benson, C. S., Sharmila, G., & Arunakaran, J., (2012). Ethanolic neem (*Azadirachta indica* A. Juss) leaf extract induces apoptosis and inhibits the IGF signaling pathway in breast cancer cell lines. *Biomedicine and Preventive Nutrition, 2*(1), 59–68. https://doi.org/10.1016/j.bionut.2011.12.008.
Forster, P., & Moser, G., (2000). *Status Report on Global Neem Usage*. GTZ, Ed. Eschborn.
Gautam, M. K., Gangwar, M., Singh, S. K., & Goel, R. K., (2015). Effects of *Azadirachta indica* on vascular endothelial growth factor and cytokines in diabetic deep wound. *Planta Medica, 81*(9), 713–721. https://doi.org/10.1055/s-0035-1545917.
Ghatule, R. R., Goel, S., Gautam, M. K., Singh, A., Joshi, V. K., & Goel, R. K., (2012). Effect of *Azadirachta indica* leaves extract on AA-induced colitis in rats: Role of antioxidants, free radicals, and myeloperoxidase. *Asian Pac. J. Trop. Dis.,* S651–657.
Hall, R., Beale, M., Fiehn, O., Hardy, N., Sumner, L., & Bino, R., (2002). Plant metabolomics: The missing link in functional genomics strategies. *The Plant Cell, 14*(7), 1437–1440. Published by: American society of plant biologists (ASPB) Stable URL: https://www.jstor.org/stable/3871617 (accessed on 16 December 2021).
Haque, E., Mandal, I., Pal, S., & Baral, R., (2006). Prophylactic dose of neem (*Azadirachta indica*) leaf preparation restricting murine tumor growth is nontoxic, hematostimulatory and immunostimulatory. *Immunopharmacol. Immunotoxicol., 28*, 33–50.
Horton, R., (2005). The neglected epidemic of chronic disease. *The Lancet, 366*(9496), 1514. doi: 10.1016/s0140-6736(05)67454-5.
Islas, J. F., Acosta, E., G-Buentello, Z., Delgado-Gallegos, J. L., Moreno-Treviño, M. G., Escalante, B., & Moreno-Cuevas, J. E., (2020). An overview of neem (*Azadirachta indica*) and its potential impact on health. *Journal of Functional Foods, 74*, 104171.
Kelsey, D. J., Nieto-Delgado, C., Cannon, F. S., & Brennan, R. A., (2015). Organic neem compounds inhibit soft-rot fungal growth and improve the strength of anthracite bricks

bound with collagen and lignin for use in iron foundry cupolas. *J. Appl. Microbiol., 1*, 65–75.

Klassen, A., et al., (2017). Metabolomics: Definitions and significance in systems biology. In: Sussulini, A., (ed.), *Metabolomics: From Fundamentals to Clinical Applications* (pp. 3–17). Springer International Publishing.

Krishnan, et al., (2012). A draft of the genome and four transcriptomes of a medicinal and pesticidal angiosperm *Azadirachta indica. BMC Genomics, 13*, 464. http://www.biomedcentral.com/1471-2164/13/464 (accessed on 16 December 2021).

Kumar, G. H., Vidya, P. R., Vinothini, G., Vidjaya, L. P., & Nagini, S., (2010). The neem limonoids azadirachtin and nimbolide inhibit cell proliferation and induce apoptosis in an animal model of oral oncogenesis. *Investigational New Drugs, 28*(4), 392–401.

Kumar, V. S., & Navaratnam, V., (2013). Neem (*Azadirachta indica*): Prehistory to contemporary medicinal uses to humankind. *Asian Pacific Journal of Tropical Biomedicine, 3*(7), 505–514. doi: 10.1016/s2221-1691(13)60105-7.

Kuravadi, N. A., & Gowda, M., (2019). Phylogeny of neem and related species in the *Meliaceae* family. In: Gowda, M., Sheetal, A., & Kole, C., (eds.), *The Neem Genome: Compendium of Plant Genomes*. Springer Nature Switzerland AG, Gewerbestrasse 11, 6330 Cham, Switzerland. https://doi.org/10.1007/978-3-030-16122-4.

Kuravadi, N. A., Yenagi, V., Rangiah, K., Mahesh, H. B., Rajamani, A., Shirke, M. D., Russiachand, H., et al., (2015). Comprehensive analyses of genomes, transcriptomes and metabolites of neem tree. *Peer J., 3*, e1066.

Luntz, A. J. M., & Nisbet, A. J., (2000). Azadirachtin from the neem tree *Azadirachta indica*: Its action against insects. *Anais da Sociedade Entomológica do Brasil, 29*, 615–632.

Luscombe, D. K., & Taha, S. A., (1974). Pharmacological studies on the leaves of *Azadirachta indica. Journal of the Pharmacy and Pharmacology, 26*, 111.

Mallick, A., Ghosh, S., Banerjee, S., Majumder, S., Das, A., Mondal, B., et al., (2013). Neem leaf glycoprotein is nontoxic to physiological functions of Swiss mice and Sprague Dawley rats: Histological, biochemical and immunological perspectives. *Int. Immunopharmacol., 15*, 73–83. doi: 10.1016/j.intimp.2012.11.006.

Mamta, S., Jyoti, S., Rajeev, N., Dharmendra, S., & Abhishek, G., (2013). Phytochemistry of medicinal plants. *J. Pharmacogn. Phytochem., 1*(6), 168–182.

Manikandan, P., Letchoumy, P. V., Gopalakrishnan, M., & Nagini, S., (2008). Evaluation of *Azadirachta indica* leaf fractions for *in vitro* antioxidant potential and *in vivo* modulation of biomarkers of chemoprevention in the hamster buccal pouch carcinogenesis model. *Food Chem. Toxicol., 46*, 2332–43; PMID:18442880; http://dx.doi.org/10.1016/j.fct.2008.03.013.

Manzoni, C., Kia, D. A., Vandrovcova, J., Hardy, J., Wood, N. W., Lewis, P. A., & Ferrari, R., (2018). Genome, transcriptome and proteome: The rise of omics data and their integration in biomedical sciences. *Briefings in Bioinformatics, 19*(2), 286–302.

Martínez-Romero, A., Ortega-Sánchez, J. L., Hernández-González, S. I., Olivas- Calderón, E. H., & Alba-Romero, J. J., (2016). Application of neem tree in agriculture, industry, medicine, and environment: A review. *African Journal of Traditional, Complementary and Alternative Medicines, 13*(2), 191. doi: 10.4314/ajtcam.v13i2.23.

Mesgari-Abbasi, M., Valizadeh, H., Mirzakhani, N., & Vahdatpour, T., (2019). Protective effects of di- and tri-peptides containing proline, glycine, and leucine on liver enzymology and histopathology of diabetic mice. *Archives of Physiology and Biochemistry*, 1–10. doi: 10.1080/13813455.2019.1662453.

Mohanty, S., Patra, A., & Chhonkar, P., (2008). Neem (*Azadirachta indica*) seed kernel powder retards urease and nitrification activities in different soils at contrasting moisture and temperature regimes. *Bioresour. Technol., 99*, 894–899. doi: 10.1016/j.biortech.2007.01.006.

Mordue, A. J., & Nisbet, A. J., (2000). Azadirachtin from the neem tree *Azadirachta indica*: Its action against insects. *An. Soc. Entomol. Brasil, 29*, 615–632.

Mukherji, B., (1957). Indigenous Indian drugs used in the treatment of diabetes. *Journal of Scientific and Industrial Research, 16A*, 1–18.

Musalia, L., Anandan, S., Sastry, V. R., & Agrawal, D., (2000). Urea-treated neem (*A. indica* A. juss) seed kernel cake as a protein supplement for lambs. *Small Rumin. Res., 35*, 107–116. doi: 10.1016/S0921-4488(99)00085-1.

Nagesh, A. K., Vijay, Y., Kannan, R., Mahesh, H. B., Anantharamanan, R., Meghana, S., Heikham, R., et al., (2015). Comprehensive analyses of genomes, transcriptomes and metabolites of neem tree. *Peer J., 3*, e1066.

Narnoliya, L. K., Rajakani, R., Sangwan, N. S., Gupta, V., & Sangwan, R. S., (2014). Comparative transcripts profiling of fruit mesocarp and endocarp relevant to secondary metabolism by suppression subtractive hybridization in *Azadirachta indica* (neem). *Molecular Biology Reports, 41*(5), R41. doi: 10.1007/s11033-014-3174-x.

Naseri, R., Navabi, S. J., Samimi, Z., Mishra, A. P., Nigam, M., Chandra, H., & Farzaei, M. H., (2020). Targeting Glycoproteins as a therapeutic strategy for diabetes mellitus and its complications. *DARU Journal of Pharmaceutical Sciences*. doi: 10.1007/s40199-020-00327-.

National Research Council (NRC), (1992). *Neem: A Tree for Solving Global Problems*. Washington, DC: The National Academies Press. https://doi.org/10.17226/1924.

Omóbòwálé, T. O., Oyagbemi, A. A., Adejumobi, O. A., Orherhe, E. V., Amid, A. S., Adedapo, A. A., & Yakubu, M. A., (2016). Preconditioning with *Azadirachta indica* ameliorates cardiorenal dysfunction through reduction in oxidative stress and extracellular signal-regulated protein kinase signaling. *Journal of Ayurveda and Integrative Medicine, 7*(4), 209–217. https://doi.org/10.1016/j.jaim.2016.08.006.

Patel, S. M., et al., (2016). Potential of neem (*Azadirachta indica* L.) for prevention and treatment of oncologic diseases. *Semin. Cancer. Biol., 40, 41*, 100–115.

Patil, P., Patil, S., Mane, A., & Verma, S., (2013). Antidiabetic activity of alcoholic extract of neem (*Azadirachta Indica*) root bark. *National Journal of Physiology, Pharmacy and Pharmacology, 3*(2), https://doi.org/10.5455/njppp.2013.3.134-138.

Radwanski, S. A., (1977). Neem tree. *World Crops, 29*, 22–66, 111–113, 167–168, 222–224.

Rajakani, R., Narnoliya, L., Sangwan, N. S., Sangwan, R. S., & Gupta, V., (2014). Subtractive transcriptomes of fruit and leaf reveal differential representation of transcripts in *Azadirachta indica*. *Tree Genetics & Genomes, 10*, 1331–1351. doi 10.1007/s11295-014-0764-7.

Rharrabe, K., Amri, H., Bouayad, N., & Sayah, F., (2008). Effects of azadirachtin on post-embryonic development, energy reserves and α-amylase activity of *Plodia interpunctella* Hübner (Lepidoptera: Pyralidae). *J. Stored Prod. Res., 44*, 290–294.

Saleem, S., Muhammad, G., Hussain, M. A., & Bukhari, S. N. A., (2018). A comprehensive review of phytochemical profile, bio-actives for pharmaceuticals, and pharmacological attributes of *Azadirachta indica. Phytotherapy Research, 32*(7), 1241–1272. doi: 10.1002/ptr.6076.

Satyanarayana, K., Sravanthi, K., Shaker, I. A., & Ponnulakshmi, R., (2015). Molecular approach to identify antidiabetic potential of *Azadirachta indica. J. Ayurveda Integr Med., 6*, 165–174. doi: 10.4103/0975-9476.157950.

Schmutterer, H., (1990). Properties and potentials of natural pesticides from neem tree. *Annual Review of Entomol., 35,* 271–298.

Shravan, K. D., Ramakrishna, R., Santhosh, K. M., & Kannappan, N., (2011). *In vivo* Antidiabetic evaluation of Neem leaf extract in alloxan-induced rats. *Journal of Applied Pharmaceutical Science, 1*, 100–105.

Singh, A., Negi, M. S., Moses, V. K., Venkateswarlu, B., Srivastava, P. S., & Lakshmikumaran, M., (2002). Molecular analysis of micropropagated neem plants using AFLP markers for ascertaining clonal fidelity. *In Vitro Cellular & Developmental Biology: Plant, 38*(5), 519–524. Published by: Society for *in vitro* biology. URL: https://www.jstor.org/stable/23321169 (accessed on 16 December 2021).

Song, X., et al., (2017). Transcriptomics and proteomic studies reveal acaricidal mechanism of octadecanoic acid-3, 4 – tetrahydrofuran diester against *Sarcoptes scabiei* var. cuniculi. *Sci. Rep., 7*, 45479. doi: 10.1038/srep45479.

Subapriya, R., & Nagini, S., (2005). Medicinal properties of neem leaves: A review. *Curr. Med. Chem. Anticancer Agents, 5,* 149–156. doi: 10.2174/1568011053174828.

Tapanelli, S., et al., (2016). Transmission blocking effects of neem (*Azadirachta indica*) seed kernel limonoids on *Plasmodium berghei* early sporogonic development. *Fitoterapia. 114*, 122–126.

Thiellement, H., Bahrman, N., Damerval, C., Plomion, C., Rossignol, M., Santoni, V., De Vienne, D., & Zivy, M., (1999). Proteomics for genetic and physiological studies in plants. *Electrophoresis, 20*, 2013–2026.

Umar, M. I., et al., (2014). Multi-constituent synergism is responsible for the anti-inflammatory effect of *Azadirachta indica* leaf extract. *Pharm. Biol., 11*, 1411–1422.

Vijayan, V., Tiwari, P. K., & Meshram, G. P., (2013). Inhibitory effects of neem seed oil and its extract on various direct-acting and activation-dependent mutagens-induced bacterial mutagenesis. *Pharm. Biol., 51*(12), 1525–1530.

Vinothini, G., Manikandan, P., Anandan, R., & Nagini, S., (2009). Chemoprevention of rat mammary carcinogenesis by *Azadirachta indica* leaf fractions: Modulation of hormone status, xenobiotic-metabolizing enzymes, oxidative stress, cell proliferation and apoptosis. *Food and Chemical Toxicology, 47*(8), 1852–1863. PMID: 19427891. https://doi.org/10.1016/j.fct.2009.04.045.

Wang, H., Lai, D., Yuan, M., & Xu, H., (2014). Growth inhibition and differences in protein profiles in azadirachtin-treated *Drosophila melanogaster* larvae. *Electrophoresis, 35*, 1122–1129.

Zhou, Y., Qin, D. Q., Zhang, P. W., Chen, X. T., Liu, B. J., Cheng, M. D., & Zhang, Z. X., (2020). Integrated LC-MS and GC–MS-based untargeted metabolomics studies of the effect of azadirachtin on *Bactrocera dorsalis* larvae. *Scientific Reports, 10*, 2306. https://doi.org/10.1038/s41598-020-58796-9.

Zhu, J., Lu, X., Fan, X., Wu, R., Diao, H., Yu, R., et al., (2017). A new cytotoxic salanninclass limonoid alkaloid from seeds of *Azadirachta indica* A. juss. *Chinese Chemical Letters, 29*(8), 17–19. https://doi.org/10.1016/j.cclet.2017.11.042.

CHAPTER 2

Gymnema sylvestre (Asclepiadaceae): Multi-Potent Medicinal Plant with Potential for Antidiabetes and Obesity

RAMACHANDRA REDDY PAMURU,[1] AMBEDKAR YARRAPALLE,[1,2] and D. VIJAYA LAKSHMI[3]

[1]*Department of Biochemistry, Yogi Vemana University, Vemanapuram, Kadapa – 516005, Andhra Pradesh, India, E-mails: reddyprbiotech@gmail.com; prcrbio@yogivemanauniversity.ac.in*

[2]*Department of Biotechnology, Yogi Vemana University, Vemanapuram, Kadapa – 516005, Andhra Pradesh, India*

[3]*Department of Microbiology, Yogi Vemana University, Vemanapuram, Kadapa – 516005, Andhra Pradesh, India*

ABSTRACT

A wonderful plant *Gymnema sylvestre* belongs to the family Asclepiadaceae holds phytochemicals with multiple medicinal properties. *G. sylvestre* has made mark in traditional medicine preparations. The most predominant medical property of *G. sylvestre* is antidiabetic and anti-obesity. Since obesity and diabetes are interlinked, *G. sylvestre* secondary metabolite gymenemic acid acts against to these metabolic disorders. A plenty of medicinal properties include anticancer, anti-inflammatory, antioxidant, antimicrobial, etc., of *G. sylvestre* have been identified. This chapter

Antidiabetic Plants for Drug Discovery: Pharmacology, Secondary Metabolite Profiling, and Ingredients with Insulin Mimetic Activity. Deepu Pandita, MPhil, Anu Pandita, MSc, & Chander Bhanu, MBBS, PGDDCE (Eds.)
© 2023 Apple Academic Press, Inc. Co-published with CRC Press (Taylor & Francis)

emphasizes the biology, traditional use, phytochemicals, anti-obese and antidiabetic properties and mechanism of action, other medicinal properties, pharmacology, and health benefits of *G. sylvestre*.

2.1 BIOLOGY OF *G. SYLVESTRE*

Gymnema sylvestre R. Br. (*G. sylvestre* R. Br.) is mostly found in western and central part of India, and in tropical regions of Australia, Japan, Vietnam, and Africa. In India, *G. sylvestre* has various names with varied (*G. sylvestre*) (Table 2.1).

TABLE 2.1 Various Common Names of *Gymnema sylvestre* in Different Indian Languages

SL. No.	Indian Language	Common Name	References
1.	Sanskrit	Madhunashini, Meshashringi	Council for Scientific and Industrial Research
2.	Hindi	Merasingi, Gur-mar	
3.	Marathi	Kalikardori, Kavali, Vakundi	
4.	Telugu	Podapatri	
5.	Kannada	Sannagerasehambu	
6.	Tamil	Cherukurinja, Adigam	
7.	Gujarathi	Mardashingi, Dhuleti	

G. sylvestre is a perennial, slow growing, pubescent, and woody climber belongs to family Asclepiadaceae. This shrub leaves are generally ovated or ellipatic in shape with a size of 0.5–1.25 inches X 1.25–2.0 inches. Inflorescence of *G. sylvestre* is umbellate cymes of upto 3 inches in size holding small yellow flowers. *G. sylvestre* is mostly grown in wild but cultivated as medicinal plant intermittently. *G. sylvestre* is medicinally important for the treatment of diabetes in Indian medicine like Ayurveda, Homeopathy, Yunani and Folk (Kanetkar et al., 2007). The common name 'gur-mar' of *G. sylvestre* in Hindi represents as 'sugar destroyer.' Besides its common use in diabetes, *G. sylvestre* has great potentiality in treating asthma, eye diseases, snake bite, inflammation, and for family planning. Moreover, *G. sylvestre* is holding activities like hepatoprotective, antimicrobial, sweet suppressing and anti-hypercholesterolemic. It possesses anti-dental caries activity against *Streptococcus mutans*. The leaves of *G.*

sylvestre also used for *Prodenia eriania* caterpillar feeding and in skin cosmetics preparations (Komalavalli and Rao, 2000).

2.2 *G. SYLVESTRE* AS TRADITIONAL MEDICINE

The traditional medicine played an important role in the ancient and middle era of human development. Especially the countries like India, China, and some other Asian and African continent are depended on traditional medicine. Out of all, India has great history of using traditional medicine with respect to Ayurveda, Homeopathy, Folk and Yunani. Most of the preparations in these traditional medicines are prepared from a source of flora. This is possible because of different medicinal vegetation spread all over in Indian sub-continent. Incredible advancement in the medical science has moving towards the development of traditional medicine with refinement against several infectious and other diseases. Modern health care system focused on quality and security of life with modified traditional medicine and should be available for all equally (World Health Organization, 1978). Traditionally, *G. sylvestre* has played great role in lowering the sugar levels of an individual. Though the traditional medicine not specifically defined the active principle and its molecular mechanism of action about the compounds of *G. sylvestre* but addressed the treatment for diabetes. As per the ancient traditional medicine book 'Shushruta' it is very clear that *G. sylvestre* was used for treating glycosuria, a urinary disease (Nadkarni, 1986). The usage of *G. sylvestre* in different Indian traditional medicine is presented in Table 2.2.

TABLE 2.2 Usage of *Gymnema sylvestre* in Different Indian Traditional Medicine

SL. No.	Type of Traditional Medicine	Usage	References
1.	Ayurveda	Treating diabetes, snake bites and malaria. Also used to treat diseases caused by phlegm and piles.	Kirtikar and Basu, (1975); Singh et al. (2008)
		Treating renal and vesicle calculi, jaundice, dyspepsia, asthma, constipation, hemorrhoids, cardiopathy, amenorrhea, bronchitis, and leukoderma.	Sastry (1994); Nadkarni (1996); Anis et al. (2000); Mathew (2004)

TABLE 2.2 *(Continued)*

SL. No.	Type of Traditional Medicine	Usage	References
		Leaves, roots, and stem, have been used as cardiotonic, digestive, diuretic, laxative, stimulant, stomachic, and uterine tonic in traditional medicine systems.	Mathew (2004)
		Obesity reduction	Ghose (1984)
2.	Folk	Roots and leaves used against asthma, eye and gastric problems, parkinsonism, urinary problems, and diabetes	Chopra et al. (2002); Anis et al. (2000); Potawale et al. (2008)
3.	Homeopathy	Roots and leaves used in the treatment of diabetes, pathogenic diseases, and snake bite.	Boericke (1996); Rakshit et al. (2013); Kishore and Singh (2017)
4.	Yunani	Roots and leaves used as anti-eczematic, antidiabetic, anti-thrust, anti-itching, anti-cough properties, anti-leprosy and eczema.	Pawan Kumar et al. (2015)

2.3 SECONDARY METABOLITES OR PHYTOCHEMISTRY OF *G. SYLVESTRE*

The chemistry of *G. sylvestre* has importance in the preparation of medicine for various ailments. Various studies used different solvents like chloroform, methanol, water, petroleum ether, and hexane for isolation of secondary metabolites from *G. sylvestre* using gas chromatography coupled with mass spectrometry (Sathya et al., 2010). The phytochemical studies revealed the chemical composition of different parts of *G. sylvestre*. Predominant chemical gymnemic acid and its derivatives are found in *G. sylvestre*. Saponins are the richest compounds in the leaves of *G. sylvestre* which comes under class oleanane and dammarane. In these, oleanane type saponins are gymnemic acid and gymnema saponins, whereas dammarane saponins are gymnemasides (Foster, 2002; Khramov et al., 2008). These compounds are coming under triterpenoid saponins family and its quantity varies with location of plant growing and time of harvesting of leaves. The concentration of gymnemic acid lies in between 0.67 to 1.06% in leaves of

G. sylvestre collected at different seasons. Novel (D)-glycoside structure of *G. sylvestre* saponins executes anti-sweet principles. The other common gymnemoside identified in *G. sylvestre* leaves is gurmarin (anti-sweetener). Besides these two major compounds also identified anthraquinones, flavones, alkaloids, hentriacontane, d-quercitol, pentatriacontane, tannic acid, phytin, α and β-chloriphylls, β-amyrin (glycosides), stigmasterol, resins, lupeols, tartaric acid, calcium oxalate, butyric acid and formic acid (Fletcher, 1999; Kanetkar, 2007; Porchezhian, 2003; Puratchimani, 2004; Raju, 2006; Yoshikawa, 1997; Zhu, 2008; Kanetkar et al., 2007). The root and leaf isolated compounds of *G. sylvestre* is presented in the Figure 2.1.

The leaves of *G. sylvestre* had been detected with anthraquinones and acidic glycosides and its derivatives. There are nine acidic glycosides of *G. sylvestre* isolated as secondary metabolites from all parts of the plant and major compounds among these are gymnemic acid A, B, C, D (Dateo and Long, 1973). A1, A2, A3 and A4 are the different forms of gymnemic acid A. It is found that the anti-sweetness of gymnemic acid is found responsible with the ester group of genin portion and this structural arrangement is present in almost all triterpene saponins. The glycosides like glucuronic acid is identified in A1 gymnemic acid, whereas A2 and A3 found with both glucuronic acid and galactose (Chakravarti and Debnath, 1981). The structure of gymnemic acid and its derivatives are holding the basic component 3*β*, 16*β*, 21*β*, 22*α*, 23, 28-hexahydroxy-olean-12-ene (3-O-*β*-glucuronide of gymnemagenin). Deacyl gymnemic acid mediates the addition of tigloyl, butyl and methyl groups to gymnemic acid. A special peptide containing 35 amino acid residues identified in *G. sylvestre* with a molecular weight of 42.09 kDa is gurmarin (Imoto et al., 1991). This peptide holding anti-sweetness function identified through testing taste response electro physiologically in rats (Gent et al., 1999). The change of pH on the tongue gurmarin as anti-sweetness polypeptide and is maximum at its isoelectric point with hydrophobic interactions (Chattopathyay, 1999; Aral et al., 1995). Yoshikawa et al. (1991) isolated another anti-sweetness compound from *G. sylvestre* belongs to gymnema saponins which hold hydroxylongi spinogenin structure with a moiety aglycone of one or two glucose at hydroxyl 23 or 28 position. The gymnemic acid, gymnemagenin, dacyl gymnemic acid, gymnestrogenin and hydroxynogispinogenin isolated from *G. sylvestre* was reviewed by Tiwari et al. (2014).

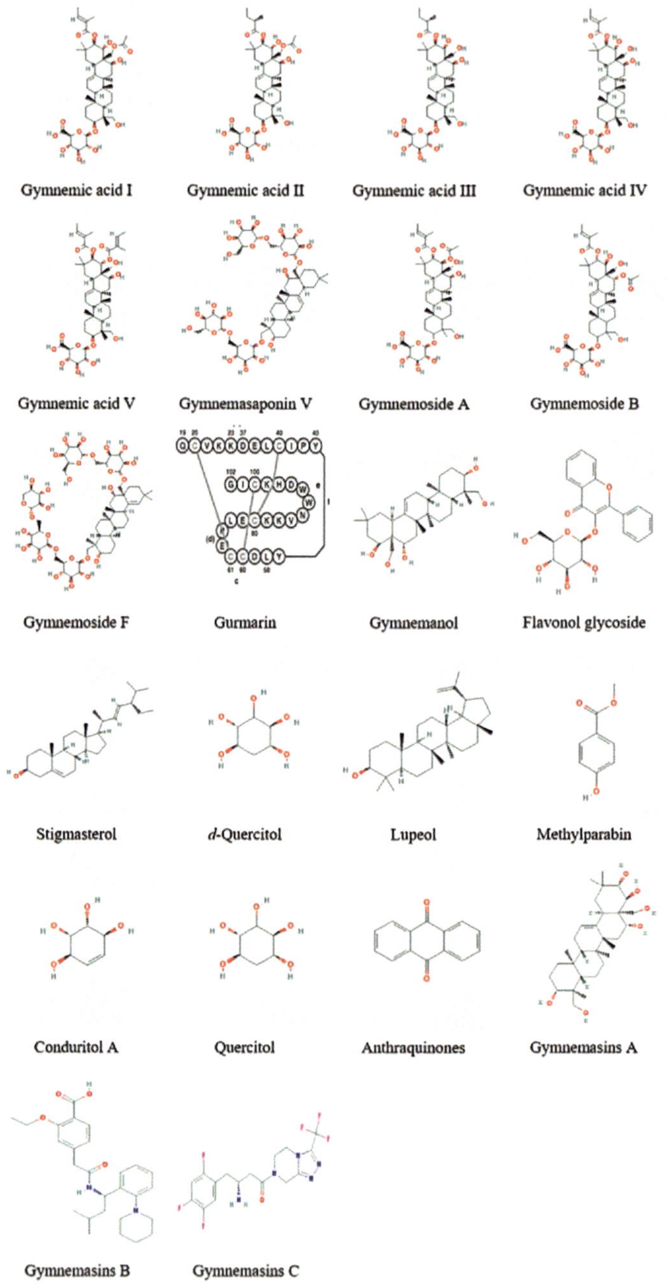

FIGURE 2.1 Structure of various phytochemicals isolated from *Gymnema sylvestre* leaves and roots (the structures are obtained from PubChem).

2.4 BIOSYNTHESIS AND MOLECULAR BIOLOGY OF *G. SYLVESTRE*

Several saponins derived from plants including saiko saponins, soyasaponins and ginsenosides show remarkable beneficiary effects on human health. Due to importance of these compounds has great potential in developing them as pharmaceutical products. The known plant-based pharmacological medicine prepares for adjuvant hemolytic, anticancer, antimicrobial, etc. (Haridas et al., 2001; Park et al., 2001). Triterpenoid and its skeleton beta-amyrin and its related glycosides are holding sugar residues are synthesized via isoprenoid pathway through 2,3-oxidosqualene cyclization (Yendo et al., 2010). Anyhow, the biosynthetic pathway of saponins is characterized but still, it is incomplete. Since the structural similarity of gymnemic acid with moiety of triterpene aglycone, knowing the biosynthetic pathway of triterpene in other plants can give a clue to know the metabolomics and genomics of G. Sylvester. In the same way, the metabolomics and genomics of metabolic profiling of *G. sylvestre* is not having very few reports. The studies in this series are conducted for *G. sylvestre* to identify the genes and its characterization through isolation and cloning. This can help to find the intermediates of metabolic pathways of phytochemical compounds in *G. sylvestre* responsible for pharmacological properties (Tiwari et al., 2014).

The genes with partial cDNAs holding molecular mechanism of *G. sylvestre* medicinal functions are isolated by Tiwari et al. (2009). Still the transcriptomic data of *G. sylvestre* is not available. Characterizing the putative pathway genes are responsible for biosynthesis of gymnemic acid, triterpene glycosides may help to improve the medicinal value of these compounds in the plant through gene manipulation studies.

2.5 ROLE OF *G. SYLVESTRE* IN CONTROLLING OBESITY AND DIABETES

Obesity and diabetes are interlinked disorders in humans. Obesity lowers the metabolism of individual, which eventually causes diabetes (type 2). It is clear from the recent research that chance of getting type 2 diabetes is 80 times more in obese people than less than 22 body mass index ones. It is believed that obesity induces prediabetes which eventually grows diabetes mellites. Obesity triggers the change in body metabolism, where fat molecules released into the blood from adipose tissue and reduces the insulin response from islets of Langerhans thereby diminished sensitivity of insulin.

G. sylvestre a potential plant medicine used to reduce obesity and diabetes. Many studies were reported the *G. sylvestre* potentiality in reducing the weight and type 2 diabetes in animal models and human subjects. In rats, long-term exposure of *G. sylvestre* showed body weight loss (Luo, 2007; Shigematsu, 2001). Ramkumar et al. (2009) were also reported weight reduction in rodents by gymnema dietary application. Clinical data suggests that a mixture of *G. sylvestre*, fenugreek, vitamin C, chitosan and niacin-chromium complex as dietary supplement reduced the body weight (Preuss, 2004; Woodgate and Conquer, 2003). Kumar (2010) also reported the reduced weight and diabetes Millets with administration of *G. sylvestre* extract. In Wistar rats Shigematsu et al. (2001) observed reduction in the body weight with administration of *G. sylvestre* ethanolic extracts. More likely, Fatani et al. (2015) reported significant ($P<0.0001$) reduction in the body weight of diabetic (Steptozotocin exposed) Albino rats with ethanolic extracts of *G. sylvestre*. Whereas Ahmed et al. (2010) reported the increase in the body weight and other organs like pancreas and liver in alloxan induced diabetic Wistar rats treated with *G. sylvestre* ethanolic extract.

G. sylvestre extracts are established medicine to reduce hyperglycemia. These extracts can reduce small intestinal absorption of glucose, and increases glucose breakdown, gluconeogenesis, and glycogen biosynthesis. It is also proved that *G. sylvestre* extracts stimulates pancreatic cells to produce insulin (Leach, 2007; Liu, 2009; Persaud, 1999; Shimizu, 1997).

Several studies revealed the effects on sugar levels in blood by *G. sylvestre* in animal models. Diabetic rats treated with *G. sylvestre* reduced blood glucose in most of the reports (Chattopadhyay, 1999; Gholap, 2003; Gupta and Seth, 1962; Gupta and Variyar, 1964; Rahman, 1989; Ramkumar et al., 2008; Ramkumar et al., 2009; Yadav, 2010). A potential extract of *G. sylvestre* called gurmarin suppressed taste response to maltose, fructose, sucrose and lactose in rats (Harada, 2000; Katsukawa, 1999). Clinically, usual medication and extracts of *G. sylvestre* administration to type 1 and 2 diabetes patients lowered the HbA1c (glycosylate hemoglobin), pre-prandial glucose in plasma and fasting glucose in blood (Baskaran, 1990a; Joffe, 2001; Shanmugasundaram, 1990a). Levels of HbA1c, fasting blood glucose, oral hyperglycemia, hypertension, secondary complications, and anthropomorphic parameters were lowered with elevated insulin in patients treated with *G. sylvestre* medication. Significant reduction in the mean post-prandial glucose in blood (55 mg/dL) and increase in insulin levels in C-peptide (149 pm/l) and serum (8 Mu/ml) were reported in patients of type 2 diabetes treated with 500 mg *G. sylvestre* leaves twice a day for a period of 60 days (Al-Romaiyan, 2010).

The daily recommended dose of *G. sylvestre* leaf extract as per Al-Romaiyan (2010) and Kumar (2010) is 500 to 1,000 mg for duration of 2 to 3 months. Combination of *G. sylvestre* extract plus hydroxycitric acid treatment resulted in reduced weight, BMI, markers of urinary fatty acid oxidation, and lipid content (Pothuraju, 2014). Joffe (2001), Preuss (2004), and Shanmugasundaram (1990a) investigated the antidiabetic role of gymnemic acid with an extract concentration of 200 to 400 mg. Low HbA1c, FBG, insulin resistance, kidney function, and apolipoprotein B were reported in the patients medicated with *G. sylvestre* (Kumar, 2010). Table 2.3 shows the antidiabetic and antioxidant compounds and their mode of action isolated from *G. sylvestre*.

TABLE 2.3 *Gymnema sylvestre* Compounds and Their Antidiabetic and Antioxidant Functions

SL. No.	Compounds of *G. sylvestre*	Activity	References
1.	Gymnemic acid homolog	Inhibits sweetness	Sugihara et al. (2000); Kurihara (1992)
2.	Oleanane	Antidiabetic	Stoecklin (1968)
3.	Alkaloid conduritol A	Antidiabetic	Kazutaka et al. (1993)
4.	Gymnemic acid	Lowers blood sugar by preventing intestinal glucose absorption	Stoecklin et al. (1967)
5.	Gymnemic acid A1 and derivatives	Anti-sweetener	Spasov et al. (2008); Kim et al. (2016)
6.	Gymnemasides A to F	Serum hypoglycemia	Yoshikawa et al. (1997a and b); Kazuko et al. (1992)
7.	Gurmarin (polypeptide)	–	Ninomiya and Imoto (1995)

TABLE 2.3 *(Continued)*

SL. No.	Compounds of *G. sylvestre*	Activity	References
8.	Gymnestrogenin (gymnemasaponin)	Antidiabetic and regeneration of B cell	Shimizu et al. (2001)
	Gymenmagenin (gymnemasaponin)	Antidiabetic by lowering glucose absorption and regeneration of B cell	
9.	Terpenoid, phenol Folic acid, tannin Flavonoid, cinnamic acid	Hypoglycemia by lowering intestine glucose absorption. Inhibits α-glucosidase and α-amylase. Stimulates glucose uptake and insulin production. Reducing power and scavenges free radicals.	Bahadoran et al. (2013); Hanhineva et al. (2010)
10.	Ascorbic acid	Elevates insulin sensitivity and glucose uptake. Scavenges hydrogen peroxide (H_2O_2), superoxide ($O^{-}\cdot$) and hydroxyl ($OH\cdot$) radicals.	Salem et al. (2009); Johtatsu et al. (2007)
11.	Anthroquinone, its derivatives Nonacosane, d-quercitol Hentriacontane, phytin, resin Penta-tri-acontane, butyric acid, formic acid, tartaric acid β-amyrene glycosides, α- and β-chlorophyll	Hypoglycemic activity and free radicals ($O^{-}\cdot$, $OH\cdot$ and H_2O_2) scavenging function.	Trease and Evans (2002); Bhansali et al. (2013)

2.6 OBESITY AND DIABETES CONTROLLING MECHANISM BY GYMNEMIC ACID AND *G. SYLVESTRE* EXTRACTS

The potential role of *G. sylvestre* in treating the diabetes is well reported in many studies. The mechanistic action of this plant against diabetes is explained differently by various authors (Figure 2.2). A mixture of saponins called Gymnemic acid is a major constituent of *G. sylvestre* structurally resembles glucose. Since its analogy with glucose binds with intestinal sugar receptors and blocks the absorption of glucose thereby lowers the blood glucose levels (Figure 2.2) (Sahu et al., 1996; Tiwari et al., 2014). Whereas Chen and Guo (2017) found inhibitors of α-glucosidase in affinity ultra-filtration high performance liquid chromatography (HPLC)-mass spectroscopic (MS) study. These α-glucosidase inhibitors suppress the insulin independent enzymes like gluconeogenic enzymes, fructose 1,6 diphosphatase, sorbitol dehydrogenase, glycogen phosphorylase and glucose 6 phosphate. On the other hand, α-glucosidase inhibitors elevate the activity of glyceraldehyde 3 phosphate dehydrogenase, glycogen synthetase, glucose 6 phosphate dehydrogenase and hexokinase which belongs to insulin dependent enzymes. However, α-glucosidase inhibitors by activating insulin dependent enzymes and lowering the insulin independent enzymes, rises the phosphorylase activity (Figure 2.2) (Chen and Guo, 2017). Moreover, *G. sylvestre* is proved plant product which induces insulin expression and secretion in β-cells in Langerhans of pancreas. Studies suggest that *G. sylvestre* has great potential to regenerate insulin producing cells in pancreas (Aralelimath and Bhise, 2012; Nakamura et al., 1999; Shanmugasundaram et al., 1990b). This was proved by Ahmed et al. (2010), where significant increase in the regeneration of β-cells in Langerhans was recorded with methanolic extract of *G. sylvestre* and is extrapolated for the treatment of type I diabetes with complete recovery of pancreatic cell activity (Patel et al., 2009) (Figure 2.2). However, the stimulated pancreatic insulin by *G. sylvestre* is found to have similar function like natural insulin and reduces the blood sugar absorption. The mechanism of gymnemic acid in lowering the glucose absorption in the mouth from food is like its function in the intestine. Gymnemic acid binds to taste buds on the tongue and blocks the absorption of food sugars thereby reduces the sweetness of glucose. The key enzyme of glycolysis glyceraldehyde 3 phosphate dehydrogenase interacts with gymnemic acid and plays anti-hyperglycemic role through the interaction of acylomoeity of gymnemic acid with this enzyme (Ishijima et al., 2008; Sugihara et al., 2000). Unlike gymnemic acid, gurmarin also reduces sweetness and bitterness on the tongue. The compounds of *G. sylvestre* lower the small intestinal assimilation of glucose and lipids. Bone (2002) compared the mechanism of incretin lowering glucose with gymnemic acids.

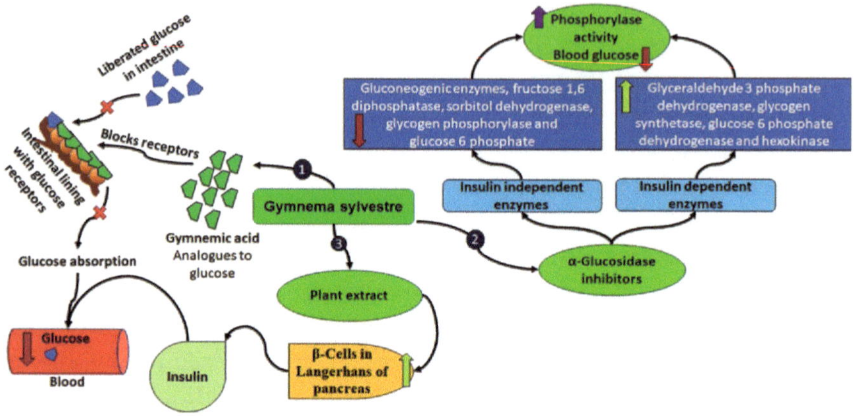

FIGURE 2.2 Mechanism of antidiabetic function of *Gymnema sylvestre*: (1) Glucose lowering function of gymnemic acid; (2) Alfa-glucosidase inhibitors induced blood glucose maintenance; (3) direct action of *G. sylvestre* on insulin production from beta-cells of Langerhans.

Note: ✖ : No binding or no absorption; ⬇ : Lowering or decrease; ⬆ : Elevate or increase

2.7 OTHER ACTIVITIES OF *G. SYLVESTRE*

The medicinal role of *G. sylvestre* not only restricted to anti-obesity and diabetes, but it also has great potentiality in correcting various ailments in humans. Reports are in connection with various biological functions of *G. sylvestre* extracts and its constituents tested in different animal systems and clinical studies. The proven functions of *G. sylvestre* are anti-cancerous, antioxidant, anti-inflammatory, lipid lowering activity, ophthalmic, antimicrobial, etc. Besides these functions *G. sylvestre* also found with adverse effects including cytotoxic effects. Various beneficiary and adverse effects of *G. sylvestre* and its isolates are discussed in this section.

2.7.1 ANTICANCER

Inflammation-induced cancer is becoming predominant in worldwide. The western countries are more affected and leading with various types of cancers. Anti-genotixic, cytotoxic and radical scavenging activity of

G. sylvestre showed anticancer activity. Anti-inflammatory and anticancer functions of Gymnema are due to the presence of specific phenolic and saponins (Ohmori, 2005; Khanna, 2009; Arunahalam et al., 2015). Ethanolic *G. sylvestre* extract of 50–555 nmol/year showed 50% inhibition on cancer cell growth in the two-stage carcinogenesis mice (Yasukawa et al., 2014). Wu et al. (2012) tested the *G. sylvestre* polysaccharides named as GSP 11, GSP 22, GSP 33, GSP 44 and GSP 55 showed anti-proliferation activity and improved immunological activity by elevating thymus and spleen index, serum hemolysin and phagocytic function. They were observed significant reduction in the proliferation of U937 cancer cells 78.6% and 83.8% with polysaccharides GSP11 and GSP33, respectively, whereas 78.2% inhibition reported in SGC cancer cells with *G. sylvestre* polysaccharides GSP22 (Wu et al., 2012). Human breast carcinoma (MCF7) and human lung adenocarcinoma (A549) cell lines were treated with *G. sylvestre* ethyl, chloroform and ethanolic extracts for testing anti-tumor activity (Srikanth et al., 2010). Srikanth et al. (2010) revealed all these three extracts are effective (anti-cancerous) against MCF7 cell lines, whereas ethyl and chloroform extracts are showed more anti-proliferative activity against A549 cells. Khanna (2010) reported the anti-tumor function of *G. sylvestre* compound gymnemagenol against HeLa cancer cells.

Moreover, Tamaki et al. (2010) found significant anti-tumor activity of *G. sylvestre* through inhibiting intestinal-BCRP (breast cancer resistance protein) in breast cancer cell lines. In dimethylbenzanthracene-induced papilloma Swiss albino mice Agarwal et al. (2016) tested the anti-tumor function of *G. sylvestre* methanolic extract and found lowered incidence of tumor with low burden of carcinoma and less papillomas in the plant extract treated group. The potential anti-tumor activity of *G. sylvestre* reported in many studies. Ethanolic *G. sylvestre* extract performed potent anticancer activity against human skin melanoma (A375) cell lines. Also reported the anti-cytotoxic function of these extract were tested with liver cell lines (WRL-68), besides its anti-proliferation function against skin papilloma cell lines (Chakraborty et al., 2013). BCRP inhibition by *G. sylvestre* flavonoid administration was reported Imai et al. (2004). They also reported the flavonoid induced BCRP substrates increased multi-drug resistance. Mao (2005) suggested the inhibitory role of *G. sylvestre* against BCRP may be due to elevated activity of its substrates (daunorubicin, epirubicin, methotrexate, topotecan, and flavopiridol) where plant components increases absorption and availability of these substrates.

2.7.2 ANTIOXIDANT

Stress release of any organism starts by initiating the antioxidant mechanism, where lipid peroxidation and antioxidant enzymes are activated to detoxify the radical scavengers generated during oxidative stress. Antioxidant role of biological compounds are placing very important role for maintenance of good health. *G. sylvestre* is a proved antioxidant by several research groups. Identified antioxidants of *G. sylvestre* are alkaloids, cinnamic acid, phenols, flavonoids, folic acid, flavonoids, tartaric acid, butyric acid, ascorbic acid, etc. Rupanar et al. (2012) tested and reported the antioxidant activity of *G. sylvestre* by comparing with DPPH (1,1-diphenyl-2-picrylhydrazyl). Significant DPPH radical scavenging activity and antioxidant role of gymnema ethanolic extract than *Capsicum frutescens* and *Averrhoa bilimbi* was proved (Rahman et al., 2014). Gunasekaran et al. (2019) demonstrated the hydroxy free radical scavenging role of *G. sylvestre* and found 87.3% of DPPH inhibition and 59.8% free radical inhibition. In the similar type of study gymnema extract showed significant radical scavenging function against super oxide, hydroxy peroxide and ferric ions (Rachh et al., 2009). Moreover, in rats gymnema showed antioxidant function against oxidative stress induced by nitric oxide, super oxide, high fat diet and hydrogen peroxide radicals (Kishore and Singh, 2015; Chakrapani and Periandavan, 2018; Arun et al., 2014).

2.7.3 ANTIMICROBIAL

The role of *G. sylvestre* in controlling the growth of microbial pathogens is tested on bacterial, viral, and fungal species. Antimicrobial activity of gymnema was tested by many research groups. The level of antimicrobial activity of *G. sylvester* varies from solvent to solvent extraction and from part of plant used for extraction. David and Sudarsanam (2013) and Tahir et al. (2017) tested antimicrobial activity of *G. sylvestre* methanolic leaf extract and reported inhibition of *Bacillus cereus, Escherichia coli, Candida kefyr*, and *Candida albicans*. They were also tested the aqueous extract and found moderate inhibitory activity against *Staphylococcus aureus, C. perfringens type-A, C. kefyr*, and *C. krusei* whereas, hexane extract potentially inhibited the growth of *C. perfringens type-A, B. cereus, S. aureus, S. enterica,* and *H. paragallinarum*. Anti-influenza viral activity of gymnemic acid A and B have tested by Porchezhian (2003). The dried leave alcoholic extract of *G. sylvestre* inhibited the growth of *Bacillus subtilis* (*B. subtilis*), *Bacillus pumilis, Staphylococcus aureus*, and *Pseudomonas aeruginosa* (Satdive, 2003). The *Streptococcus aureus, Escherichia Coli, Klebsiella pneumoniae (K. pneumoniae),* and *Pneumococcus aeruginosa* (Gram '+ve' and '-ve' bacteria) inhibited by *G. sylvestre* (Swami and Prabakaran, 2012). The antimicrobial function of *G. sylvestre* extracted using various solvents is presented in Table 2.4.

TABLE 2.4 Antimicrobial Activity of Various Solvent Gymnema Extracts Against Selected Microbes

SL. No.	Solvent Used for Extraction	Inhibitory Activity Against Microbes	References
1.	Aqueous	*Staphylococcus aureus, C. perfringens type-A, C. kefyr* and *C. krusei*	David and Sudarsanam (2013); Tahir et al. (2017)
	Hexane	*C. perfringens type-A, B. cereus, S. aureus, S. enterica* and *H. paragallinarum*	
	Methanol	*Bacillus cereus, Escherischia coli, Candida kefyr* and *Candida albicans*	
	Ethanolic and aqueous	*Salmonella typhi, S. Typhimurium*, and *S. paratyphi*	
2.	Ethanol	*Bacillus subtilis, Bacillus pumilis, Staphylococcus aureus* and *Psedomonas aeruginosa*	Satdive (2003)
3.	Ethanol, chloroform, and ethyl acetate	*P.* vulgaris, *E. coli, P.* aeroginosa, K. pneumoniae, and *S. aureus*	Pasha et al. (2009); Paul and Jayapriya (2009)
4.	Ethanol	Gram-positive (*Bacillus cereus, Bacillus subtilis, Streptococcus cremoris, Streptococcus fecalis, Staphylococcus aureus*) and five gram-negative (*Klebsiella pneumoniae, Proteus vulgaris, Salmonella paratyphi, Shigella boydi, Shigella dysentriae*)	Senthil Kumar (2015)

TABLE 2.4 *(Continued)*

SL. No.	Solvent Used for Extraction	Inhibitory Activity Against Microbes	References
5.	Petroleum ether, chloroform, ethanol, methanol, and acetone	Gram-positive *(Staphylococcus aureus*) and three Gram-negative *(Escherichia Coli, Klebsiella pneumoniae,* and *Pseudomonas aeroginosa*) bacteria	Swami and Prabakaran (2012)
6.	Ethanol	*E. coli, Pseudomonas aeruginosa, Enterococcus hirae, Bacillus subtilis, Micrococcus luteus, Staphylococcus aureus*	Thanwar et al. (2016)
7.	Chloroform and methanol	*Escherichia coli* and *Klebsiella pneumoniae*	Murugan et al. (2012)
8.	Ethanol	*Bacillus pumilis, B. subtilis, Pseudomonas aeruginosa* and *Staphylococcus aureus*	Satdive et al. (2003)
9.	Methanol	Bacteria *P. aeruginosa, S. aureus,* and the fungus *F. oxysporum*	Gunasekaran et al. (2019)
10.	Aqueous	*Bacillus subtilis, Staphylococcus aureus, Escherichia coli, Klebsiella aerogenes,* and *Aspergillus niger*	Shirish et al. (2017)
11.	Aqueous and ethanol	*Staphylococcus aureus, Escherichia coli,* and *Candida albicanse*	Gupta and Singh (2014)

A variety of microorganisms (*E. hirae, E. coli, B. subtilis, P. aeruginosa, S. aureus, M. luteus* and *C. albicans*) used for test inhibitory function of *G. sylvestre* (Gunasekaran et al., 2019; Arora and Sood, 2017; Thanwar et al., 2016). Disk diffusion method also revealed the antimicrobial activity of Gymnema compounds. Gymnemic acid showed inhibitory zones with *A. niger* (9.25 mm), *V. cholera* (8.65 mm), *E. coli* (8.65 mm), *S. aureus* (7.12 mm), *C. albicans* (6.43 mm), and *S. mutans* (6.00 mm) (Gupta and Singh, 2014). Antimicrobial activity against B. cereus and E. coli of gymnemic acid was reported by Shivanna and Raveesha (2009). For enhanced anti-microbial activity *G. sylvestre* extract added with nanoparticles of Zno and nanofibers of poly-ε-capro lactone (Karthikeyan et al., 2019; Rama-lingam et al., 2019). Gram '+ve' bacteria like *Streptococcus pneumoniae* and *Staphylococcus aureus,* and Gram '-ve' bacteria *Escherichia coli, Proteus vulgaris, K. pneumoniae, Pseudomonas aeruginosa,* and *Shigella dysenteriae* are effectively inhibited by ZnO nanoparticles with Gymnema (Karthikeyan et al., 2019). Whereas Gymnema with poly-ε-capro lactone effective against *S. epidermidis, S. aureus, Escherichia coli* and *Pseudomonas aeruginosa* (Ramalingam et al., 2019).

2.7.4 LIPID LOWERING ACTIVITY

Lipid accumulation in the body is responsible for obesity and diabetes. Reduction of fat content normalizes the body metabolism. The hypolipidemic role is potentially identified in the leaf extract of *G. sylvestre*. In rats, lipid peroxidation reduced due to *G. sylvestre* administration further it is lowered oxidative stress (Ananthan, 2004; Ramkumar et al., 2004, 2008, 2009). Recently, hypolipidemia activity of *G. sylvestre* is tested in high fat diet induced hyperlipidemia female Wistar rats where they found significant effect on reducing the triglycerides, cholesterol and low-density lipoprotein, and elevating high density lipoproteins (Singh et al., 2017). Shigematsu (2001) observed decreased serum triglycerides without altering serum cholesterol after 10 weeks of gymnema leaf extract administration. Observed lowered lipid content in diabetic rats exposed with *G. sylvestre* extract and gymnemic triacetate (Ramkumar et al., 2008; Luo, 2007; Daisy, 2009). Nakamura (1999) demonstrated the excretion of cholesterol and cholic acid derived bile through feces in the rats fed with gymnema extract in a dose-dependent manner. The hypolipidemia in high cholesterol exposed rats treated with *G. sylvestre* leaf extract is reported by Dholi and Raparla (2014), and Rachh et al. (2010). In rats, they found elevated levels of low-density lipoproteins, cholesterol and triglycerides, and low high-density lipoproteins due to one week high cholesterol treatment is reverted with exposure to *G. sylvestre* extract and suggested that hypolipidemia function of this plant is due to holding an acidic components (tannins, flavonoids, saponins, etc.) (Dholi and Raparla, 2014; Rachh et al., 2010). Hypolipidemic function of Gymnema is also tested in diabetic rats. Moreover, lowered cholesterol, low-density lipoproteins, triglycerides, and very low-density lipoproteins are tested in these rats (Kumar et al., 2013).

Clinical studies suggest that Gymnema supplements to diabetic patients who are in regular antidiabetic medication lowered the plasma triglycerides, free fatty acids and cholesterol (Baskaran, 1990b; Kumar,

2010; Shanmugasundaram, 1990a). Low lipid content is also observed in these reports.

Clinical manifestations of *G. sylvestre* well reported in many studies. It has potential to reduce total cholesterol, triglycerides, and sugar levels with elevated blood insulin C-peptide. Shanmugasundaram et al. (1990a) reported significant reduction in levels of plasma glucose and HbA1c in 64 diabetic subjects treated with *G. sylvestre* extract and reduced dose of insulin injection. Reduction in the triglycerides, fasting and post prandial glucose levels in blood were reported in 58 type-2 diabetic subjects treated with *G. sylvestre* for 90 days (Kumar et al., 2010). Li et al. (2015) treated hard capsule of gelatin with leaf powder of *G. sylvestre* to 32 type-2 diabetic patients for 30 days and reported low levels of low-density lipids, cholesterol, and triglycerides.

2.7.5 ANTI-INFLAMMATORY ACTIVITY

Inflammation is an initial condition of chronic disorders. The anti-inflammatory role of *G. sylvestre* is tested in many studies. Porchezhian (2003) reported that gymnema extracts reduce mast cell histamine release in vitro. Carrageenan induced inflammation found to be reduced significantly with the treatment of *G. sylvestre* methanolic extract (Kumar et al., 2012). Diwan (1995) observed lowered rat paw edema induced by carrageenan, but it is less than naproxen induced suppression of edema. Anyhow, in the peritoneal ascites mice, the efficacy of edema inhibition for gymnema and naproxen are almost similar. The only difference between these two treatments is naproxen irritates gastric mucosa and prevents formation of beneficial granuloma. Similarly, Yasukawa et al. (2014) reported the anti-inflammatory function of *G. sylvestre* ethanolic extract in mice suffering with TPA-induced inflammation. About 50% of the inhibitory function of inflammation-induced second stage carcinogenesis was reported in this study.

2.7.6 OPHTHALMIC

Diabetes in humans not only alters the carbohydrate metabolism, but also causes glaucoma in the eyes. Hyperglycemia is a condition of diabetes that induces cataracts in the eyes. These sugar-induced cataracts are get reduced and eyes are protected by a product in combination with *G. sylvestre* extract (Moghaddam, 2005).

2.7.7 WOUND HEALING FUNCTION

The potential role of *G. sylvestre* tested against many disorders. The wound healing role of *G. sylvestre* is tested in rats. At first, Malik et al. (2009) found significant wound healing activity of *G. sylvestre* alcoholic extract in wounded rats. Later Kiranmai et al. (2010) discussed the wound healing function of hydro alcoholic *G. sylvestre* extract. This activity is explained with phytochemicals especially flavonoids initiated free radical scavenging mechanism.

2.7.8 HEPATO PROTECTIVE FUNCTION

Additional role of *G. sylvestre* teste is hepatoprotection. This was first evaluated by Srividy et al. (2010) *in vitro* and *in vivo*. They tested different concentrations of hot macerated *G. sylvestre* hydro alcoholic extracts (200, 400, 600 and 800 μg/ml) incubated with freshly prepared hepatocytes in vitro. Found significant reduction in D-galactosamine induced hepatotoxicity at 200, 400 and 600 μg/ml, whereas 800 μg/ml showed cytotoxicity of hepatocytes. The restoration of D-galactosamine induced damage in hepatocytes treated with 200, 400, 600 and 800 μg/ml hydro alcoholic extract of *G. sylvestre* in a dose dependent manner *in vivo*.

2.7.9 IMMUNOMODULATORY FUNCTION

Immunity an ultimate function of immune system maintains good health of an individual. During attack of pathogens or illness immunomodulators helps to increase the immunity thereby lowers the pathogenicity of pathogen or illness. Gupta et al. (2010) reported the immunomodulatory role of *G. sylvestre*. They tested the immunostimulatory function of Gymnema leaves. It is found that movement of neutrophils in different assays like nitroblue tetrazolium, pahagocytosis and chemotaxis tests in killed *Candida albicans*. Malik et al. (2009) tested remarkable immunostimulatory function by incubating human neutrophils with *G. sylvestre* leaf water extract at different concentrations (10, 25, 50, 100, and 1,000 μg/ml *in vitro*.

2.7.10 ANTI-ARTHRITIC FUNCTION

Arthritis is one of the leading disease-causing severe health problems and its treatment leads to several other side effects. Finding natural drugs against arthritis has great importance at this point of time. Because of this the anti-arthritic function of *G. sylvestre* leaf extract played great importance. Malik et al. (2010) tested the anti-arthritic role of *G. sylvestre* water and petroleum ether extracts. This activity of this plant may be due to the occurrence of saponin glcosides, steroids and triterpenoids. Eric and Lawrence (1996) explained the anti-inflammatory role of this plant at bone and cartilage. The plant extract can cause reduction in paw swelling by blocking pain responsible cytokines TNF-α and IL-1β, and interferons. Further studies need to be established the anti-arthritic activity of *G. sylvestre*.

2.7.11 ADVERSE FUNCTIONS

Nonetheless the *G. sylvestre* has great potential to show beneficiary effects in protecting human health; it has its own limitations and shows adverse effects too. However, no change in behavior, neurological and autonomic effects were observed with *G. sylvestre* administration, but in general Asclepiadaceae (milkweed) family members are allergic to humans and animals. They show allergic cross reactivity in individuals. Ogawa et al. (2004) found no toxic effects in Wistar male and female rats treated with *G. sylvestre* for about 52 weeks. The taste of gymnema is bitter and creates loss of sweetness in the mouth (Ulbricht, 2011). Nevertheless, this plant causes hypoglycemia in both diabetic and non-diabetic individuals (Khare et al., 1983). Furthermore, *G. sylvestre* effect on hypoglycemia is persistent even after the treatment stopped (Srivastava et al., 1986). Hepatotoxicity was reported in patients treated with *G. sylvestre* as tea. This was evidenced with laboratory and hepato histological studies (Shiyovich, 2010). The increase in spontaneous blood pressure was reported in rats fed with *G. sylvestre*. But this study has no further support to explain its clinical importance (Preuss, 1998). The LD50 value of *G. sylvestre*, i.e.,

3,990 mg/kg found in Albino mice showed toxicological effects in normal and diabetic mice with a safety ratio of 11.08 and 16.03 respectively. Whereas LD50 of *G. sylvestre* water and ethanolic extract with 375 mg/kg treated intra-peritoneally in mice was also reported by Bhakuni and Dhar (1971).

2.8 HEALTH BENEFITS

G. sylvestre a plant of holding number of health benefits has lot of importance in pharmaceuticals. Several health benefits were notified with this plant and are mentioned beneath:

- Reduction in the food sweetness (sugar carving) in the mouth by blocking the taste bud sugar receptors.
- Lowers blood hyperglycemia by reducing the sugar absorption in the small intestine through blocking its receptors and this is a major health benefit of this plant.
- The extracts of this plant improves the insulin production in diabetic patients and lowers blood glucose.
- Reduces the risk of heart diseases through reducing the levels of cholesterol, triglycerides, and lipid contents of the blood.
- Dietary inclusion of this plant helps in body weight control and acts as anti-obesity plant by lowering apatite thereby reduced intake of food.
- Saponins and tannins of this plant are responsible for anti-inflammation, which helps in healing the injury or infection caused by hyperglycemia.
- Anti-cancerous function of this plant can act against the tumor growth and ceases the malignancy.
- Plant extract also controls several pathogens through its antimicrobial function.
- Dietary inclusion of this plant (flavonoids and tannins) may help to reduce oxidative stress by elevating antioxidant levels.
- This plant acts against diabetic neuropathy through neuroprotection.
- The extracts of this plant acts as anti-arthritics, hepatoprotective and wound healing medicine in humans.
- Ophthalmic and immune modulatory function of this plant has potential health benefit.

Besides above beneficiary role of *G. sylvestre* many other unknown functions are still under discovery.

2.9 CONCLUSION

Several beneficiary roles are attributed to *G. sylvestre* and are in practice through Ayurveda, Yunani, Folk and Homeopathy medicine but are restricted only to few parts of the globe. It is essential to develop pharmaceutical medicine for the treatment of various ailments in humans and animals. This chapter emphasized various health benefits of *G. sylvestre* and its derived compounds. The mechanistic actions of these compounds against various ailments are not yet completely done. Revolutionized studies need to be initiated to develop *G. sylvestre* based medicine pharmacologically. Hopefully anticipating new findings for various ailments from this plant in the near future.

KEYWORDS

- **breast cancer resistance protein**
- ***Gymnema sylvestre***
- **high-performance liquid chromatography**
- **mass spectroscopic**
- **pharmacology**
- **phytochemistry**

REFERENCES

Agrawal, R. C., Soni, S., Jain, N., Rajpoot, J., & Maheshwari, S. K., (2016). Chemopreventive effect of *Gymnema sylvestre* in Swiss albino mice. *Int. J. Sci. Res. Publications, 6,* 78–83.

Ahmed, A. B., Rao, A., & Rao, M., (2010). *In vitro* callus and *in vivo* leaf extract of *Gymnema sylvestre* stimulate β-cells regeneration and antidiabetic activity in Wistar rats. *Phytomed., 17*(13), 1033–1039.

Al-Romaiyan, A., Liu, B., Asare-Anane, H., Maity, C. R., Chatterjee, S. K., Koley, N., Biswas, T., et al., (2010). A novel *Gymnema sylvestre* extract stimulates insulin secretion from human islets *In vivo* and *In vitro*. *Phytother. Res., 24*(9), 1370–1376.

Ananthan, R., Latha, M., Ramkumar, K. M., Pari, L., Baskar, C., & Narmatha, B. V., (2004). Modulatory effects of *Gymnema montanum* leaf extract on alloxan-induced oxidative stress in Wistar rats. *Nutrition, 20*(3), 280–285.

Anis, M., Sharma, M. P., & Iqbal, M., (2000). Herbal ethnomedicine of the Gwalior Forest division in Madhya Pradesh, India. *Pharm. Biol., 38*(4), 241–253.

Arai, K., Ishima, R., Morikawa, S., Miyasaka, A., Imoto, T., Yoshimura, S., Aimoto, S., & Akasaka, K., (1995). Three-dimensional structure of gurmarin, a sweet taste-suppressing polypeptide. *J. Biomol. NMR, 5*(3), 297–305.

Aralelimath, V. R., & Bhise, S. B., (2012). Antidiabetic effects of *Gymnema sylvestre* extract on streptozotocin induced diabetic rats and possible b-cell protective and regenerative evaluations. *Dig. J. Nanomater. Bios., 7*(1), 135–142.

Arora, D. S., & Sood, H., (2017). *In vitro* antimicrobial potential of extracts and phytoconstituents from *Gymnema sylvestre* R Br. leaves and their biosafety evaluation. *AMB Express, 7*(1), 115.

Arun, L. B., Arunachalam, A. M., Arunachalam, K. D., Annamalai, S. K., & Kumar, K. A., (2014). *In vivo* anti-ulcer, anti-stress, anti-allergic, and functional properties of gymnemic acid isolated from *Gymnema sylvestre* R Br. *BMC Complement Altern. Med., 14*(70).

Arunachalam, K. D., Arun, L. B., Annakalai, S. K., & Arunachalam, A. M., (2015). Potential anticancer properties of bioactive compounds of *Gymnema sylvestre* and its biofunctionalized silver nanoparticles. *Int. J. Nanomed., 10,* 31–41.

Bahadoran, Z., Mirmiran, P., & Aziz, F., (2013). Dietary polyphenols as potential nutraceuticals in management of diabetes: A review. *J. Diabetes Metab. Disord., 12*(1), 1–9.

Baskaran, K., Ahamath, B. K., Shanmugasundaram, K. R., & Shanmugasundaram, E. R. B., (1990b). Antidiabetic effect of a leaf extract from *Gymnema sylvestre* in non-insulin-dependent diabetes mellitus patients. *J. Ethnopharmacol., 30*(3), 295–305.

Baskaran, K., KizarAhamath, B., Radha. S. K., & Shanmugasundaram, E. R., (1990a). Antidiabetic effect of a leaf extract from *Gymnema sylvestre* in non-insulin-dependent diabetes mellitus patients. *J. Ethnopharmacol., 30*(3), 295–300.

Bhakuni, D. S., & Dhar, M. L., (1971). Screening of Indian plants for biological activity. Part III. *Indian J. Exp. Biol.,* 9, 91–102.

Bhansali, S., Shafiq, N., Pandhi, P., et al., (2013). Effect of a deacyl gymnemic acid on glucose homeostasis and metabolic parameters in a rat model of metabolic syndrome. *Indian J. Med. Res., 137*(6), 1174–1179.

Boericke, W., (1996). *Pocket Manual of Homoeopathic Materia Medica and Repertory* (pp. 338–339). Rep. Ed. New Delhi. B. Jain publishers (P) Ltd.

Bone, K., (2002). *Gymnema: A Key Herb in the Management of Diabetes – in Phytotherapy Review and Commentary*, National Institute of Herbalists. National Herbalists Association of Australia.

Chakraborty, D., Samrat, G., Kausik, B., Avinaba, M., Sourav, S., & Khuda-Bukhsh, A. R., (2013). Antihyperglycemic drug *Gymnema sylvestre* also shows anticancer potentials in human melanoma a375 cells via reactive oxygen species generation and mitochondria-dependent caspase pathway. *Integr. Cancer Therapies., 12*(5), 433–441.

Chakrapani, L. N., & Periandavan, K., (2018). Protective role of gymnemic acid in curbing high fat diet and high fructose induced pancreatic oxidative stress mediated type-2 diabetes in Wistar rats. *Int. J. Pharm. Sci. Res., 9*(5), 2130–2139.

Chakravarti, D., & Debnath, N. B., (1981). Isolation of gymnemagenin the sapogenin from *Gymnema sylvestre* R.Br. (*Asclepiadaceae*). *J. Inst. Chemists., 53,* 155–158.

Chattopadhyay, R. R., (1999). A comparative evaluation of some blood sugar lowering agents of plant origin. *J. Ethnopharmacol., 67*(3), 367–372.

Chen, G., & Guo, M., (2017). Rapid screening for α-glucosidase inhibitors from *Gymnema sylvestre* by affinity ultrafiltration–HPLC-MS. *Front. Pharmacol., 8,* 228.

Chopra, R. N., Nayar, S. L., & Chopra, I. C., (2002). *Glossary of Indian Medicinal Plants.* National institute of science communication and information resources, New Delhi, India.

Daisy, P., Eliza, J., & Farook, K. A., (2009). A novel dihydroxy gymnemic triacetate isolated from *Gymnema sylvestre* possessing normoglycemic and hypolipidemic activity on STZ-induced diabetic rats. *J. Ethnopharmacol., 126*(2), 339–344.

Dateo, G. P. Jr., & Long, L. Jr., (1973). Gymnemic acid, the anti-saccharine principle of *Gymnema sylvestre*. studies on the isolation and heterogeneity of gymnemic acid A1. *J. Agricult. Food Chem., 21*(5), 899–903.

David, B. C., & Sudarsanam, G., (2013). Antimicrobial activity of *Gymnema sylvestre* (*Asclepiadaceae*). *J. Acute. Dis., 2*(3), 222–225.

Dholi, S. K., & Raparla, R. K., (2014). *In vivo* antidiabetic evaluation of gymnemic acid in streptozotocin induced rats. *J. Pharm. Innov., 3*(7), 82–86.

Diwan, P. V., Margaret, I., & Ramakrishna, S., (1995). Influence of *Gymnema sylvestre* on inflammation. *Inflammopharmacol., 3*(3), 271–277.

Eric, G. B., & Lawrence, J. L., (1996). *Rheumatoid Arthritis and Its Therapy—The Text Book of Therapeutics: Drug and Disease Management* (16th edn.). Williams and Wilkins Company, Baltimore, Md, USA.

Eric, G. B., & Lawrence, J. L., (1996). *Rheumatoid Arthritis and its Therapy-the Text Book of Therapeutics: Drug and Disease Management.* Baltimore, Md, USA. Williams and Wilkins Company. (16th edn, Vol. 16).

Fatani, A. J., Al-Rejaie, S. S., Abuohashish, H. M., Al-Assaf, A., Parmar, M. Y., Ola, M. S., & Ahamed, M. M., (2015). Neuroprotective effects of *Gymnema sylvestre* on streptozotocin-induced diabetic neuropathy in rats. *Exp. Ther. Med., 9*(5), 1670–1678.

Fletcher, J. I., Dingley, A. J., Smith, R., Connor, M., Christie, M. J., & King, G. F., (1999). High-resolution solution structure of gurmarin, a sweet-taste-suppressing plant polypeptide. *Eur. J. Biochem., 264*(2), 525–533.

Foster, S., (2002). *Gymnema sylvestre – in Alternative Medicine Reviews Monographs* (pp. 205–207). Thorne Research Inc.

Gent, J. F., Hettinger, T. P., Frank, M. E., & Marks, L. E., (1999). Taste confusions following gymnemic acid rinse. *Chem. Sens., 24*(4), 393–403.

Gholap, S., & Kar, A., (2003). Effects of *Inular acemosa* root and *Gymnema sylvestre* leaf extracts in the regulation of corticosteroid induced diabetes mellitus: Involvement of thyroid hormones. *Pharmazie., 58*(6), 413–415.

Ghose, S. C., (1984). *Drugs of Hindoosthan* (8th edn.). Calcutta: Hahnemann P publishing Co. Pvt. Ltd.

Gunasekaran, V., Srinivasan, S., & Rani, S. S., (2019). Potential antioxidant and antimicrobial activity of *Gymnema sylvestre* related to diabetes. *J. Med. Plants, 7*(2), 5–11.

Gupta, R. B. P., & Singh, P., (2014). Antimicrobial activity of gymnemic acid on pathogens-*Gymnema sylvestre*. *Int. J. Curr. Microbiol. App. Sci., 3*(5), 40–45.

Gupta, S. N., Pramanik, S., Tiwari, O. P., Thacker, N., Pande, M., & Upmanyu, N., (2010). immunomodulatory activity of *Gymnema sylvestre* leaves. *Inter. J. of Pharmacol., 8*(2), 1531–2976.

Gupta, S. S., & Seth, C. B., (1962). Experimental studies on pituitary diabetes. II. Comparison of blood sugar level in normal and anterior pituitary extract induced hyperglycemic rats treated with a few Ayurvedic remedies. *Ind. J. Med. Res., 50,* 708–714.

Gupta, S. S., & Variyar, M. C., (1964). Experimental studies on pituitary diabetes. IV. Effect of *Gymnema sylvestre* and *Coccinia indica* against the hyperglycaemic response of somatotropin and corticotropin hormones. *Ind. J. Med. Res., 52,* 200–207.

Hanhineva, K., Torronen, R., Bondia-Pons, I., Pekkinen, J., Kolehmainen, M., Mykkanen, H., & Poutanen, K., (2010). Impact of dietary poly-phenols on carbohydrate metabolism. *Int. J. Mol. Sci., 11*(4), 1365–1402.

Harada, S., & Kasahara, Y., (2000). Inhibitory effect of gurmarin on palatal taste responses to amino acids in the rat. *Am. J. Physiol. Regul. Integr. Comp. Physiol., 278*(6), R1513–R1517.

Haridas, V., Higuchi, M., Jayatilake, G. S., Bailey, D., Mujoo, K., Blake, M. E., Arntzen, C. J., & Gutterman, J. U., (2001). Avicins: Triterpenoid saponins from *Acacia victoriae* (Bentham) induce apoptosis by mitochondrial perturbation. *PNAS, 98*(10), 5821–5826.

Imai, Y., Tsukahara, S., Asada, S., & Sugimoto, Y., (2004). Phytoestrogens/flavonoids reverse breast cancer resistance protein/ABCG2-mediated multidrug resistance. *Cancer Res., 64*(12), 4346–4352.

Imoto, T., Miyasaka, A., Ishima, R., & Akasaka, K., (1991). Anovel peptide isolated from the leaves of *Gymnema sylvestre*-I. Characterization and its suppressive effect on the neural responses to sweet taste stimuli in the rat. *Comp. Biochem. Physiol., 100*(2), 309–314.

Ishijima, S., Takashima, T., Ikemura, T., & Izutani, Y., (2008). Gymnemic acid interacts with mammalian glycerol-3-phosphate dehydrogenase. *Mole. Cellu. Biochem., 310*(1, 2), 203–208.

Joffe, D. J., & Freed, S. H., (2001). Effect of extended release *Gymnema sylvestre* leaf extract (*Beta Fast G-XR*) alone or in combination with oral hypoglycemics or insulin regimens for type 1 and type 2 diabetes. *Diabet. Cont. Newslett., 76*(1), 23–24.

Johtatsu, T., Andoh, A., Kurihara, M., Iwakawa, H., Tsujikawa, T., Kashiwagi, A., Fujiyama, Y., et al., (2007). Serum concentrations of trace elements in patients with Crohn's disease receiving enteral nutrition. *J. Clin. Biochem. Nutr., 19,* 137–139.

Kanetkar, P., Singhal, R., & Kamat, M., (2007). *Gymnema sylvestre*: A memoir. *J. Clin. Biochem. Nutr., 41*(2), 77–81.

Karthikeyan, M., Jafar, A. A., Karthikeyan, C., & Vijaya, K. P., (2019). Enhancement of antibacterial and anticancer properties of pure and REM doped ZnO nanoparticles synthesized using *Gymnema sylvestre* leaves extract. *SN Appl. Sci., 1,* 355.

Katsukawa, H., Imoto, T., & Ninomiya, Y., (1999). Induction of salivary gumarin-binding proteins in rats fed *gymnema*-containing diets. *Chem. Sens., 24*(4), 387–392.

Kazuko, Y., Yuuji, K., Shigenobu, A., Kouji, M., Toshio, M., Liu, H. M, et al., (1992). Dammarane saponins from *G. sylvestre*. *Phytochem., 31*(1), 237–241.

Kazutaka, M., Shigeo, T., Tadashi, F., Goro, K., Shanti, P. U., Mitsunori, K., Ichimoto, I., & Nkano, Y., (1993). Isolation of conduritol A from *Gymnema sylvestre* and its effect against intestinal glucose absorption in rats. *Biosc. Bitechnol. Biochem., 57*(12), 2184–2185.

Khanna, G., (2010). Non-proliferative activity of saponins isolated from the leaves of *Gymnema sylvestre* and *Eclipta prostrata*on HepG2 cells-*in vitro* study. *Int. J. Pharm. Sci. Res., 1*(8), 38–42.

Khanna, V., & Kannabiran, K., (2009). Anticancer-cytotoxic activity of saponins isolated from the leaves of *Gymnema sylvestre* and *Eclipta prostrata* on *HeLa* cells. *Int. J. Green Pharm., 3*(3), 227–229.

Khare, A. K., Tondon, R. N., & Tewari, J. P., (1983). Hypoglycaemic activity of an indigenous drug (*Gymnema sylvestre*, "Gurmar") in normal and diabetic persons. *Ind. J. Physiol. Pharm., 27*, 257–258.

Khramov, V. A., Spasov, A. A., & Samokhina, M. P., (2008). Chemical composition of dry extracts of *Gymnema sylvestre* leaves. *Pharmaceut. Chem. J., 42*(1), 30–32.

Kim, H. J., Hong, S. H., Chang, S. H., Kim, S., Lee, A. Y., Jang, Y., Davaadamdin, O., et al., (2016). Effects of feeding a diet containing *Gymnema sylvestre* extract: Attenuating progression of obesity in C57BL/6J mice. *Asian Pac. J. Trop. Med., 9*(5), 437–444.

Kiranmai, M., Kazim, S. M., & Ibrahim, M., (2011). Combined wound healing activity of *Gymnema sylvestre* and *Tagetes erectalinn. Int. J. Pharmaceut. Appl., 2,* 135–140.

Kirtikar, K. R., & Basu, B. D., (1975). *Indian Medicinal Plants* (Vol. III). In Periodical Experts D-42. Vivek Vihar Publish, Delhi.

Kishore, L., & Singh, R., (2015). Protective effect of *Gymnema sylvestre* L. Against advanced glycation end-product, sorbitol accumulation and aldose reductase activity in homoeopathic formulation. *Ind. J. Res. Homoeopat., 9*(4), 240–248.

Kishore, L., & Singh, R., (2017). Preventive effect of *Gymnema sylvestre* homeopathic preparation on streptozotocin-nicotinamide induced diabetic nephropathy in rats. *Orient. Pharm. Exp. Med., 17,* 223–232.

Komalavalli, N., & Rao, M. V., (2000). *In vitro* micropropagation of *Gymnema sylvestre*: Multipurpose medicinal plant. *Plant Cell Tiss. Org. Cult., 61,* 97–105.

Kumar, A. R., Rathinam, K. S., & Kumar, C. A., (2012). Evaluation of anti-inflammatory activity of some selected species of *Asclepiadaceae* family. *Int. J. Chem. Sci.,10*(1), 548–556.

Kumar, S. N., Mani, U. V., & Mani, I., (2010). An open label study on the supplementation of *Gymnema sylvestre* in type 2 diabetics. *J. Die. Suppl, 7*(3), 273–282.

Kumar, V., Bhandari, U., Tripathi, C., & Khanna, G., (2013). Anti-obesity effect of *Gymnema sylvestre* extract on high fat diet-induced obesity in Wistar rats. *Drug Res., 63*(12), 625–632.

Kurihara, Y., (1992). Characteristics of anti-sweet substances, sweet proteins and sweetness inducing protein. *Crit. Rev. Food Sci. Nutr., 32*(3), 231–252.

Leach, M. J., (2007). *Gymnema sylvestre* for diabetes mellitus: A systematic review. *J. Altern. Comple. Med., 13*(9), 977–983.

Li, Y., Zheng, M., Zhai, X., Huang, Y., Khalid, A., Malik, A., Shah, P., et al., (2015). Effect of *Gymnema sylvestre*, *Citrullus colocynthis* and *Artemisia absinthium* on blood glucose and lipid profile in diabetic human. *Acta. Pol. Pharm., 72,* 981–985.

Liu, B., Asare-Anane, H., Al-Romaiyan, A., Huang, G., Amiel, S. A., Jones, P. M., & Persaud, S. J., (2009). Characterization of the insulinotropic activity of an aqueous extract of *Gymnema sylvestre* in mouse beta-cells and human islets of Langerhans. *Cell Physiol. Biochem., 23*(1–3), 125–132.

Luo, H., Kashiwagi, A., Shibahara, T., & Yamada, K., (2007). Decreased body weight without rebound and regulated lipoprotein metabolism by gymnemate in genetic multifactor syndrome animal. *Mol. Cell Biochem., 299*(1, 2), 93–98.

Malik, J. K., Manvi, F. V., Nanjware, B. R., Dwivedi, D. K., Purohit, P., & Chouhan, S., (2009). Wound healing properties of alcoholic extract of *Gymnema sylvestre* R.Br. leaves in rats. *J. Pharm. Res., 2,* 1029–1030.

Malik, J. K., Manvi, F. V., Nanjware, B. R., Dwivedi, Purohit, P., & Chouhan, S., (2010). Anti-arthritic activity of leaves of *Gymnema sylvestre* R. Br. leaves in rats. *Der. Pharmacia. Lettr., 2,* 336–341.

Mao, Q., (2005). Role of the breast cancer resistance protein (ABCG2) in drug transport. *AAPS J., 7*(1), E118–E133.

Mathew, M., (2004). Aromatic and medicinal plants research station, odakkali-A centre for promoting medicinal and aromatic plants. *Ind. Coconut J. Cochin, 34*(10), 10–15.

Moghaddam, M. S., Kumar, P. A., Reddy, G. B., & Ghole, V. S., (2005). Effect of diabecon on sugar-induced lens opacity in organ culture: Mechanism of action. *J. Ethnopharmacol., 97*(2), 397–403.

Murugan, M., Mohan, V. R., & Thamodharan, V., (2012). Phytochemical screening and antibacterial activity of *Gymnema sylvestre* (Retz) R. Br ex. Schultes and *Morinda pubescens* J.E. Smith var. *pubescens. J. Appl. Pharmaceut. Sci., 2*(2), 73–76.

Nadkarni, K. M., (1986). *Gymnema Sylvestre, in Indian Materia Medica with Ayurvedic Unani* (Vol. I, pp. 596–599) (Bombay, India: Popular Prakashan).

Nadkarni, K. M., (1996). *Indian Materia medica, in Dr. KM Nadkarni's Indian Material Medica: With Ayurvedic, Unani-Tibbi, Siddha, Allopathic, Homeopathic, Naturopathic & Home Remedies, Appendices & Indexes* (Vol. 1). Popular Prakashan.

Nakamura, Y., Tsumura, Y., Tonogai, Y., & Shibata, T., (1999). Fecal steroid excretion is increased in rats by oral administration of gymnemic acids contained in *Gymnema sylvestre* leaves. *J. Nutr., 129*(6), 1214–1222.

Ninomiya, Y., & Imoto, T., (1995). Gurmarin inhibition of sweet taste responses in mice. *Am. J. Physiol., 268*(4), R1019-25.

Ogawa, Y., Sekita, K., Umemura, T., Saito, M., Ono, A., Kawasaki, Y., Uchida, O., Matsushima, Y., Inoue, T., & Kanno, J., (2004). *Gymnema sylvestre* leaf extract: A 52-week dietary toxicity study in Wistar rats. *Shokuhin Eiseigaku Zasshi., 45*(1), 8–18.

Ohmori, R., Iwamoto, T., Tago, M., Takeo, T., Unno, T., Itakura, H., & Kondo, K., (2005). Antioxidant activity of various teas against free radicals and LDL oxidation. *Lipids, 40*(8), 849–853.

Park, H. J., Kwon, S. H., Lee, J. H., Lee, K. H., Miyamoto, K. I., & Lee, K. T., (2001). Kalopanaxsaponin A is a basic saponin structure for the anti-tumor activity of hederagenin monodesmosides. *Planta Medica., 67*(2), 118–121.

Pasha, C., Sayeed, S., Ali, M. S., & Khan, M. Z., (2009). Anti-salmonella activity of selected medicinal plants. *Turkish J. Biol., 33,* 59–64.
Patel, S. S., Shah, R. S., & Goyal, R. K., (2009). Antihyperglycemic, antihyperlipidemic and antioxidant effects of Dihar, a polyherbal ayurvedic formulation in streptozotocin induced diabetic rats. *Ind. J. Experiment. Biol., 47*(7), 564–570.
Paul, J., & Jayapriya, K. P., (2009). Screening of antibacterial effects of *Gymnema sylvestre* (L.) R.Br. – A medicinal plant. *Pharmacol. Online., 3,* 832–836.
Pawan, K. S., Munawwar, H. K., JavedInam, S., & Aslam, S., (2015). Pharmacognostical and physicochemical standardization of multiple samples of gurmar Buti leaves (*Gymnema sylvestre* R.Br.) having immense therapeutic values. *Int. J. Curr. Res. Biosci. Plant Biol., 2*(10), 9–17.
Persaud, S. J., Al-Majed, H., Raman, A., & Jones, P. M., (1999). *Gymnema sylvestre* stimulates insulin release *in vitro* by increased membrane permeability. *J. Endocrinol., 163*(2), 207–212.
Porchezhian, E., & Dobriyal, R. M., (2003). An overview on the advances of *Gymnema sylvestre*: Chemistry, pharmacology and patents. *Pharmazie., 58*(1), 5–12.
Potawale, S. E., Shinde, V. M., Libi, A., Boradem, S., Dhalawatm, H., & Deshmukh, R. S., (2008). *Gymnema sylvestre*: A comprehensive review. *Pharmacol. Online, 2,* 144–157.
Pothuraju, R., Sharma, R. K., Chagalamarri, J., Jangra, S., & Kumar, K. P., (2014). A systematic review of *Gymnema sylvestre* in obesity and diabetes management. *J. Sci. Food Agric., 94*(5), 834–840.
Preuss, H. G., Bagchi, D., Bagchi, M., Rao, C. V., Dey, D. K., & Satyanarayana, S., (2004). Effects of a natural extract of hydroxycitric acid (HCA-SX) and a combination of HCA-SX plus niacin-bound chromium and *Gymnema sylvestre* extract on weight loss. *Diabetes Obes. Metab., 6,* 171–180.
Puratchimani, V., & Jha, S., (2004). Standardization of *Gymnema sylvestre* r. Br. with reference to gymnemagenin by high-performance thin-layer chromatography. *Phytochem. Anal., 15*(3), 164–166.
Rachh, P. R., Patel, S. R., Hirpara, H. V., Rupareliya, M. T., Rachh, M. R., Bhargava, A. S., Patel, N. M., & Modi, D. C., (2009). In-vitro evaluation of antioxidant activity of *Gymnema sylvestre* R.Br. leaf extract. *Rom. J. Biol. Plant Biol., 54*(2), 141–148.
Rachh, P. R., Rachh, M. R., Ghadiya, N. R., Modi, D. D., Modi, K. P., Patel, N. M., & Rupareliya, M. T., (2010). Antihyperlipidemic activity of *Gymnema sylvestre* R. Br. leaf extract on rats fed with high cholesterol diet. *Int. J. Pharmacol., 6,* 138–141.
Rahman, A. U., & Zaman, K., (1989). Medicinal plants with hyphoglycemic activity. *J. Ethnopharmacol., 26*(1), 1–55.
Rahman, M. M., Habib, M. R., Hasan, M. A., Saha, A., & Mannan, A., (2014). Comparative assessment on *in vitro* antioxidant activities of ethanol extracts of *Averrhoa bilimbi, Gymnema sylvestre* and *Capsicum frutescens. Pharmacog. Res., 6,* 36–41.
Raju, V. S., Kannababu, S., & Subbaraju, G. V., (2006). Standardization of *Gymnema sylvestre* R.Br. by high-performance thin-layer chromatography: An improved method. *Phytochem. Anal., 17*(3), 192–196.
Rakshit, G., Singh, J. P., Pathak, S. D., Banoth, C. K., Chandra, P. K., Rajpal, Choudhury, S. K., & Singh, V. K., (2013). A multi-centric double-blind randomized homoeopathic pathogenetic trial of *Gymnema sylvestre. Ind. J. Res. Homoeopathy., 7,* 9–21.

Ramalingam, R., Dhand, C., Leung, C. M., Ong, S. T., Annamalai, S. K., Kamruddin, M., Verma, N. K., et al., (2019). Antimicrobial properties and biocompatibility of electrospun poly-ε-caprolactone fibrous mats containing *Gymnema sylvestre* leaf extract. *Mat. Sci. Eng., 98,* 503–514.

Ramkumar, K. M., Latha, M., Venkateswaran, S., Pari, L., Ananthan, R., & Bai, V. N., (2004). Modulatory effect of *Gymnema montanum* leaf extract on brain antioxidant status and lipid peroxidation in diabetic rats. *J. Med. Food., 7*(3), 366–371.

Ramkumar, K. M., Ponmanickam, P., Velayuthaprabhu, S., Archunan, G., & Rajaguru, P., (2009). Protective effect of *Gymnema montanum* against renal damage in experimental diabetic rats. *Food. Chem. Toxicol., 47*(10), 2516–2521.

Ramkumar, K. M., Rajaguru, P., Latha, M., & Ananthan, R., (2008). Effect of *Gymnema montanum* leaves on red blood cell resistance to oxidative stress in experimental diabetes. *Cell Biol. Toxicol., 24*(3), 233–241.

Rupanar, S. V., Pingale, S. S., Dandge, C. N., & Kshirsagar, D., (2012). Phytochemical screening and *In vitro* evaluation of antioxidant antimicrobial activity of *Gymnema sylvestre. Int. J. curr. Res., 8*(12), 43480–43486.

Sahu, N. P., Mahato, S. B., Sarkar, S. K., & Poddar, G., (1996). Triterpenoid saponins from *Gymnema sylvestre. Phytochem., 41*(4), 1181–1185.

Salem, S. A., Hassan, D. R., & Mowafy, A. R., (2009). A comparative impact of different types of a single antioxidant supplementation (B-carotene, alfa tocopherols and ascorbic acids) on lipid profile in hyperlipidemic rats. *Middle-east J. Sci. Res., 4,* 354–360.

Sastry, B. S., (1994). *Gymnema sylvestre. Chaukhambha.* Varanasi, India: Bhav Prakash Nighantu.

Satdive, R. K., Abhilash, P., & Fulzele, D. P., (2003). Antimicrobial activity of *Gymnema sylvestre* leaf extract. *Fitoterapia., 74*(*7,* 8), 699–701.

Sathya, A., Ramasubramaniaraja, R., & Brindha, P., (2010). Pharmacognostical, phytochemical and GC-MS investigation of successive extract of *Gymnema sylvestre* R. Br. *J. Pharma. Res., 3,* 984–987.

Senthilkumar, M., (2015). Phytochemical screening and antibacterial activity of *Gymnema sylvestre*r br. ex schult. *IJPSR., 6*(6), 2496–2503.

Shanmugasundaram, E., Gopinath, K., Shanmugasundaram, K., & Rajendran, V., (1990b). Possible regeneration of the islets of Langerhans in streptozotocin diabetic rats given *Gymnema sylvestre* leaf extracts. *J. Ethnopharmacol., 30*(3), 265–279.

Shanmugasundaram, E., Rajeswari, G., Baskaran, K., Kumar, B., Shanmugasundaram, K., & Ahmath, B., (1990a). Use of *Gymnema sylvestre* leaf extract in the control of blood glucose in insulin-dependent diabetes mellitus. *J. Ethnopharmacol., 30*(3), 281–294.

Shigematsu, N., Asano, R., Shimosaka, M., & Okazaki, M., (2001). Effect of administration with the extract of *Gymnema sylvestre* R. Br leaves on lipid metabolism in rats. *Biol. Pharm. Bull., 24*(6), 713–717.

Shigematsu, N., Asano, R., Shimosaka, M., & Okazaki, M., (2001b). Effect of long term-administration with *Gymnema sylvestre* R. BR on plasma and liver lipid in rats. *Biol. Pharm. Bull., 24*(6), 643–649.

Shimizu, K., Iino, A., Nakajima, J., Nakajyo, S., Urakawa, N., Atsuchi, M., Wada, T., & Yamashita, C., (1997). Suppression of glucose absorption by some fractions extracted from *Gymnema sylvestre* leaves. *J. Vet. Med. Sci., 59*(4), 245–251.

Shimizu, K., Ozeki, M., Iino, A., Nakajyo, S., Urakawa, N., & Atsuchi, M., (2001). Structure-activity relationships of triterpenoid derivatives extracted from *Gymnema inodorum* leaves on glucose absorption. *Japane. J. Pharmacol., 86*(2), 223–229.

Shirish, S., Pingale, Shobha, V. R., & Manohar, G. C., (2017). Evaluation of antimicrobial activity of *Gymnema sylvestre. Int. Res. J. Pharm., 8*(3), 10–12.

Shivanna, Y., & Raveesha, K. A., (2009). *In vitro* antibacterial effect of selected medicinal plant extracts. *J. Nat. Prod., 2,* 64–69.

Shiyovich, A., Nesher, L., & Sztarkier, I., (2010). Toxic hepatitis induced by *Gymnema sylvestre*, a natural remedy for type 2 diabetes mellitus. *Am. J. Med. Sci.,* 340(6), 514–517.

Singh, D. K., Kumar, N., Sachan, A., Lakhani, P., Tutu, S., Nath, R., Sachan, A. K., & Dixit, R. K., (2017). Hypolipidaemic effects of *Gymnema sylvestre* on high fat diet induced dyslipidaemia in Wistar rats. *J. Clin. Diagn, Res., 11*(5), FF01–FF05.

Singh, V. K., Umar, S., Ansari, S. A., & Iqbal, M., (2008). *Gymnema sylvestre* for diabetics. *J. Herbs. Spices Med. Plants, 14*(1, 2), 88–106.

Spasov, A. A., Samokhina, M. P., & Bulanov, A. E., (2008). Antidiabetic properties of *Gymnema sylvestre* (a review). *Pharm. Chem. J., 42*(11), 626–629.

Srikanth, A. V., Sayeeda, M., Lakshmi, N., Ravi, M., Kumar, P., & Madhava, R. B., (2010). Anticancer activity of *Gymnema sylvestre* R.Br. *Int. J. Pharm. Sci.Nanotech., 3,* 897–899.

Srivastava, Y., Venkatakrishna-Bhatt, H., Jhala, C. I., Nigam, S. K., Kumar, A., & Verma, Y., (1986). Oral *Gymnema sylvestre* R. Br. leaf extracts inducing protracted longevity and hypoglycemia in alloxan diabetic rats: Review and experimental study. *Int. J. Crude Drug Res., 24*(4), 171–176.

Srividya, A. R., Varma, S. K., Dhanapal, S. P., Vadivelan, R., & Vijayan, P., (2010). *In vitro* and *in vivo* evaluation of hepato protective activity of *Gymnema sylvestre. Int. J. Pharmaceut. Sci. Nanotechnol., 2,* 768–773.

Stoecklin, W., (1968). Glycosides and aglycons. CCCIX. Gymnestrogenin. A new pentahydroxy triterpene from *Gymnema sylvestre* leaves. *Helvetica Chimica. Acta., 51*(6), 1235–1242.

Stoecklin, W., Weiss, E., & Reichstein, T., (1967). Glycosides and aglycons. CCLXXXVIII. Gymnemic acid, the antisaccharin principle of *Gymnema sylvestre* – Isolation and identification. *Helvetica Chimica. Acia., 50*(2), 474–490.

Sugihara, Y., Nojima, H., Matsuda, H., Murakami, T., Yoshikawa, M., & Kimura, I., (2000). Antihyperglycemic effects of gymnemic acid IV, a compound derived from *Gymnema sylvestre* leaves in streptozotocin-diabetic mice. *J. Asian Nat. Prod. Res., 2*(4), 321–327.

Swami, J., & Prabakaran, (2012). Studies on antibacterial activity of *Gymnema sylvestre* against respiratory infection causing bacteria. *Int. J. Curr. Advan. Res., 1,* (1), 1–4.

Tahir, M., Rasheed, M. A., Niaz, Q., Ashraf, M., Anjum, A. A., & Ahmed, M. U., (2017). Evaluation of antibacterial effect of *Gymnema sylvestre* R. Br. species cultivated in Pakistan. *Pak. Vet. J., 37*(3), 245–250.

Tamaki, H., Satoh, H., Hori, S., Ohtani, H., & Sawada, Y., (2010). Inhibitory effects of herbal extracts on breast cancer resistance protein (BCRP) and structure inhibitory potency relationship of isoflavonoids. *Drug Metab. Pharmacok., 25*(2), 170–179.

Thanwar, M., Dhananjay, D., Gharia, A. K., & Supriya, C., (2016). Antibacterial study of *Gymnema sylvestre* plant. *Int. J. Chem. Stud., 4*(3), 80–83.

The Wealth of India: A Dictionary of Indian Raw materials and Industrial products. Raw materials, vol. IV. Council of Scientific and Industrial Research, New Delhi, pp. 276–277, 1956.

Tiwari, P., Mishra, B. N., & Sangwan, N. S., (2014). Phytochemical and pharmacological properties of *Gymnema sylvestre*: An important medicinal plant. *BioMed. Res. Int.* Article ID 830285.

Tiwari, P., Sabir, F., Asha, & Sangwan, N. S., (2009). *Gymnema Sylvestre Glycosyl Transferase mRNA, Partial CDS NCBI Database* (accession no. GU181368).

Trease, G. E., & Evans, W. C., (2002). *Pharmacognosy* (15th edn., pp. 42–44, 221–229, 246–249, 304–306, 331, 332, 391–393). Saunders Publishers, London.

Ulbricht, C., Abrams, T. R., Basch, E., Davies-Heerema, T., Foppa, I., Hammerness, P., Rusie, E., et al., (2011). An evidence-based systematic review of gymnema (*Gymnemasylvestre* R. Br.) by the natural standard research collaboration. *J. Diet Suppl., 8*(3), 311–330.

Woodgate, D., & Conquer, J. A., (2003). Effects of a stimulant –free dietary supplement on body weight and fat loss in obese adults: A six-week exploratory study. *Curr. Ther. Res. Clin. Exp., 64*(4), 248–262.

World Health Organization., (1978). *Primary Health Care*. Geneva: Alma-Ata.

Wu, X., Mao, G., Fan, Q., Zhao, T., Zhao, J., Li, F., & Yang, L., (2012). Isolation, purification, immunological and anti-tumor activities of polysaccharides from *Gymnema sylvestre*. *Food. Res. Int., 48*(2), 935–939.

Yadav, M., Lavania, A., Tomar, R., Prasad, G. B., Jain, S., & Yadav, H., (2010). Complementary and comparative study on hypoglycemic and antihyperglycemic activity of various extracts of *Eugenia jambolana seed, Momordica charantia* fruits, *Gymnema sylvestre*, and *Trigonella foenum graecum* seeds in rats. *Appl. Biochem. Biotechnol., 160*(8), 2388–2400.

Yasukawa, K., Okuda, S., Nobushi, Y., (2014). Inhibitory effects of *Gymnema* (*Gymnema sylvestre*) leaves on tumor promotion in two-stage mouse skin carcinogenesis. *J. Evid. Based Complement. Altern. Med.* Article ID, 328684.

Yoshikawa, K., Arihara, S., & Matsuura, K., (1991). A new type of antisweat principles occurring in *Gymnema sylvestre*. *Tetrahedron Let., 32*(6), 789–792.

Yoshikawa, K., Murakami, T., & Matsuda, H., (1997). Medicinal food stuffs X1structures of new triterpene glycosides, gymnemosides- c, -f, -e and -f, from the leaves of the *Gymnema sylvestre* R.Br.: Influence of gymnema glycosides on glucose uptake in rat's small intestinal fragments. *Chem. Pharm. Bull., 45*(12), 2034–2038.

Yoshikawa, M., Murakami, T., Kadoya, M., Li, Y., Murakami, N., Yamahara, J., & Matsuda, H., (1997). Medicinal foodstuffs. IX. The inhibitors of glucose absorption from the leaves of *Gymnema sylvestre* R. Br. (*Asclepiadaceae*): Structures of gymnemosides a and b. *Chem. Pharm. Bull., 45*(10), 1671–1676.

Zhu, X. M., Xie, P., Di, Y. T., Peng, S. L., Ding, L. S., & Wang, M. K., (2008). Two new triterpenoid saponins from *Gymnema sylvestre*. *J. Integr. Plant Biol., 50*(5), 589–592.

CHAPTER 3

Perspectives of *Syzygium cumini* (L.) Skeels, as a Potential Rational Herbal Medicine for Diabetes

LEPAKSHI MD. BHAKSHU,[1] K. VENKATA RATNAM,[2] DEEPU PANDITA,[3] ANU PANDITA,[4] PULALA RAGHUVEER YADAV,[5] and R. R. VENKATA RAJU[6]

[1]*Department of Botany, PVKN Government College (Autonomous), Chittoor, 517 002, Andhra Pradesh, India, E-mails: lmbakshu@gmail.com; bakshu@pvkngcchittoor.info*

[2]*Department of Botany, Rayalaseema University, Kurnool – 518 007, Andhra Pradesh, India*

[3]*Government Department of School Education, Jammu, Jammu and Kashmir, India*

[4]*Vatsalya Clinic, Krishna Nagar, New Delhi, India*

[5]*Department of Biotechnology, Indian Institute of Technology Hyderabad, Kandi – 502 285, Telangana, India*

[6]*Department of Botany, Sri Krishnadeveraya University, Ananthapuramu – 515003, Andhra Pradesh, India*

ABSTRACT

A systematic revisionary study was performed with antidiabetic properties of *Syzygium cumini* (SC) using regular indexing webpages such as

Antidiabetic Plants for Drug Discovery: Pharmacology, Secondary Metabolite Profiling, and Ingredients with Insulin Mimetic Activity. Deepu Pandita, MPhil, Anu Pandita, MSc, & Chander Bhanu, MBBS, PGDDCE (Eds.)
© 2023 Apple Academic Press, Inc. Co-published with CRC Press (Taylor & Francis)

Scopus, Science Direct, PubMed, Google Scholar, NCBI and online journals, monographs and books, etc., and enumerated the antidiabetic properties scientifically along with evidence in different experimental models. *S. cumini* has enormous application in the management of diabetes and its complications solely or in combination proved to be safe medicament and it possess additional benefits such as anti-inflammatory, antioxidant, anorexigenic, anti-HIV, antimicrobial, and antifertility effect. The antidiabetic (AD) efficiency of *S. cumini* different part extracts/ formulations prepared using fruit, fruit pulp, seeds, stem bark, and leaves were discussed. The phytoconstituents isolated from the various parts of *S. cumini* are flavonoids (isoquercetin, myricetin, kaempferol), phenolics (ellagic acid, anthocyanins), glucosides (jambolin or antimellin) and alkaloid (jambosine) might be responsible for the biological activities. The study also emphasized the biochemical and molecular mechanism of prevention of diabetes in addition to controlling the lipid profile and beneficial role in the protective role of vital organs in diabetes and its complications. The extensive research on the *S. cumini* has appreciated the beneficial roles and discussed in view of diabetes and its multiple roles in maintenance of hyper-glycemic conditions.

3.1 INTRODUCTION

3.1.1 ABOUT THE PLANT

Syzygium cumini (L.) Skeels, (Common name: black Jamun, java plum, XXXyricetin, black plum, Portuguese plum, purple plum, Malabar plum, Indian blackberry, Jamaica and damson plum, Blackberry, Family: Myrtaceae), widely distributed in India, eastern Africa, and Southeast Asia (Srivastava and Chandra, 2013). The medicinal or botanical properties of this plant have been published with different synonyms like *Eugenia jambolana* Lam., *Myrtus cumini* Linn., *Syzygium jambolana* DC., *S. jambolanum* (Lam.) DC., *E. djouant* Perr., *Calyptranthes jambolana* Willd., *E. cumini* (Linn.) Druce. And *E. caryophyllifolia* Lam (Ayyanar and Subash-Babu, 2012). The fruits are delicious (pericarp) and seasonal with conspicuous purple color which is commercially grown as a horti-cultural tree crop in many states of India. Sagrawat et al. (2006) reviewed about the medicinal properties and biological activities and revealed its multiple benefits from its various parts.

The alcoholic or aqueous extracts *S. cumini* has been extensively investigated for the antidiabetic effects using different models such as Streptozotocin/Alloxan induced rats or rabbits (Rafiullah et al., 2006; Kar et al., 2003; Baliga et al., 2012; Maiti et al., 2013). The seeds provided a novel therapeutic agent as antihyperglycemic agents since the phytochemicals which are proved for its effects (Kumari et al., 2016; Zulcafli et al., 2020). The phytochemical components reported from the various parts of *S. cumini* are flavonoids (isoquercetin, myrecitin, kaempferol), phenolics (ellagic acid, anthocyanins), glucosides (jambolin or antimellin) and alkaloid (jambosine) might be involved in the biological activities (Ayyanar and Suresh-Babu, 2012).

3.1.2 TRADITIONAL CLAIMS

The traditional claims (Figure 3.1) on the usage of this plant for diabetes or its complications as follows (Bijauliya et al., 2018; Chhetri et al., 2005; Ayyanar and Suresh-Babu, 2012):

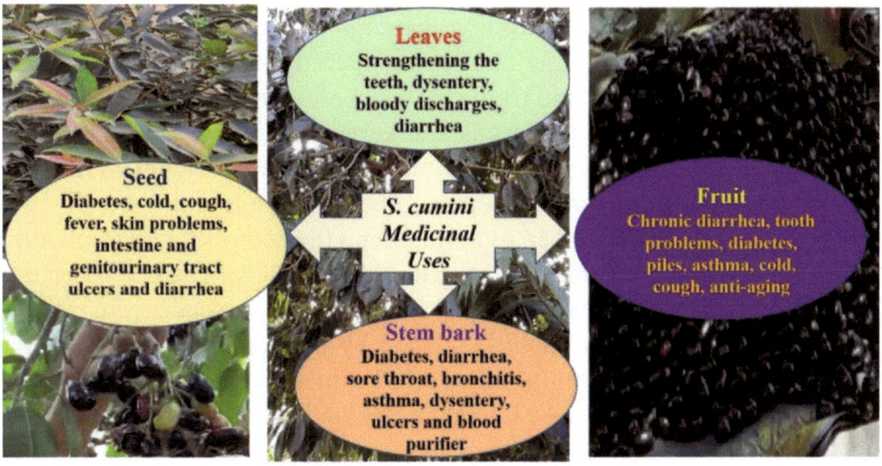

FIGURE 3.1 Medicinal properties of different parts of *S. cumini*.

- The tribes of Sikkim and Darjeeling Himalaya (India) were using the decoction obtained from the Stem bark for the treatment of diabetes;

- The fruits were used to treat digestive disorders including diarrhea;
- The fruit juice along with curd is good against digestive problems;
- The fruit juice is effectively used in Teeth problems;
- Jamun juice is beneficial in treating piles;
- The fruit juice helps in cold, cough and asthma;
- Jamun juice believed to improve the immune system;
- The fruit juice acts as an anti-aging agent.

3.2 EFFECT OF EXTRACTS ON ANIMAL MODELS

The AD (antidiabetic) efficiency of the fruit pulp, seeds, stem bark and leaves of *S. cumini* were evaluated scientifically using different animal models such as rats and rabbits. The extracts, formulations, fractions or individual phytochemical constituents were proved as effective against the artificially induced diabetes mellitus. The research work has been systematically discussed along with the additional beneficial effects exerted by plant samples hereunder.

The significant glucose reduction in blood samples has been evaluated by ethanol extract of seeds in hyperglycemic normal, rats 250 mg/kg bodyweight concentration. Further, it altered the higher concentrations of glucose, cholesterol, and triglycerides in alloxan-evoked diabetic (DB) rats. The ethanol extract of seeds and kernels when administered orally, remarkably reduced glucose level and urea or sugar level in urine of STZ- induced experimental rats than glibenclamide found to be effective. Further, the extract is proved as normalizing the activities of carbohydrate metabolizing enzymes significantly (Ravi et al., 2003). The glucose and fat lowering effects of alcoholic extract of *E. jambolana* seeds were reported significantly in diabetic rabbits induced by alloxan, and the assessment has done in fasting blood glucose at 90 min in addition to fall in peak blood glucose in the glucose tolerance test (sub and mild DB rabbits). The tested extract exerted elevation in the serum insulin level at 32.8% mild diabetics and 26.9% severe diabetic rabbits. The glycogen levels of liver and muscle remarkably enhanced which indicate the beneficial role of the extract in the management of diabetes (Sharma et al., 2003). The ethanolic extract of seeds exhibited the improvement in the pancreatic tissues when compared to the tissue damage in the control correlated with the hypoglycemic and histopathological examinations of diabetic rats, induced by alloxan (Singh and Gupta, 2007). Mycaminose isolated and characterized as active principle from the seed extracts which showed significant effect in streptozotocin-induced rat models when compared to the glibenclamide (Kumar et al., 2008).

A comparative antidiabetic efficiency of fresh cow urine and herbal formulation consisting of seven medicinal herbs prepared in cow urine was evaluated. It is revealed that the formulation exhibited concentration-dependent glucose-lowering activity in alloxan treated rat models (Jarald et al., 2008). *E. jambolana* seed extract possesses high flavonoid concentration displayed an improvement in glucose and lipid profile in the diabetic rats induced by streptozotocin. The short-term studies on the glucose-lowering property in fasting blood samples of DB rats reported that a significant fall after 60 minutes in the glucose tolerance test.

The assessment of glucose controlling role of the extract in the mild and severe diabetic (MD and SD, respectively) rats, has been proved in glucose tolerant test and effectively broke the glucose peak in one hour. In addition, the extracts also effected the glucose metabolizing enzymes (hexokinase, glucose-6-phosphatase) glycogen biosynthesis, and found improvement in a significant level. The flavonoids stimulated about 16% increased insulin levels when tested in the isolated pancreatic islets. In the same way the extracts affected the lipid metabolism significantly decreases the levels of triglycerides, low density lipid (LDL) and enhances the high density lipid (HDL) as in the normal rats indicating its preventing role of coronary heart diseases (Sharma et al., 2008). Whereas a similar effect with the use of methanol extract was demonstrated by Mastan et al. (2009) and proved that the seed extract prolonged the effect of gliclazide (a standard antidiabetic medicament) without any hypoglycemic shock when used in combination.

The filtered fruit pulp juice demonstrated a noteworthy activity on glucose concentration in blood in STZ-induced DB rats and showed normalized lipid profile and body weight, and the results were compared with the 5 mg/kg glyburide (standard) found equivalent effect with 200 mg/kg in the experimental models. In addition, the antioxidant enzyme levels were elevated and involving the antioxidant mechanism (Rekha et al., 2008). The AD activity of sequential fractions from the petroleum ether, chloroform, acetone, methanol and water from the seeds and the methanol fraction at 100 mg/kg b.wt. concentration reported as effective fraction and yielded Cuminoside evaluated as effective at 50 milligram/kg, and remarkably controlled the FBG (fasting blood glucose) concentration in the DB rat models. Further cuminoside strongly reduced the harmful lipid biomarkers, hepatic enzyme levels such as alanine aminotransferase, lactate dehydrogenase, aspartate aminotransferase and enhanced the levels of beneficial lipids (Farswan et al., 2009).

S. cumini leaf ethanolic extract also proved as effective on the STZ or nicotinamide-STZ-provoked DB rabbits at 100 milligrams/kg b.wt. on fasting blood glucose levels over the aqueous extract significantly (Qadir et al., 2009). Pandey (2010) investigated antidiabetic effects of mixed water extract prepared from S. *cumini* and *Aegle marmelos* (Bilva) in alloxan treated rats through elevating the insulin levels in serum and strongly controlled the excessive glucose and lipid markers and normalized equal to untreated animals whereas the flower extracts shown not effective on diabetes. In addition, the leaf alcoholic extract exhibited a noteworthy beneficial effect on the glucose/lipid profile in alloxan induced DB rats (Schoenfelder et al., 2010).

The alcoholic extract and ash obtained from the flowers exhibited significant antidiabetic effect (fasting blood glucose) on the streptozotocin or nicotinamide-streptozotocin induced diabetic rabbits with the tested concentrations while the aqueous extract not affected (Nisa et al., 2010). The partially purified principle from the seed extracts were proved as effective on the alloxan induced diabetes in various intensities (mild or severe diabetics) in rabbits and observed that a significant fall in fasting glucose levels in blood in addition to a significant decrease in the glycosylated hemoglobin. The extract also elevated levels of serum insulin, indicating rejuvenation of beta-cell function (Sharma et al., 2011; Subramoniam, 2016).

Eight-week treatment of the ethanolic extract of seeds at 20 mg/kg b.wt. studied in detail and investigated for antidiabetic property in the diabetic mice induced by alloxan. The extract also characterized and isolated the phyto-constituent "morroniside (iridoid glycoside)" and proved as highly effective antidiabetic principle. The ethanol extract investigated for the levels of patho-markers of diabetes such as glucose, glucokinase, fructosamine, glycosylated hemoglobin and the expression of protein markers like insulin, Glucose transporters (GLUT2 and GLUT4) and reported to controlled by the tested extract to normal. The antioxidant biomarkers also (catalase, lipid peroxidase, superoxide dismutase and thiols) found increased significantly in the target animal groups (Figure 3.2). The production of Nitric oxide and oxygen free radicals also effected and controlled by the treated mice significantly. The inflammatory cytokines and signal proteins of apoptosis such as iNOS, IFNγ, NFkB, STAT1, Caspase3 and Bcl2 and Bax expression were significantly affected and normalized in the tissues (liver and pancreas) of the treated mice. In addition, the extracts also influenced the DNA damage and normalized indicating its protective role and beneficial in diabetes (Samadder et al., 2011). The treatment with Homeopathic herbal formulation prepared using *S. cumini* and *Cephalandra indica*, for a month showed significant enhancement in the serum insulin level and reduced blood glucose quantity. Further, it enhanced the low levels of glucose transporter-4 (GLUT-4), v-akt (murine thymoma viral oncogene homolog), insulin receptor, Akt, and p-Aktser473 markers expression (Sampath et al., 2013).

The seed aqueous extract (100, 200 or 400 mg/kg) of *S. cumini* evaluated for insulin release effect on diabetic rats induced with high fat diet – Streptozotocin showed normalized the glucose and insulin in the serum, Tumor Necrosis Factor-α, lipid profile, and oxidation status of pancreas with a normalization of β-cell function. The extract also exhibited the protective activity against the damage of islets and β-cells at 400 mg/kg in addition to enhanced marker protein expression PPARγ and PPARα in the liver (Sharma et al., 2012). The fraction of the ethyl acetate from the methanol extract exhibited additional beneficial effect normalization of oxidative status of hemoglobin and testicular impairment and testosterone levels in the serum in the streptozotocin-induced diabetic rats indicates gender potentiality through the rejuvenating the germ cells even proved recovered at gene expression of Bax and Bcl-2 (Ghosh et al., 2014; Figure 3.3 and Table 3.1).

The extract of fruit with the combination of roots of banana, seeds of tamarind and Ivy gourd (*Coccinia grandis*) reported for hypoglycemic and hypolipidemic effect with enhanced profile of antioxidant enzymes (Chatterjee et al., 2013). *S. cumini* SB (stem-bark) extract also affected the lipid metabolism and reduced the total cholesterol in the alloxan induced DB rats (Tripathi and Kohli, 2014). The volatile oil components of the raw fruit pulp and paste of the leaves were reported the inhibition activity of the α-amylase, indicating its mild antidiabetic activity (Nishandhini et al., 2015). The water extract (WE) prepared from SC seed tested on the DB rats (in streptozotocin-induced a short-term model) and evaluated on adenosine deaminase (ADA) activity and lipoperoxidation (kidney, liver, cerebral cortex and pancreas) revealed that, significant reversion of ADA activity and ameliorated the lipoperoxidation. Even the treated rats were exhibited normalized lipid profile in liver (Bitencourt et al., 2015; Figure 3.2). The fraction of ethyl acetate (seeds) demonstrated for the glucose reducing effect in STZ- induced DB rats and showed inhibition on the carbohydrate digesting enzymes (Jana et al., 2015).

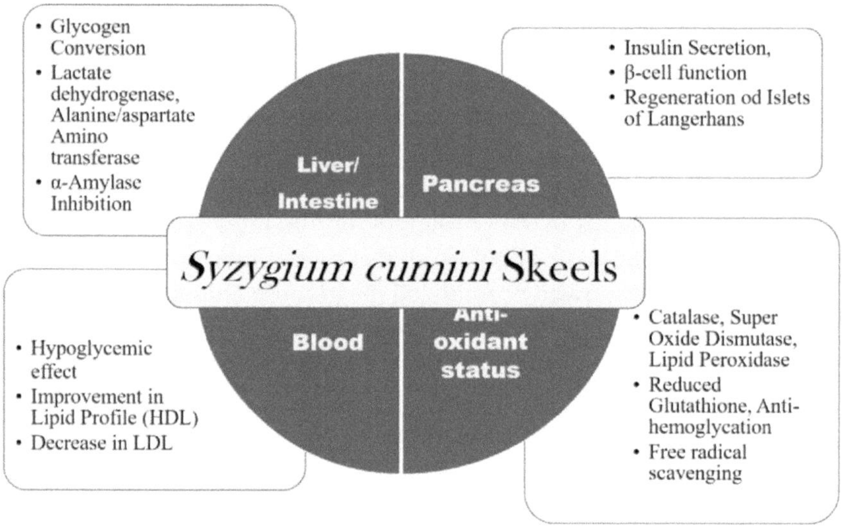

FIGURE 3.2 Biochemical effect of *S. cumini* extracts on diabetic profile.

TABLE 3.1 Antidiabetic Effect of Different Parts of *S. cumini*

Part Used	Mode of Extraction	Mechanism of Antidiabetic Effect	Reference(s)
Flowers	Ethanol, aqueous and ash	Decreased elevated FBG levels	Nisa and Malik (2010)
Fruit pulp	Aqueous	Reduced elevated glucose concentrations in the blood and enhanced antioxidant enzymes	Rekha et al. (2008)
Fruit and seed	Ethanol	Reduced elevated serum glucose (SG) levels and enhanced insulin levels	Raza et al. (2017)
Leaf	Ethanol/aqueous	Reduced elevated FBG levels	Qadir et al. (2009)

TABLE 3.1 *(Continued)*

Part Used	Mode of Extraction	Mechanism of Antidiabetic Effect	Reference(s)
Leaf	Hydro-alcoholic	Hypoglycemic/hypo-lipidemic	Baldissera et al. (2016)
Leaf	Hydro-ethanolic	Reduced triglyceride levels in serum, liver, and skeletal muscle Enhanced insulin secretion in INS-1E β cells	Sanches et al. (2016)
Leaf	Ethanol	Glucose/lipid-lowering effect	Schoenfelder et al. (2010)
Leaf	Ethanol/aqueous	Reduced oxidative stress	Chages et al. (2018)
Leaf	Ethanol/aqueous	Glucose transporter	Ajiboye et al. (2018)
Leaf	Ethanol/aqueous	α-amylase (AAL) and α-glucosidase (AGL)	Poongunran et al. (2017)
Seed kernel	Ethanol	Reduced elevated glucose concentrations in the blood and urea/sugar concentrations in urine	Ravi et al. (2003)
Seed	Ethanol	Reduced elevated glucose/lipid concentrations and enhanced insulin levels	Sharma et al. (2003)
Seed	Ethanol	Improved damaged pancreatic tissue	Singh and Gupta (2007)
Seed	Petroleum ether, chloroform, acetone, methanol and water	Reduced lipid profile/hepatic enzymes	Farswan et al. (2009)
Seed	Ethanol	Reduced blood glucose/ glycosylated hemoglobin and inhibited inflammatory markers	Samadder et al. (2011)
Seed	Water	Reduced glucose /lipid profile and TBARS and regulation of PPARγ and PPARα	Sharma et al. (2012)
Seed	Water	Inhibited adenosine deaminase and lipid peroxidation	Bitencourt et al. (2015)
Seed	Ethyl acetate	Glucose lowering effect and inhibition of carbohydrate digestion enzymes	Jana et al. (2015)

TABLE 3.1 *(Continued)*

Part Used	Mode of Extraction	Mechanism of Antidiabetic Effect	Reference(s)
Seed	Decoction/ herbal drink	Glucose lowering effect	Perera et al. (2017)
Seed	Aqueous	Reduced bad lipid profile, enhanced insulin secretion, and activated PPARγ and PPARα	Sharma et al. (2017)
Seed	Petroleum ether	Hypoglycemic effect	Proma et al. (2018), Kandan and Govindan (2017)
Seed	Ethereal fraction of ethanol extract	Protected pancreatic tissue/islets of Langerhans	Mandal et al. (2008)
Seed	EtoAc fraction of methanol extract	Suppressed aldose reductase and protein tyrosine phosphatase activities	Sawant et al. (2015)
Seed	Ethanol	Decreased BG concentration	Shankar et al. (2007)
Seed	Water	Enhanced G6PD, glucose levels in hepatic/muscle tissues and increased C-peptide/insulin level in plasma	Dusane and Joshi (2011)
Seed kernel	Acetone	Suppression of AGL enzyme	Shinde et al. (2008)
Seed	Ethanol	Inhibition of human pancreatic α-amylase	Ravi Kumar et al. (2011)
Seed	Water	Suppression of AAL enzyme	Kumar et al. (2013)
Seed	Methanol	Suppression of AGL enzyme	Middha et al. (2012)
Seed kernel	Water and ethanol	Reduced elevated glucose levels and altered hepatic/skeletal muscle glycogen	Grover et al. (2000)
Seed	Powder	Reduced fasting blood glucose levels	Sridhar et al. (2005)
Seed	Ethanol/aqueous	Regulation of PPARγ and PPARα in 3T3-L1 cell lines	Sharma et al. (2008)
Stem bark	Water	Lowered glucose levels in blood/urine samples and recovered body weight, intake of food and fluids	Saravanan and Pari (2008)
Stem bark	Water	Reduced BG levels and bad cholesterol levels	Tripathi and Kohli (2014)

Whereas hydroalcoholic extract of *S. cumini* leaf exhibited good hypoglycemic and hypolipidemic effects, attributed to the presence of myricetin and phenolic content (Baldissera et al., 2016). Antidiabetic efficiency of water decoction and herbal drink prepared from the seeds was investigated in a clinical experiment. The potential glucose reducing effect of extract and drink may be attributed to the presence of active constituents such as gallic acid, ellagic acid and umbelliferone (phenolics). The presence of these components was confirmed by chromatographic techniques such as TLC, HPLC and LC (Perera et al., 2017). Antioxidant and antidiabetic efficiency of α-hydroxy succinamic acid, purified from fraction (IIC) of fruit pulp was determined during the oral administration at 15 mg/kg significantly controlled the glucose and lipid contents (Tanwar et al., 2017). Sharma et al. (2017) reported antidiabetic role of the aqueous extract (400 mg/kg/day) of seeds in the high fat diet rats induced with the streptozotocin normalized the production of insulin, lipid profile and anti-inflammatory markers levels along with the expression of PPARγ and PPARα activity and confirmed its antidiabetic effects when compared with the standard drug (metformin) (Figure 3.3).

Fruit and seed extracts of *S. cumini* demonstrated for hyperglycemic activity in the male Sprague Dawley rats and the ethanolic extracts attenuated glucose in serum to 5.35% and 12.29% in control as well as experimental rats, including the insulin level indicating its preventive role in hyperglycemia (Raza et al., 2017). The polyphenol rich portion prepared from *S. cumini* leaves was assessed in oxidative stress induced diabetes in rat models (Chagas et al., 2018). The results state that fruit pulp oil significantly inhibited α-amylase enzyme activity and leaf oil enhanced antioxidant potential in the *in vitro* studies. Majeed et al. (2018) investigated the antidiabetic potential of herbal formulation prepared from seven active herbs in alloxan induced diabetes in rat models. The formulation significantly controlled the serum glycosylated hemoglobin and glucose concentration levels and found enhanced insulin concentrations in serum samples of alloxan induced DB animals (Figure 3.2). The fraction obtained from petroleum ether from the seeds also reported for significant hypoglycemic effect (Proma et al., 2018; Kandan and Govindan, 2017). Ajiboye et al. (2018) demonstrated the antidiabetic effect of treated with the leaf extract containing high polyphenols showed the significant preventive role with the normalized profile including glucose transporter levels in the alloxan-induced rats for two weeks.

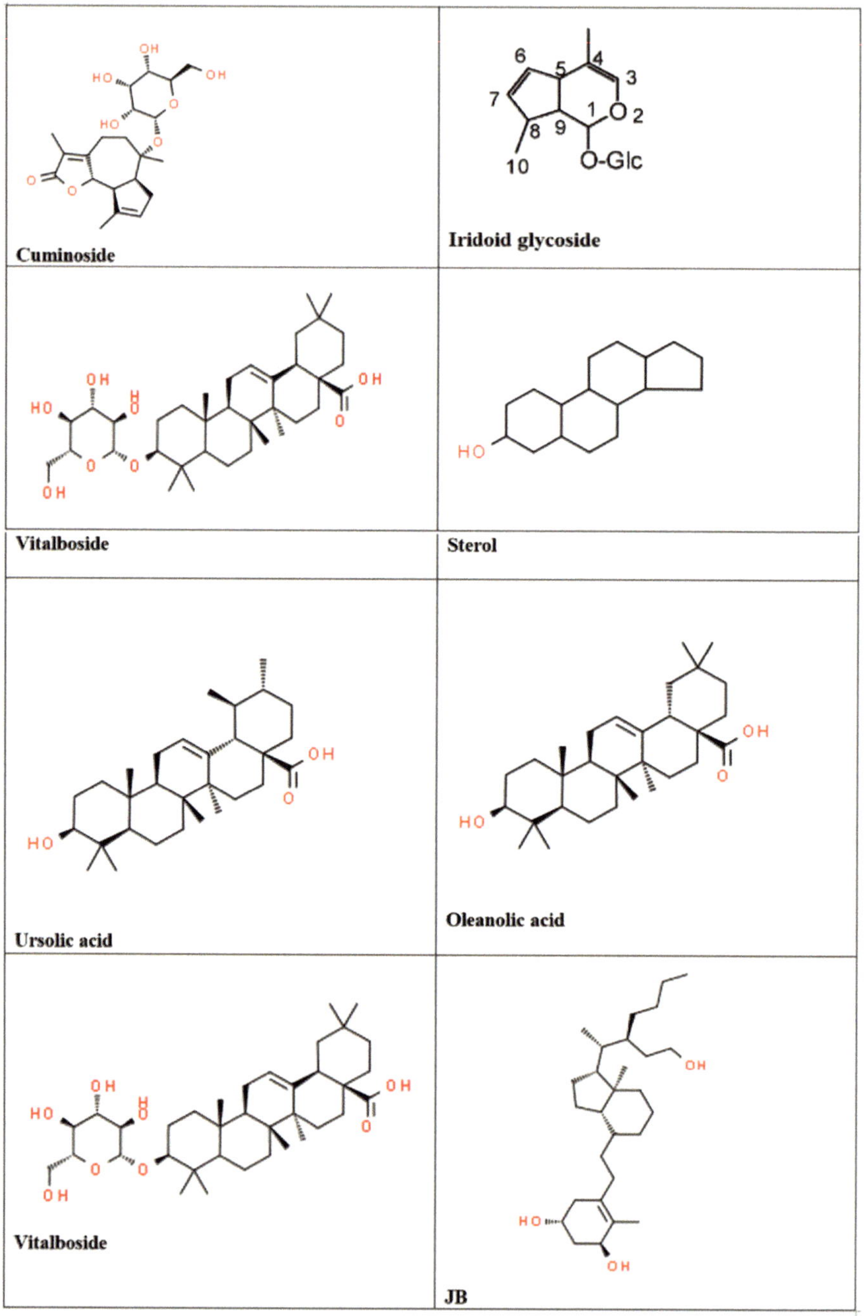

FIGURE 3.3 Important active principles of *S. cumini.*

3.3 EFFECT OF EXTRACTS/POLY HERBAL FORMULATIONS IN *IN-VITRO* CELL-LINES

The flavonoid-rich portion of seeds was discovered to be effective through the activation of the peroxisome proliferators-activated receptors (PPARα and PPARγ) up to 3–4 folds (over control) and their capacity to change into 3T3-L1 preadipocytes suggesting the preventing role and control of diabetes (Sharma et al., 2008). DB 14201 is an Ayurvedic formulation demonstrated paucible antidiabetic mechanism of action by regulation of glucose uptake, TNF-alpha and free fatty acid release in 3T3-L1 cell lines (murine adipocytes) and insulin release in β TC-6 cell lines (murine pancreatic beta cell) (Gopalakrishna et al., 2017). The antidiabetic activity of jamun (SC) along with five medicinal plants was assessed in *in-vitro* cell lines models. It revealed that jamun extract exhibited an inhibitory effect on glucose diffusion across the dialysis membrane assay. Ajebli and Eddouks (2019) reviewed that 41 crude extracts of medicinal plants rich in tannins and isolated 19 compounds reported to have a promising blood-glucose-lowering effect. They concluded that these components could act as promising active principles for the discovery of new antidiabetic drugs.

The effect of *S. cumini* leaf hydro-ethanolic extract on metabolic changes in lean and monosodium L-glutamate (MSG)-induced obese rats exhibited pronounced control on the weight gain and formation of adipose tissues and lipid profile. Further, the tested extracts were showed beneficial effects in insulin resistance and partial reversal of glucose intolerance in affected rats and INS-1E β-cells (Sanches et al., 2016).

3.4 EFFECT OF PHYTOCONSTITUENTS (TABLE 3.2)

Thyagarajan (et al., 2016) demonstrated that Vitalboside A, a triterpenoid isolated from SC seed methanolic extract intensify transportation of glucose through "IRb-triggered PI3K/Akt activation and allosteric suppression of PTP1B in 3T3-L1 adipocytes and L6 myotube cell lines." Oral administration of saturated fatty acid, i.e., "Δ5 lipid, and sterol rich LH II" portion, obtained from *E. jambolana* (EJ) seed ethanol extract for a period of 21 days, exhibited significant changes in the higher concentration of glucose in alloxan-challenged mild diabetic/severely diabetic rabbits. It also exhibited significant lipid-lowering activity in both types of animal models tested (Sharma et al., 2011). The ethereal fraction separated from ethanolic extract of the *S. cumini* seeds, oral administration to the diabetic rats exhibited significant protection to the pancreatic tissues and recovery of the islets of Langerhans which identified as ferulic acid. Whereas normal did not show any significant change in the concentration of glycogen in the skeletal and liver tissues, activities of G6PG catalase and peroxidase enzymes in liver. The protective mechanisms suggested that the ferulic acid for the therapeutic effect against diabetes (Mandal et al., 2008). Ethyl acetate fraction separated from methanol fractions of *S. cumini* seeds strongly inhibited aldose reductase and protein-tyrosine phosphatase enzymes in in-vitro studies. Phenolic components such as ruby phenol, gallic acid, ellagic acid, and valoneic acid dilactone isolated from ethyl acetate fraction significantly inhibited aldose reductase by expressing lower IC_{50} values and ellagic acid, valoneic acid dilactone, and ruby phenol inhibited protein-tyrosine phosphatase (Sawant et al., 2015). Compounds JB1 (identified as "3,15-dihydroxy Δ 3 androstene [16,17-C] (6′ methyl, 2′-1,3-dihydroxy-1-propene) 4H pyran," and JB2 (identified as "3-hydroxy androstane [16,17-C] (6′ methyl, 2′-1-hydroxy-isopropene-1-yl) 4,5,6 H pyran") characterized from ethanol extract of SC seed by chromatography techniques. A comparative antidiabetic efficiency of both principles and ethanol extract was evaluated in alloxan – challenged animal models. Ethanol extract significantly lowered elevated glucose concentration in blood samples of treated animals in single-dose and prolonged treatments. JB1 showed more potent activity than JB2, which was comparable to ethanolic extract (Shankar et al., 2007). Antidiabetic potential of organic solvent extracts, i.e., methanol, ether, chloroform, acetone, and aqueous extracts of SC was carried out in normal and non-insulin-dependent diabetes mellitus Wistar rat models (Farswana et al., 2009) for 21 days experimental schedule. Of the tested five extracts, methanol extract strongly reduced blood sugar levels in fasting samples. The active principles characterized from methanol extract was cuminoside also studied for its glucose-lowering and antioxidant potential (Farswana et al., 2009). The fraction purified form of SC seeds was assessed for its glucose-lowering effects in STZ-challenged Swiss mice for 21 days treatment schedule. The results indicated that the fraction at 2g/L significantly restored glucose concentration levels to normal, enhanced G6PD, glucose levels in hepatic and muscle tissues. Further it increased C-peptide and insulin levels in plasma. Histological studies revealed that it enhanced growth of neo-islets in treated rats (Dusane and Joshi, 2011). The glucose/lipid-lowering effect of water extract prepared from SC bark was investigated in STZ-induced rat models. The extract at 300 mg kg^{-1} significantly lowered sugar levels in both blood and urine samples than other concentrations tested. It further strongly restored STZ-induced physiological changes, i.e., daily intake of food and fluids, body weight in treated animals. During oral glucose tolerance test, the extract strongly reduced glucose concentration levels in blood samples at 30 to 120 min than the positive control animals (Saravanan and Pari, 2008).

TABLE 3.2 Antidiabetic Effect of Phyto-Constituents Isolated from *S. cumini*

Active Principle	Part Used (Extract)	Mode of Action	Reference(s)
Cuminoside	Seed (methanol)	Reduced FBG level	Farswan et al. (2009)
Iridoid glycoside	Seed (ethanol)	Reduced generation of ROS, down-regulated NFkB, IFNγ, iNOS, Bcl 2	Samadder et al. (2011)
Essential oil	Fruit pulp/leaf paste	α-amylase inhibition	Nishandhini et al. (2015)
α-Hydroxy succinic acid	Fruit pulp	Glucose/lipid-lowering effect and antioxidant effect	Tanwar et al. (2017)
Essential oil	Fruit pulp/leaf	α-amylase inhibition	Chagas et al. (2018)
Vitalboside	Seed	Augmented glucose transport and inhibited PTP 1B in 3T3 L1 and L6 myotubes	Thiyagarajan et al. (2013)
Δ5 Lipid, sterol	Seed (ethanol)	Glucose/lipid-lowering effect	Sharma et al. (2011)
JB1 and JB2	Seed	Reduced elevated glucose levels	Shankar et al. (2007)
Ursolic and oleanolic acid	Leaf	α-amylase inhibition	Poongunran et al. (2017)

The structures of some important active principles of *S. cumini* are given in Figure 3.3.

3.5 EFFECT ON THE CARBOHYDRATE METABOLISM

3.5.1 EFFECT OF SJ EXTRACTS ON KEY ENZYMES OF CARBOHYDRATE METABOLISM (FIGURES 3.2 AND 3.4)

Herbal formulation, prepared from *S. jambolanum* (SJ), significantly lowered glucose concentration in fasting blood samples and enhanced key marker enzymes of carbohydrate metabolism, i.e., glucose -6- phosphate, hexokinase and glucose -6- phosphate dehydrogenase in STZ induced diabetes in rats. Further, the formulation decreased the bad lipid marker levels such as cholesterol, triglycerides, LDL and VLDL and enhanced the good cholesterol (HDL) levels (Maiti et al., 2013). Antidiabetic efficiency of aqueous leaf extract of *S. cumini* along with phenolic compounds, i.e., rutin, gallic acid, and chlorogenic acid was assessed by using adenosine deaminase activity in erythrocytes challenged with elevated glucose concentrations *in vitro* methods. The extract in combination with caffeine and insulin exhibited potent Antidiabetic activity than phenolic compounds (De Bona et al., 2014).

3.5.2 EFFECT OF SJ EXTRACTS ON CARBOHYDRATE DIGESTING ENZYMES (IN VITRO)

The acetone extracts of seed kernels were reported to inhibit the α-glucosidase activity of rat, Bacterial or yeast and found significant when compared to the acarbose. These results were correlated in the *in vivo* models suggesting as one of the mechanisms of action (Shinde et al., 2008). In addition, the seed extract also proved as a potential inhibitor of human pancreatic α-amylase even at low concentrations (Ravi Kumar et al., 2011). SJ seed aqueous extract strongly inhibited α-amylase by expressing low IC_{50} value (1.33mg/ml), in comparison with acarbose (Kumar et al., 2013). Whereas methanolic extract strongly inhibited α-glucosidase enzyme *in vitro* assay (Middha et al., 2012). Antidiabetic efficiency of hydro-alcoholic extract of MD1 formulation was assessed in an in vitro study (Telapolu et al., 2018). It revealed that the extract exhibited noteworthy suppression activity against key carbohydrate digestive marker enzymes, i.e., α-glucosidase and α-amylase. Further, the extract significantly modulated messenger RNA (mRNA) expression levels of PPARγ and GLUT4 in 3T3L1 cells (Figures 3.2 and 3.4).

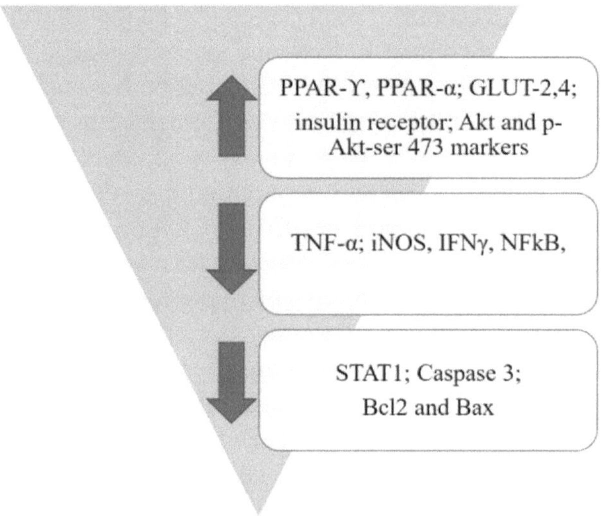

FIGURE 3.4 Effect of *Syzygium cumini* on inflammatory markers and gene expression in diabetes.

Effect of *S. cumini* leaf extract on α-amylase was studied in *in vitro* studies. The active fraction separated from the leaf extract was determined to be a 3:1 (ursolic acid and oleanolic acid) using spectral/chromatography techniques. The fraction exhibited significant inhibitory activity on porcine pancreatic alpha-amylase by expressing low IC_{50} value (39.9 μg/ml). Further, the fraction strongly inhibited alpha-glucosidase enzyme with an IC_{50} of 28.2 μg/ml. Two terpenoids, i.e., ursolic acid and oleanolic acid expressed less IC_{50} values of 6.7 and 57.4 μg/ml against α-amylase and 3.1 and 44.1 μg/ml against α-glucosidase respectively (Poongunran et al., 2017).

3.6 SYNERGETIC EFFECT IN THE FORMULATIONS

The glucose-lowering effect of aqueous decoction, ethanol extract and lyophilized powder obtained from kernels of *E. jambolana* (EJ) and Guduchi were evaluated in animal models for a duration of 21 to 120 days experimental schedule. The results indicated that lyophilized powder of SJ seed kernels at 200 mg/kg per day and water extract of guduchi at 400 mg/kg per day reduced ~70% glucose concentrations in treated diabetic animals in 3 and 15 weeks of experimental schedule. Further, the extract modulated glycogen content in liver and skeletal muscle and key marker enzymes were partially restored by the tested both plant extracts (Grover et al., 2000). Administration of EJ seed powder with 500 and 1,000 mg/kg exhibited notable improvement in body weight and decrease in glucose concentration in fasting animal group (Sridhar et al., 2005). Glucose lowering activity and antioxidant effect of hyponidd, an herbal-mineral formulation mixer of 10 herbal extracts, was assessed in STZ-induced animal models. Oral administration of hyponidd at 100 and 200 mg/kg, exhibited noteworthy glucose-lowering activity and enhanced glycogen content in liver and total hemoglobin in treated animals in 45 days experimental duration. It further reduced elevated levels of glycosylated hemoglobin, TBARS, hydro-peroxides, α-tocopherol and ceruloplasmin in diabetic rats. It significantly enhanced the antioxidant enzyme levels, i.e., glutathione and vitamin C concentrations. The results were compared with glibenclamide (Babu and Prince, 2004). Dianex herbal formulation prepared from water extracts of gurmar, jamun, bitter guard, neem, *Cassia auriculata,* bilvam, ashwagandha and turmeric was assessed for glucose-lowering effect in normal and STZ-challenged mice models. The formulation exhibited notable glucose-lowering activity at 250–500 mg/kg concentration in both treated and untreated mice in acute and long-term studies. Further, it reduced elevated levels of biochemical markers such as lipid profile, ALT, AST, urea and creatinine levels to normal levels. Histopathological studies results revealed that the formulation enhanced tissue regeneration in both animals. *In vitro* study results revealed that it exhibited strong antioxidant activity against the tested models (Mutalik et al., 2005). The methanol extracts of *Aegle marmelos* and *S. cumini* were cumulatively analyzed for glucose absorption activity in L6 myotubes. The extracts showed noteworthy glucose absorption activity at 100 ng/ml concentration when compared to insulin and rosiglitazone. The extracts enhanced the expression levels of glucose transporting molecules, i.e., Glut-4, PPARγ and PI3 kinase and suppressed the expression levels of cycloheximide, which interfere in protein synthesis (Anandharajan et al., 2006). Antidiabetic efficacy of "GSPN kwath" a polyherbal formulation of *Gymnema sylvestre* consisting 10 antidiabetic medicinal plants was investigated (Mahajan et al., 2015) in type- II diabetic patients. Daily administration of 10 grams of the formulation significantly reduced glucose and glycosylated hemoglobin concentrations in blood samples of patients for a period of six months. It also altered higher levels of bad cholesterol levels and enhanced good cholesterol levels. "Diakyur" polyherbal formulation consists of about seven herbs, significantly decreased plasma glucose levels and inhibited TBARS production in erythrocyte membranes, liver and kidney tissue at 1,600 mg/kg in alloxan-induced diabetic rats and rabbits (Joshi et al., 2007). The glucose/lipid-lowering effect and antioxidant activities of Dihar herbal formulation consisting of eight medicinal plants, i.e., jamun, bitter guard, amla, gurmar, elachi, neem, guduchi and turmeric in STZ-challenged type-1 diabetic animal models. The formulation at 100 mg/kg, significantly altered STZ- induced serum glucose/lipid levels, creatinine and urea levels in diabetic animals. It further, enhanced antioxidant enzyme levels and TBARS mediated free radical formation in STZ-induced diabetic animal models (Patel et al., 2009). Polyherbal formulation (MTEC) constituted with the hydromethanol extract of banana root, seeds of Tamarind and leaves of jamun and Ivy gourd at specific ratio was carried out in STZ-challenged rat models. The formulation significantly normalized the fasting blood glucose level, glycogen contents of liver and skeletal muscle along with said carbohydrate metabolic enzymes such as glucose-6-phosphatase and lactate dehydrogenase and decreases the activity of hexokinase as well as liver glucose-6-phosphate dehydrogenase, skeletal and cardiac muscles towards the control levels (Chatterjee et al., 2009).

3.7 MULTIFACETED BENEFICIAL EFFECTS

Since diabetes mellitus is a multi-syndromic and effects all vital organs and functional proteins or various biomolecules and leading to many complications. *S. cumini* has studied extensively and revealed the following beneficial effects and impact the role of human health even in the combination with other effective antidiabetic plants (Table 3.3):

1. Leaf extract reported for normalized antioxidant profile in the stress related parameters (RBC hemolysates and erythrocyte membranes using isolated system) and inhibited the marker enzymes such as adenosine deaminase, acetylcholinesterase and the antioxidant enzymes (De Bona et al., 2011).
2. The jamun extracts controlled the gastric ulcer related issues and enhanced the glutathione, mucin and mucosal glycoprotein in the

experimental rats indicated the anti-ulcer in diabetics (Chaturvedi et al., 2009).
3. The fruit pulp inhibited the free radicals and improved the antioxidant activity and strongly affected the lipid markers including the bodyweight in the treated diabetic rat liver (Rekha et al., 2008). The leaf extracts strongly reduced the triglycerides and prevented the DNA damage in the vital organs such as lung, gastrocnemius, kidney, liver and heart and improved the antioxidant status in addition to control over the bodyweight (Baldissera et al., 2016).
4. The renal protective capacity of FIIc (purified compound) from water extract of EJ fruit pulp was investigated in STZ-induced severe diabetic rats in *in-vivo* methods. Oral administration of FIIc at 15 mg/kg b. wt./day for two months exhibited noteworthy glucose-lowering effect in fasting blood samples in a concentration-dependent manner. It further showed notable improvement in body weight, blood urea, plasma creatinine levels, urine volume, urinary sugar and microalbuminuria (Tanwar et al., 2010).
5. The hydroethanolic extract *S. cumini* leaf showed a significant role in the control of monosodium L-glutamate induced obesity in the newborn rats for eight weeks treatment. In addition, Triton and WR1339-induced hypertriglyceridemia also being studied and demonstrated its role in controlling overweight and restored the total cholesterol, triglycerides and free fatty acids, fasting serum glucose, as well as insulin sensitivity, to levels similar to lean rats. It also halved the triglyceride content into very-low-density lipoprotein particles along with healing of liver steatosis These effects were proved up to the molecular level in the hepatic protein expression of the endoplasmic reticulum chaperone GRP94, downregulation of hepatic triglyceride secretion pathway by reducing the splicing of X-box binding protein 1 (XBP-1s), as well as protein disulfide isomerase (PDI) and microsomal triglyceride transfer protein (MTP) translational levels supporting the downregulation of the XBP-1s/PDI/MTP axis in the liver of MSG-obese rats as a novel feasible mechanism for the antihypertriglyceridemic effect promoted by the polyphenolic phytoconstituents in the leaf (França et al., 2019).

TABLE 3.3 Antidiabetic Effect of Polyherbal Formulations or Mixed Extracts of *S. cumini*

Formulation/ Mixed Extract	Number of Plants Used	Mechanism of Action	Reference(s)
DB 14201	–	Inhibition of glucose uptake Inhibited TNF-α production in 3T3-L1 and β-TC6 cell lines	Gopalakrishna et al. (2017)
Mixed extract	7 plants	Glucose lowering activity	Jarald et al. (2008)
Mixed extract	2 plants	Reduced higher levels of glucose and lipid levels	Pandey (2010)
Mixed extract	2 plants	Enhanced serum insulin levels, GLUT-4 levels, and reduced blood glucose levels	Sampath et al. (2013)
Mixed extract	3 plants	Hypoglycemic and hypolipidemic activity	Chatterjee et al. (2013)
Herbal formulation	7 plants	Lowered serum glucose and glycosylated hemoglobin levels and enhanced serum insulin levels	Proma et al. (2018); Kandan and Govindan (2017)
Herbal formulation	Single plant	Reduced FBG/lipid marker levels and enhanced carbohydrate metabolism enzymes	Maiti et al. (2013)
MD1	–	Inhibition of carbohydrate digestive marker enzymes	Telapolu et al. (2018)
Hyponidd	10 plants	Lowered blood glucose levels and increased hepatic glycogen and total hemoglobin	Sridhar et al. (2005)
Dianex	8 plants	Reduced elevated levels of glucose, lipid markers, ALT, AST, urea, and creatinine	Mutalik et al. (2005)
Mixed extract	2 plants	Regulation of PPARγ and PPARα, GLUT-4 in L6 myotubes	Anandharajan et al. (2006)
GSPN kwath	10 plants	Lowered blood glucose levels and glycosylated hemoglobin	Mahajan et al. (2015)
Diakyur	7 plants	Reduced plasma glucose levels and inhibited TBARS production in erythrocyte membranes	Joshi et al. (2007)

TABLE 3.3 *(Continued)*

Formulation/ Mixed Extract	Number of Plants Used	Mechanism of Action	Reference(s)
Dihar	8 plants	Reduced elevated levels of glucose, lipid markers, TBARS and enhanced antioxidant enzyme levels	Patel et al. (2009)
MTEC	4 plants	Reduced FBG levels and glycogen contents in liver and skeletal muscle tissues	Chatterjee et al. (2009)

3.8 FUTURE PERSPECTIVES

The extensive research on the effect of *S. cumini* on diabetes has a clear role in the prevention and treatment of diabetes. The introduction of phyto-constituents in the regular treatment of diabetes in the preliminary stages awaiting the human trials since it is natural and even it is not reported any toxic effects at the tested doses in various systems as far as the research studies.

3.9 CONCLUSION

The present study on the role of Jamun and its parts on the maintenance of normal health and preventing diabetes and its complications is very much appreciated by the scientific world. The extracts of various parts were proved to control the glucose, insulin, fat and other metabolic pathways in the normalizing the health as well as provide protection to the vital organs and biomolecules including DNA. Owing to its multifarious pharmaco-logical properties such as antioxidant, anti-inflammatory, antibacterial, antifungal, anti-HIV, antifertility, anorexigenic, etc., the scientists paved carpet towards the development of safe medicaments to control the onset diabetes. Moreover, the usage of fruits of *S. cumini* is having strong rational support to control diabetes. However, further research is required to introduce *S. cumini* as a medicament in the treatment of diabetes and related complications and even in large-scale production of active fractions or principles. The fruits of *S. cumini* were available at affordable cost and for its delicious taste they might be used as functional food for overall improvement of human health. However, the preparation of modern medicine for diabetes by formulation with effective combination and use as a medicament has to be taken for the task in the scientific world and to develop as a safe drug.

KEYWORDS

- **antidiabetic activity**
- **glucose/lipid profile**
- **molecular mechanism**
- **phytoconstituents**

REFERENCES

Ajebli, M., & Eddouks, M., (2019). The promising role of plant tannins as bioactive antidiabetic agents. *Current Medicinal Chemistry, 26*(25), 4852–4884.

Ajiboye, B. O., Ojo, O. A., Akuboh, O. S., Abiola, O. M., Idowu, O., & Amuzat, A. O., (2018). Anti-hyperglycemic and anti-inflammatory activities of polyphenolic-rich extract of *Syzygium cumini* Linn. leaves in alloxan-induced diabetic rats. *J. Evidence-Based Integra. Med., 23.* 2515690X18770630.

Anandharajan, R., Jaiganesh, S., Shankernarayanan, N. P., Viswakarma, R. A., & Balakrishnan, A., (2006). *In vitro* glucose uptake activity of *Aegles marmelos* and *Syzygium cumini* by activation of glut-4, PI3 kinase and PPARγ in L6 myotubes. *Phytomedicine, 13*(6), 434–441.

Ayyanar, M., & Subash-Babu, P., (2012). *Syzygium cumini* (L.) skeels: A review of its phytochemical constituents and traditional uses. *Asian Pac. J. Trop. Biomed., 2*(3), 240–246.

Babu, P. S., & Prince, P. S. M., (2004). Antihyperglycaemic and antioxidant effect of hyponidd, an ayurvedic herbomineral formulation in streptozotocin-induced diabetic rats. *J. Pharm. and Pharmacol., 56*(11), 1435–1442.

Baldissera, G., Sperotto, N. D. M., Rosa, H. T., Henn, J. G., Peres, V. F., Moura, D. J., Roehrs, R., et al., (2016). Effects of crude hydroalcoholic extract of *Syzygium cumini* (L.) skeels leaves and continuous aerobic training in rats with diabetes induced by a high-fat diet and low doses of streptozotocin. *J. Ethnopharmacol., 194,* 1012–1021.

Baliga, M. S., Fernandes, S., Thilakchand, K. R., D'souza, P., & Rao, S., (2013). Scientific validation of the antidiabetic effects of *Syzygium jambolanum,* DC. (Black plum), a traditional medicinal plant of India. *J. Altern. Complement Med., 19*(3), 191–197.

Bijauliya, R. K., Alok, S., Sabharwal, M., & Chanchal, D. K., (2018). *Syzygium cumini* (Linn.)-an overview on morphology, cultivation, traditional uses and pharmacology. *Inter. J. Pharmaceu. Sci. Res., 9*(9), 3608–3620.

Bitencourt, P. E. R., De Bona, K. S., Cargnelutti, L. O., Bonfanti, G., Pigatto, A., Boligon, A., Athayde, M. L., et al., (2015). *Syzygium cumini* seed extract ameliorates adenosine deaminase activity and biochemical parameters but does not alter insulin sensitivity and pancreas architecture in a short-term model of diabetes. *J. Comple. Integra. Med., 12*(3), 187–193.

Chagas, V. T., Coelho, R. M. R. D. S., Gaspar, R. S., Da Silva, S. A., Mastrogiovanni, M., Mendonça De, C. J., Ribeiro De, M. N. S., et al., (2018). Protective effects of a polyphenol-rich extract from *Syzygium cumini* (L.) skeels leaf on oxidative stress-induced diabetic rats. *Oxid. Med. and Cell. Long., 2018*.

Chatterjee, K., Ali, K. M., & Ghosh, D., (2009). Anti-hyperglycaemic, antioxidative activities of a formulated polyherbal drug MTEC (Modified) in streptozotocin-induced diabetic rat. *Journal of Medicinal Plants Research, 3*(6), 468–480.

Chatterjee, K., Ali, K. M., De, D., Bera, T. K., Jana, K., Maiti, S., Ghosh, A., & Ghosh, D., (2013). Hyperglycemia-induced alteration in reproductive profile and its amelioration by the polyherbal formulation MTEC (modified) in streptozotocin-induced diabetic albino rats. *Biomarkers and Genomic Medicine, 5*(1, 2), 54–66.

Chaturvedi, A., Bhawani, G., Agarwal, P. K., Goel, S., Singh, A., & Goel, R. K., (2009). Ulcer healing properties of ethanolic extract of *Eugenia jambolana* seed in diabetic rats: study on gastric mucosal defensive factors. *Ind. J. Physiol. Pharmacol., 53*(1), 16–24.

Chhetri, D. R., Parajuli, P., & Subba, G. C., (2005). Antidiabetic plants used by Sikkim and Darjeeling Himalayan tribes, India. *Journal of Ethnopharmacology, 99*(2), 199–202.

De Bona, K. S., Bellé, L. P., Bittencourt, P. E. R., Bonfanti, G., Cargnelluti, L. O., Pimentel, V. C., Ruviaro, A. R., et al., (2011). Erythrocytic enzymes and antioxidant status in people with type 2 diabetes: Beneficial effect of *Syzygium cumini* leaf extract *in vitro*. *Diabetes Research and Clinical Practice, 94*(1), 84–90.

De Bona, K. S., Bonfanti, G., Bitencourt, P. E. R., Cargnelutti, L. O., Da Silva, P. S., Da Silva, T. P., Zanette, R. A., et al., (2014). *Syzygium cumini* is more effective in preventing the increase of erythrocytic ADA activity than phenolic compounds under hyperglycemic conditions *in vitro*. *Journal of Physiology and Biochemistry, 70*(2), 321–330.

Dusane, M. B., & Joshi, B. N., (2011). Seeds of *Syzygium cumini* (L.) skeels: Potential for islet regeneration in experimental diabetes. *Zhong xi yi jie he xue bao [Journal of Chinese Integrative Medicine], 9*(12), 1380.

Farswan, M., Mazumder, P. M., & Parcha, V., (2009). Modulatory effect of an isolated compound from *Syzygium cumini* seeds on biochemical parameters of diabetes in rats. *International Journal of Green Pharmacy, 3*(2).

Farswan, M., Mazumder, P. M., Parcha, V., & Upaganlawar, A., (2009). Modulatory effect of *Syzygium cumini* seeds and its isolated compound on biochemical parameters in diabetic rats. *Pharmacognosy Magazine, 5*(18), 127.

França, L. M., Coêlho, C. F. F., Freitas, L. N. C., Souza, I. L. S., Chagas, V. T., Debbas, V., De Lima, T. M., et al., (2019). *Syzygium cumini* leaf extract reverts hypertriglyceridemia via downregulation of the hepatic XBP-1s/PDI/MTP axis in monosodium L-glutamate-induced obese rats. *Oxid Med Cell Longev., 2019,* 9417498.

Ghosh, A., Jana, K., Ali, K. M., De, D., Chatterjee, K., & Ghosh, D., (2014). Corrective role of *Eugenia jambolana on testicular imp*airment in streptozotocin-induced diabetic male albino rat: An approach through genomic and proteomic study. *Andrologia, 46*(3), 296–307.

Gopalakrishna, P. G. K., Bharate, S. S., Awasthi, A., Verma, R., Mishra, G., Singh, A.T., Jaggi, M., et al., (2017). Antidiabetic potential of polyherbal formulation DB14201: Preclinical development, safety and efficacy studies. *Journal of Ethnopharmacology, 197,* 218–230.

Grover, J. K., Vats, V., & Rathi, S. S., (2000). Anti-hyperglycemic effect of *Eugenia jambolana and Tinospora cordifolia* in experimental diabetes and their effects on key metabolic enzymes involved in carbohydrate metabolism. *J. Ethnopharmacol., 73*(3), 461–470.

Jana, K., Bera, T. K., & Ghosh, D., (2015). Antidiabetic effects of *Eugenia jambolana in the streptozot*ocin-induced diabetic male albino rat. *Biomark. and Genom. Med., 7*(3), 116–124.

Jarald, E. E., Edwin, S., Tiwari, V., Garg, R., & Toppo, E., (2008). Antidiabetic activity of cow urine and a herbal preparation prepared using cow urine. *Pharmaceutical Biology, 46*(10, 11), 789–792.

Joshi, C. S., Priya, E. S., & Venkataraman, S., (2007). Hypoglycemic and anti-lipidperoxidative effects of a polyherbal formulation, diakyur, in experimental animal models. *Journal of Health Science, 53*(6), 734–739.

Kandan, P., & Govindan, S., (2017). Antidiabetic activity and phytochemical constituents of *Syzygium cumini* seeds in Puducherry region, South India *Inter. J. Pharmacog. Phytochem. Res., 9*(7), 985–989.

Kar, A., Choudhary, B. K., & Bandyopadhyay, N. G., (2003). Comparative evaluation of hypoglycemic activity of some Indian medicinal plants in alloxan diabetic rats. *Journal of Ethnopharmacology, 84*(1), 105–108.

Kumar, A., Raju, I., Jayachandran, T., Deecaraman, M., Aravindan, P., Padmanabhan, N., & Krishan, M., (2008). Antidiabetic activity of *Syzygium cumini* and it's isolated compound against streptozotocin-induced diabetic rats. *J. Med. Plant. Res., 2,* 246–249.

Kumar, P., Mehta, M., Satija, S., & Garg, M., (2013). Enzymatic *in vitro* antidiabetic activity of few traditional Indian medicinal plants. *Journal of Biological Sciences, 13*(6), 540–544.

Kumari, B., Sharma, V., & Yadav, S., (2016). The therapeutic potential of *Syzygium cumini* seeds in diabetes mellitus. *Journal of Medicinal Plants Studies, 5,* 212–218.

Mahajan, S., Chauhan, P., Subramani, S. K., Anand, A., Borole, D., Goswamy, H., & Prasad, G., (2015). Evaluation of "GSPF Kwath": A *Gymnema sylvestre*-containing polyherbal formulation for the treatment of human type 2 diabetes mellitus. *European Journal of Integrative Medicine, 7*(3), 303–311.

Maiti, S., Ali, K. M., Jana, K., Chatterjee, K., De, D., & Ghosh, D., (2013). Ameliorating effect of mother tincture of *Syzygium jambolanum* on carbohydrate and lipid metabolic disorders in streptozotocin-induced diabetic rat: Homeopathic remedy. *J. Nat. Sci. Bio. and Med., 4*(1), 68.

Majeed, W., Khaliq, T., Aslam, B., & Khan, J. A., (2018). Polyherbal formulation prevents hyperglycemia by modulating the biochemical parameters and upregulating the insulin signaling cascade in alloxan induced hyperglycemic rats. *Pakistan Veterinary Journal, 38*(2), 121–126.

Mandal, S., Barik, B., Mallick, C., De, D., & Ghosh, D., (2008). Therapeutic effect of ferulic acid, an ethereal fraction of ethanolic extract of seed of *Syzygium cumini* against streptozotocin-induced diabetes in male rat. *Methods and Findings in Experimental and Clinical Pharmacology, 30*(2), 121–128.

Mastan, S. K., Latha, T. B., Latha, T. S., Srikanth, A., Chaitanya, G., & Kumar, K. E., (2009). Influence of methanolic extract of *Syzygium cumini* seeds on the activity of gliclazide in normal and alloxan-induced diabetic rats. *Pharmacol Online, 3,* 845–850.

Middha, S. K., Usha, T., Tripathi, P., Marathe, K. Y., Jain, T., Bhatt, B., Masurkar, Y. P., & Pande, V., (2012). An *in vitro* studies on indigenous ayurvedic plants, having hypoglycemic activity. *Asian Pacific Journal of Tropical Disease, 2,* S46–S49.

Mutalik, S., Chetana, M., Sulochana, B., Devi, P. U., & Udupa, N., (2005). Effect of dianex, a herbal formulation on experimentally induced diabetes mellitus. *Phytotherapy Research: An International Journal Devoted to Pharmacological and Toxicological Evaluation of Natural Product Derivatives, 19*(5), 409–415.

Nisa, T. M., & Malik, A. S., (2010). Antidiabetic activity of inorganic metals of *Eugenia jambolana Lam. (Myrtaceae)* flowers. *Pharmacologyonline, 2,* 979–985.

Nishandhini, S., Sudha, V., Mallavarapu, G. R., & Murugan, R., (2015). Chemical compositions, α-Amylase inhibitory and antioxidant activities of the essential oils from unripe fruit pulp and leaves of *Syzygium Cumini. Int. J. Pharm. Pharm. Sci., 7*(2), 515–519.

Pandey, S., (2010). Treatment of diabetes and hyperlipidemia with extracts of *Eugenia jambolana seed and Aegle marmelos* leaf extracts in alloxan-induced diabetic rats. *International Journal of Pharma and Bio Sciences, 1*(2).

Patel, S. S., Shah, R. S., & Goyal, R. K., (2009). Antihyperglycemic, Antihyperlipidemic and Antioxidant Effects of Dihar, a Polyherbal Ayurvedic Formulation in Streptozotocin-Induced Diabetic Rats. *Ind. J. Exp. Biol., 47*(7), 564–570.

Perera, P. R. D., Ekanayake, S., & Ranaweera, K., (2017). Antidiabetic compounds in *Syzygium cumini* decoction and ready to serve herbal drink. *Evidence-Based Comple. and Alter. Med., 2017.*

Poongunran, J., Perera, H. K. I., Jayasinghe, L., Fernando, I. T., Sivakanesan, R., Araya, H., & Fujimoto, Y., (2017). Bioassay-guided fractionation and identification of α-amylase Inhibitors from *Syzygium cumini* leaves. *Pharmaceutical Biology, 55*(1), 206–211.

Proma, N. M., Naima, J., Islam, M. R., Papel, J. A., Rahman, M. M., & Hossain, M. K., (2018). Phytochemical constituents and antidiabetic properties of *Syzygium cumini* Linn. seed. *Int. J. Pharm. Sci. Res., 9*(5), 1806–1814.

Qadir, M. I., & Malik, S. A., (2009). Effect of *Eugenia jambolana leaves extracts o*n blood glucose levels of experimental diabetic rabbits. *Pharmacologyonline, 3,* 829–835.

Rafiullah, M. R. M., Siddiqui, A. W., Mir, S. R., Ali, M., Pillai, K. K., & Singh, S., (2006). Antidiabetic activity of some Indian medicinal plants. *Pharmaceutical Biology, 44*(2), 95–99.

Ravi, K. A., Ponnusamy, S., Ravindran, R., Zinjarde, S., & Bhargava, S., (2011). Evaluation of traditional Indian antidiabetic medicinal plants for human pancreatic amylase inhibitory effect *in vitro. Evidence-based Complementary and Alternative Medicine.* Art. no., 515647.

Ravi, K., Rajasekaran, S., & Subramanian, S., (2003). Hypoglycemic effect of *Eugenia jambolana seed kernels on s*treptozotocin-induced diabetes in rats. *Pharmaceutical Biology, 41*(8), 598–603.

Raza, A., Butt, M. S., & Suleria, H. A. R., (2017). Jamun (*Syzygium cumini*) seed and fruit extract attenuate hyperglycemia in diabetic rats. *Asian Pacific Journal of Tropical Biomedicine, 7*(8), 750–754.
Rekha, N., Balaji, R., & Deecaraman, M., (2008). Effect of aqueous extract of *Syzygium cumini* pulp on antioxidant defense system in streptozotocin-induced diabetic rats. *Iranian Journal of Pharmacology and Therapeutics, 7*(2), 137–140.
Sagrawat, H., Mann, A. S., & Kharya, M. D., (2006). Pharmacological potential of *Eugenia jambolana:* A review. *Pharmacogn Mag., 2,* 96–104.
Samadder, A., Chakraborty, D., De, A., Bhattacharyya, S. S., Bhadra, K., & Khuda-Bukhsh, A. R., (2011). Possible signaling cascades involved in attenuation of alloxan-induced oxidative stress and hyperglycemia in mice by ethanolic extract of *Syzygium jambolanum*: Drug-DNA interaction with calf thymus DNA as target. *Euro. J. Pharmaceu. Sci., 44*(3), 207–217.
Sampath, S., Narasimhan, A., Chinta, R., Nair, K. J., Khurana, A., Nayak, D., Kumar, A., & Karundevi, B., (2013). Effect of homeopathic preparations of *Syzygium jambolanum* and *Cephalandra indica* on gastrocnemius muscle of high fat and high fructose-induced type-2 diabetic rats. *Homeopathy, 102*(3), 160–171.
Sanches, J. R., França, L. M., Chagas, V. T., Gaspar, R. S., Dos, S. K. A., Gonçalves, L. M., Sloboda, D. M., et al., (2016). Polyphenol-rich extract of *Syzygium cumini* leaf dually improves peripheral insulin sensitivity and pancreatic islet function in monosodium L-glutamate-induced obese rats. *Front Pharmacol., 10*(7), 48.
Saravanan, G., & Pari, L., (2008). Hypoglycaemic and anti-hyperglycemic effect of *Syzygium cumini* bark in streptozotocin-induced diabetic rats. *J. Pharmacol. and Toxicol., 3*(1), 1–10.
Sawant, L., Singh, V. K., Dethe, S., Bhaskar, A., Balachandran, J., Mundkinajeddu, D., & Agarwal, A., (2015). Aldose reductase and protein tyrosine phosphatase 1B inhibitory active compounds from *Syzygium cumini* seeds. *Pharmaceutical Biology, 53*(8), 1176–1182.
Schoenfelder, T., Warmlin, C. Z., Manfredini, M. S., Pavei, L. L., Réus, J. V., Tristão, T. C., Fernandes, M. S., & Costa-Campos, L., (2010). Hypoglycemic and Hypolipidemic effect of leaves from *Syzygium Cumini* (L.) skeels, myrtaceae. in diabetic rats. *Revista Brasileira de Farmacognosia, 20*(2), 222–227.
Shankar, M. B., Parikh, J. R., Geetha, M., Mehta, R. S., & Saluja, A. K., (2007). Antidiabetic activity of novel androstane derivatives from *Syzygium cuminii* Linn. *Journal of Natural Remedies, 7*(2), 214–219.
Sharma, A. K., Bharti, S., Kumar, R., Krishnamurthy, B., Bhatia, J., Kumari, S., & Arya, D. S., (2012). *Syzygium cumini* ameliorates insulin resistance and β-cell dysfunction via modulation of PPARγ, dyslipidemia, oxidative stress, and TNF-α in type 2 diabetic rats. *J. Pharmacol. Sci., 119*(3), 205–213.
Sharma, B., Balomajumder, C., & Roy, P., (2008). Hypoglycemic and hypolipidemic effects of flavonoid rich extract from *Eugenia jambolana seeds on* streptozotocin induced diabetic rats. *Food and Chemical Toxicology, 46*(7), 2376–2383.
Sharma, S. B., Nasir, A., Prabhu, K. M., Murthy, P. S., & Dev, G., (2003). Hypoglycaemic and hypolipidemic effect of ethanolic extract of seeds of *Eugenia jambolana in alloxan-induce*d diabetic rabbits. *J. Ethnopharmacol., 85*(2, 3), 201–206.

Sharma, S. B., Rajpoot, R., Nasir, A., Prabhu, K. M., & Murthy, P. S., (2011). Ameliorative effect of active principle isolated from seeds of *Eugenia jambolana on* carbohydrate metabolism in experimental diabetes. *Evidence-Based Complementary and Alternative Medicine, 2011.*

Sharma, S., Pathak, S., Gupta, G., Sharma, S. K., Singh, L., Sharma, R. K., Mishra, A., & Dua, K., (2017). Pharmacological evaluation of aqueous extract of *Syzigium cumini* for its antihyperglycemic and anti-dyslipidemic properties in diabetic rats fed a high cholesterol diet—role of PPARγ and PPARα. *Biomedicine & Pharmacotherapy, 89,* 447–453.

Shinde, J., Taldone, T., Barletta, M., Kunaparaju, N., Hu, B., Kumar, S., Placido, J., Zito, S. W., (2008). α-Glucosidase inhibitory activity of *Syzygium cumini* (Linn.) skeels seed kernel *in vitro* and in goto–kakizaki (GK) rats. *Carbohydrate Research, 343*(7), 1278–1281.

Singh, N., & Gupta, M., (2007). Effects of ethanolic extract of *Syzygium cumini* (Linn.) seed powder on pancreatic islets of alloxan diabetic rats. *Indian J. Exp. Biol., 45*(10), 861–867.

Sridhar, S. B., Sheetal, U. D., Pai, M., & Shastri, M. S., (2005). Preclinical evaluation of the antidiabetic effect of *Eugenia jambolana* Seed Powder in streptozotocin-diabetic rats. *Braz. J. Med. Biol. Res., 38*(3), 463–468.

Srivastava, S., & Chandra, D., (2013). Pharmacological potentials of *Syzygium cumini*: A review. *Journal of the Science of Food and Agriculture, 93*(9), 2084–2093.

Subramoniam, A., (2016). *Plants with Antidiabetes Mellitus Properties*. Taylor and Fransis Group, CRC Press, Boca Raton.

Tanwar, R. S., Sharma, S. B., & Prabhu, K. M., (2017). *In vivo* assessment of antidiabetic and antioxidative activity of natural phytochemicals isolated from fruit-pulp of *Eugenia jambolana in Streptozotocin*-induced diabetic rats. *Redox Report, 22*(6), 301–307.

Tanwar, R. S., Sharma, S. B., Singh, U. R., & Prabhu, K. M., (2010). Attenuation of renal dysfunction by anti-hyperglycemic compound isolated from fruit pulp of *Eugenia jambolana in* streptozotocin-induced diabetic rats. *Indian J. Biochem. Biophys., 47*(2), 83–89.

Telapolu, S., Kalachavedu, M., Punnoose, A. M., & Bilikere, D., (2018). MD-1, a polyherbal formulation indicated in diabetes mellitus ameliorates glucose uptake and inhibits adipogenesis–an *in vitro* study. *BMC Complemen. and Altern. Med., 18*(1), 113.

Thiyagarajan, G., Muthukumaran, P., Sarath, K. B., Muthusamy, V. S., & Lakshmi, B. S., (2016). Selective inhibition of PTP 1B by vitalboside A from *Syzygium cumini* enhances insulin sensitivity and attenuates lipid accumulation via partial agonism to PPAR γ: *In vitro* and in silico investigation. *Chem. Bio. & Drug Desig., 88*(2), 302–312.

Tripathi, A. K., & Kohli, S., (2014). Pharmacognostical Standardization and antidiabetic activity of *Syzygium cumini* (Linn.) barks (Myrtaceae) on streptozotocin-induced diabetic rats. *J. Comple. Integra. Med., 11*(2), 71–81.

Zulcafli, A. S., Lim, C., Ling, A. P., Chye, S., & Koh, R., (2020). Antidiabetic potential of *Syzygium* sp.: An overview. *Yale. J. Biol, Med., 93*(2), 307–325.

CHAPTER 4

Cinnamomum zeylanicum: Nature's Wonder Plant with Antidiabetic Prominence

SASIREKA RAJENDRAN,[1] ANU PANDITA,[2] VINOTH RATHINAM,[3] SUGUMARI VALLINAYAGAM,[4] DEEPU PANDITA,[5] and MADHESWARAN MUTHUSAMY[6]

[1]*Department of Biotechnology, Mepco Schlenk Engineering College, Sivakasi, Tamil Nadu, India*

[2]*Vatsalya Clinic, Krishna Nagar, New Delhi, India*

[3]*Department of Electronics and Communication Engineering, P.S.R. Engineering College, Sivakasi, Tamil Nadu, India*

[4]*Department of Biotechnology, Mepco Schlenk Engineering College, Sivakasi, Tamil Nadu, India*

[5]*Government Department of School Education, Jammu, Jammu and Kashmir, India, E-mail: deepupandita@gmail.com*

[6]*Muthayammal Engineering College, Rasipuram, Tamil Nadu, India, E-mail: prof.madheswaran@gmail.com*

ABSTRACT

Diabetes mellitus (DM) is a condition wherein the human body cannot secrete adequate insulin or a state where produced insulin is not utilized which leads to abnormally higher levels of blood sugar (glucose). The

Antidiabetic Plants for Drug Discovery: Pharmacology, Secondary Metabolite Profiling, and Ingredients with Insulin Mimetic Activity. Deepu Pandita, MPhil, Anu Pandita, MSc, & Chander Bhanu, MBBS, PGDDCE (Eds.)
© 2023 Apple Academic Press, Inc. Co-published with CRC Press (Taylor & Francis)

metabolism of carbohydrates, proteins, and fats is getting affected due to prolonged hyperglycemia Prevalence of this disease is very common among adults nowadays. Treatment is also challenging even though drugs are used. The need for an efficient alternative drug is vital to manage the disease. Cinnamon with secondary metabolites of camphor eugenol, trans-alpha bergamotene, oxygenated terpenoids, terpene hydrocarbons, α-copaene, α-bergamotene, β-caryophyllene, and trans-cinnamyl acetate especially cinnamaldehyde, is one of the efficient Wonder Plants of Nature with anti-diabetic potential that can be used as a better substitute to cure diabetes. The current chapter will discuss these properties of *Cinnamomum zeylanicum.*

4.1 INTRODUCTION

In the 16th century, cinnamon emerged as an age-old spice and it was identified by the Portuguese people in Sri Lanka. The plant has been used for thousands of years since it enhances the taste in the foods, and also it holds potential medicinal benefits. Cinnamon is the most widely used aromatic component worldwide, following pepper. The taxonomy for *Cinnamomum zeylanicum* comes from Sri Lanka's previous term, Ceylon, which identifies the plant's origin. The genus Cinnamomum refers to the family "Lauraceae" and it was found that nearly 250 aromatic and flavoring species scattered among Australia, China, and Southeast Asia. Normally, dehydrated core bark of the plant, *Cinnamomum zeylanicum* Blume (syn. *Cinnamomum verum*) commonly known as "true" cinnamon or Ceylon cinnamon, native to Sri Lanka and the Malabar Coast of India, is used for a commercial purpose. Cinnamon is found to be pale tan and slightly sweet in the aroma. Cassia, native to Myanmar (Burma) is another closely related spice but of having substandard quality. Cassia cinnamon is originated as *C. cassia* (syn. *C. aromatica*) according to Vietnam and China. Similarly, *Cinnamomum burmanii*, Indonesian cinnamon originated from the Java region and Sumatra, and the *Cinnamomum tamala*, Indian cassia originates from the region of Myanmar (northeastern part) and India, and *Cinnamomum loureiroi* (syn. *Saigon cassia*) especially from Vietnam (Mbaveng and Kuete, 2017).

FIGURE 4.1 Parts of *Cinnamomum* spp.

Source: (a) https://www.kindpng.com/imgv/hwomwoR_cinnamomum-zeylanicum-botanical-illustration-hd-png-download/; (b) https://images.app.goo.gl/PGtb9tQvWQt5-kwDk6; (c) https://images.app.goo.gl/TjB12NgZ45hYLZTa6; (d) https://images.app.goo.gl/yXNEtK3AHLiH64ms8; (e) https://images.app.goo.gl/fWgKesQCWiLqcyNF6.

The Cinnamon species is native to Sri Lanka, and it is considered as one of the important manufacturers and sellers of leaf oil and bark oil with 24,000 ha in farming. In addition to Sri Lanka, small quantities of cinnamon bark are produced by Seychelles, Madagascar, and India with greater quality. In India, true cinnamon is manufactured in Kerala and coastal parts of Karnataka, also in several places of Tamil Nadu (Krishnamoorthy et al., 1996). Prominent Islands for cinnamon are the Andaman, Nicobar, Lakshadweep, and the Maldives. Based on the International Herald Tribune in 2006, nearly 90% of the entire world's cinnamon is manufactured by Sri Lanka, followed by Vietnam, India, and China. According to the FAO, 40% of the whole world's Cassia genus of cinnamon is by Indonesia. The major importers of cinnamon are France, Mexico, Singapore, Saudi Arabia, Hong Kong, West Germany, Taiwan, the USA, and the UK (Thomas, 2012).

Figure 4.1 represents classic cinnamon plant parts. Usually, the bark occurs in golden brown subsequently crushed into powdered form. The cinnamon spice cultivated in Sri Lanka holds superiority because of its organoleptic properties and even within the plantations, there are noteworthy compositional distinctions. The occurrence of diverse cultivars has been recorded on the island (Senanayake et al., 1989). They are familiarized by the nature of sensory evaluation and are represented by taxonomy like 'honey,' 'camphoraceous,' 'sweet,' and also 'bloom,' 'wild,' and 'mucilaginous.' The honey and sweet cultivars are widely produced (Senanayake et al., 1997).

The whole plant is found to be useful for many applications. However, the anti-diabetic effect of this plant is found to be explored nowadays by many studies. There are several reports which exhibit anti-diabetic nature of cinnamon in animal models. This chapter presently focuses on the anti-diabetic effect of cinnamon and its importance.

4.2 NATURE OF CINNAMON

The cinnamon tree prefers humid perennial conditions that may enlarge to 7 m in its normal condition. On the other hand, under farming, the plant habitually grew like a bush, restricting the plant's tallness to below 3 m, ever since the stems are constantly cut back to produce new shoots for bark/quills extraction. The plant possesses thick scabrous bark and strong branches. The cinnamon bark is found to have a sweet aromatic taste in its ordinary or boiled state and possess a beautiful smell and hot nature. The cinnamon leaves are present in an ovate-oblong shape and it grows up to a height of 7 to 18 cm. The plants to some extent are bitter and found to have an appealing fragrance when it is baked or deep-fried. The flowers are organized in panicles, greenish with a distinctive odor. It is noted that the fruit occurs in purple color, with a 1 cm elongated berry with one seed. The plant prospers greatest in a warm, damp frigid environment at little elevation and is typically complete to reap after three years. The preferred soil would not be drenched as this creates a bitter bark. 8 or 10 side branches may arise on the plant and these can be collected after 3 years to obtain quality cinnamon bark. Cinnamon is primarily acquired by exposing the middle region of the bark to air and made dried, which is sold as spines or fine particles. The dried-out leaves may develop a component of spice combinations also. The essential oil pulled out from the cinnamon bark is utilized as a dense basis of cinnamon flavor, especially in the tinned foodstuff and confectionery industries. Cinnamon gets a unique smell and fragrance because of the volatile oils present in the bark or leaves (Wang et al., 2014).

4.3 PROPERTIES OF CINNAMON

The properties of cinnamon include aromatic stimulant, anticancer, nematicidal, antioxidant, sweet, hypoglycemic, insecticidal, anodyne, warming, anti-inflammatory, astringent, anti-mycotic, demulcent, and antimicrobial (Figure 4.2).

4.4 CHEMICAL COMPOSITION OF CINNAMON AND STRUCTURE OF METABOLITES

The composition of cinnamon varies in the parts such as top, middle, and bottom segments, and at distinct development phases (Geng et al., 20110). Cinnamon bark top and center segments are highly resourceful for extraction of volatile oils when compared with whole bark and especially of 12 years old. The oil present in the cinnamon bark is found to be rich in trans-cinnamaldehyde (Singh et al., 2007). The compound's presence is minimum at the initial years of growth (33.95%) and maximum at 6-years of growth (76.4%). Composition of Cinnamon bark are as follows: vitamin B1 (0.14 mg/100 g), moisture 5.4–11.4%, iron 0.04%, carbohydrates 16.6–22.6%, potassium 0.4%, protein 3.0–4.65%, fiber 25.6–30.5%, total ash 3.55%, phosphorous 0.05%, sodium 0.01%, vitamin B2 (0.21 mg/100 g), calcium 1.6%, volatile oil 0.3–2.8%, fixed oil 0.3–1.9%, niacin (1.9 mg/100 g), and vitamin C (39.8 mg/100 g). Cinnamon possesses important constituents but its occurrence varies considerably with plant location, climatic condition, age of the tree, harvest time, season, and duration of packing (Figure 4.3) (Shumaila Gul and Mahpara Safdar, 2009).

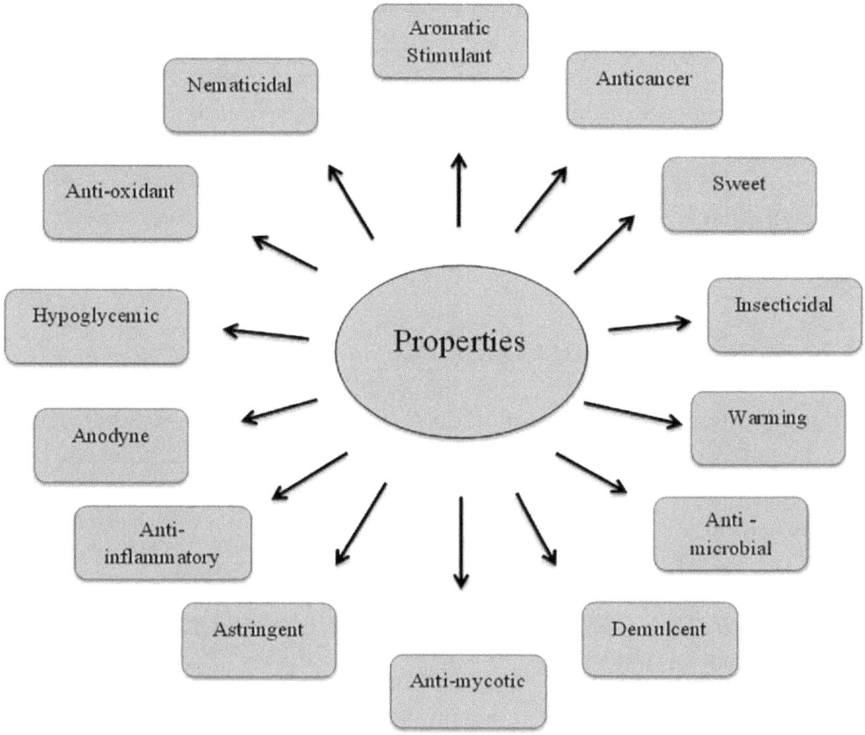

FIGURE 4.2 Different properties of cinnamon.

The components in the cinnamon plant parts are extracted by different methodologies and are identified by various analytical methods like HPLC (high-performance liquid chromatography), LC-MS (liquid chromatography-mass spectrometry, and GC-MS (gas chromatography-mass spectrometry) are listed in Table 4.1. The structure of various compounds present in Cinnamon are listed in Figures 4.3 and 4.4.

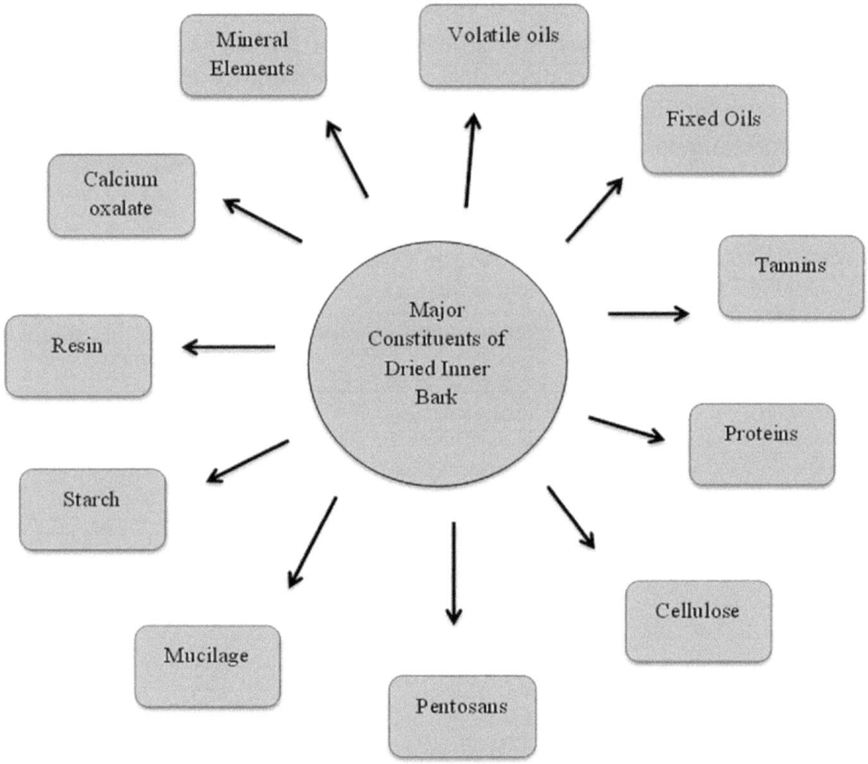

FIGURE 4.3 Major constituents of cinnamon bark.

The major constituents present in the cinnamon plant parts along with their composition are highlighted in Table 4.2. Leaves and bark have eugenol and cinnamaldehyde as major components; root bark generally holds camphor while fruits contain β-caryophyllene and trans-cinnamyl acetate. Besides buds of *C. zeylanicum* contains α-copaene, oxygenated terpenoids, and α-bergamotene, terpene hydrocarbons, as major components however flowers have constituent trans-cinnamyl acetate, α-bergamotene, and caryophyllene oxide. Other compounds such as cinnamyl alcohol, borneol, cinnamyl acetate, cinnamic acid, coumarin, hydroxyl cinnamaldehyde, etc., are present in lesser quantities (Vangalapati et al., 2012). But, the large quantity of coumarin differs in Cinnamomum species. Amongst the compound coumarin can bring damage to the liver. Amongst the leading varieties of cinnamon, *C. zeylanicum* is found to have a little amount of coumarin (17 mg/kg), whereas varieties, like *C. loureiroi* (6,970 mg/kg), *C. cassia* (310 mg/kg), and *C. burmannii* (2,150 mg/kg), to some extent have more amount of coumarin. *C. cassia* comprises nearly 1 to 2% of cassia oil (volatile oil). 65% to 80% cinnamaldehyde and Eugenol are the chief compounds in the essential oils (Figure 4.4) (Pasupuleti Visweswara Rao and Siew Hua Gan, 2014).

TABLE 4.1 Key Components of Diverse Cinnamon Species Based on the Different Types of Extract and Analytical Approaches

Cinnamon Species	Type of Extracts	Components Explored	Analytical Approaches	References
C. cassia	Aqueous-methanolic	Cinnamaldehyde, eugenol, cinnamic acid, protocatechuic acid, and (-)-epicatechin	HPLC	Yang et al. (2007)
	Methanolic extract	Eugenol, cinnamaldehyde, coumarin, and cinnamic acid	HPLC	He et al. (2005)
	Oil	2-methoxycinnamaldehyde (1.85%), gamma-muurolene (1.56%), trans-cinnamaldehyde (60%), cinnamyl alcohol acetate (1.18%), beta-bisabolene (1.25%), trans-alpha-bisabolene (2.54%), copaene (7.37%), tetradecanal (0.83 and (1)-delta-cadinene (2.92%), Cis-cinnamaldehyde (1.43%),%), alpha-muurolene (1.92%)	GC-MS analysis	Geng et al. (2011)
C. zeylanicum	Aqueous extract	Cinnamyl alcohol, cinnamaldehyde	HPLC	Shen et al. (2010)

TABLE 4.1 *(Continued)*

Cinnamon Species	Type of Extracts	Components Explored	Analytical Approaches	References
	Oil	α-amorphene (0.5%), cadinene (0.9%), α-copaene (0.8%), and Cinnamaldehyde (97.7%)	GC-MS analysis	Singh et al. (2007)
C. wilsoni	Methanol extract	Cinnamaldehyde, EUGENOL	HPLC	He et al. (2005)
C. mairei	Methanol extract	Cinnamaldehyde, eugenol	HPLC	He et al. (2005)
C. loureirii	Methanol extract	Cinnamic acid, cinnamaldehyde, coumarin	HPLC	He et al. (2005)
C. burmannii	Aqueous extract	Vanillin azine, ferulic acid, cinnamyl alcohol, guiacol, o-coumaric acid, eugenol, p-coumaric acid, cinnamic acid methyl ester, 2-methoxy-cinnamaldehyde, cinnamide, chlorogenic acid, t-cinnamic acid, clove oil, homovanillic acid, isovanillic acid, 3-methoxyl-tyrosine, resveratrol, vanillic acid, curcumin, 4-oxo-4h-1-benzopyran-carboxylic acid, and procyanidin type-A oligomers	HPLC	Anderson et al. (2004)
	Methanol extract	Procyanidin tetramer (2.35%), procyanidin trimer (13.76%), (S)-cinnamaldehyde (1.95%), (E)-cinnamic acid (2.69%), (E)-cinnamaldehyde (62.18%), Procyanidin B1 (0.7%), procyanidin dimer (1.27%), (-)-epicatechin (1.02%), procyanidin B2 (5.12%), and (1)-catechin (2.61%)	GC-MS and LC-MS analysis	Shan et al. (2007)

TABLE 4.1 *(Continued)*

Cinnamon Species	Type of Extracts	Components Explored	Analytical Approaches	References
	Oil	Cyclopentadecane (1.29%), T-cadinol (1.15%), δ-elemene (2.32%), cinnamaldehyde (2.70%), α-ylangene (0.52%), camphor (1.79%), 4-terpineol (0.50%), α-cubebene (1.56%), pentadecanol (0.96%), α-amorphene (5.39%), α-calacorene (1.71%), caryophylene (1.23%), calarene (0.94%), aromadendrene (1.47%), α-murolene (2.40%), epibicyclo-sesquiphellandrene (0.44%), β-guaiene (2.14%), patchoulene (0.25%), γ-eudesmol (1.28%), α-humulene (0.43%), santalene (0.77%), valencene (0.60%), γ-cadinene (0.50%), elaidinsaeure (7.71%), isocurcumenol (0.55%) cariophylnenyl alcohol (0.49%), δ-cadinene (5.98%), α-eudesmol (1.63%), and palmitic acid (2.48%)	GC-MS analysis	Thantsin et al. (2008)
C. osmophloeum	Oil	3-phenyl-2-propenal (4.06%), cinnamaldehyde (6.87%), cinnamyl acetate (11.71%), linalool (40.24%), trans-camphor (9.38%), coumarin (2.13%), and caryophyllene (2.65%)	GC-MS analysis	Lee et al. (2013)

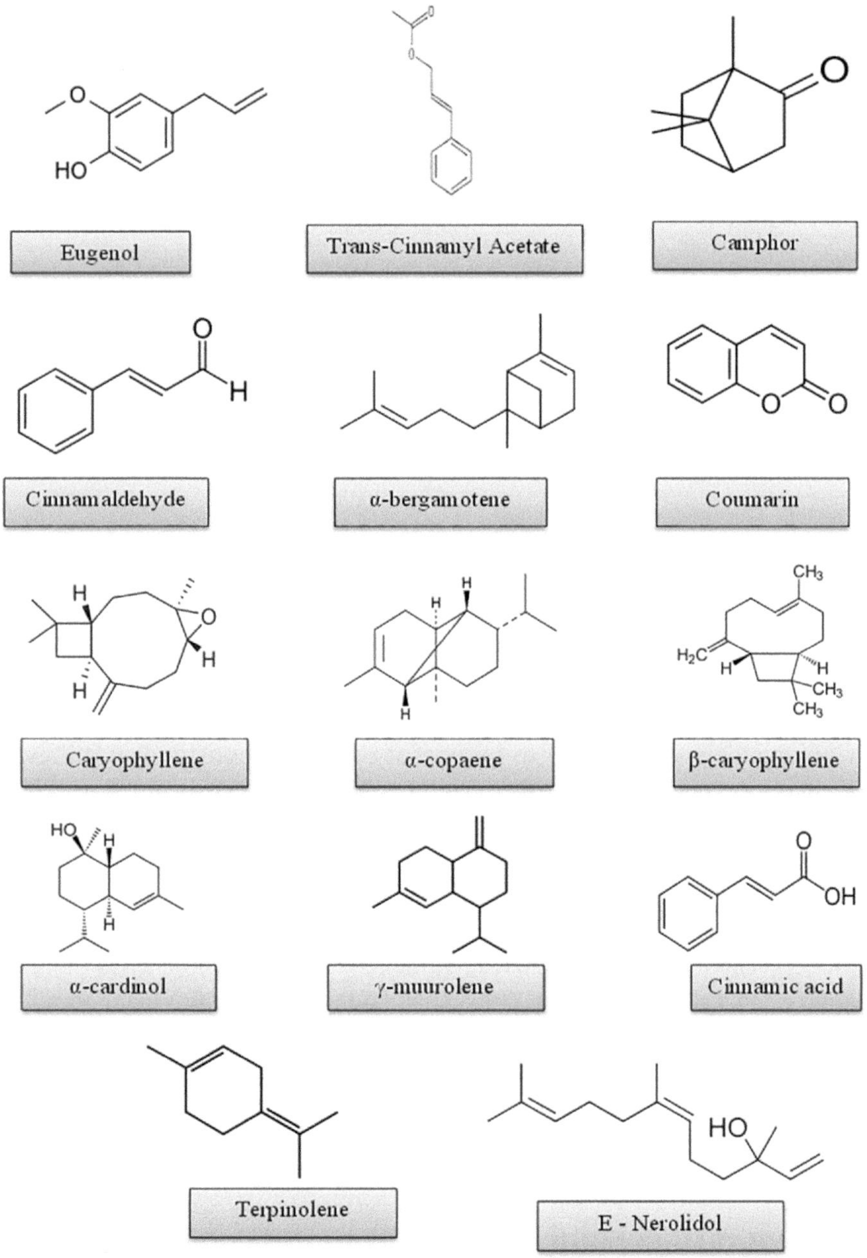

FIGURE 4.4 Chemical structures of major compounds present in cinnamon.

TABLE 4.2 Major Compounds Found in Cinnamon Plant Parts

SL. No.	Parts of *Cinnamomum zeylanicum*	Major Compound	Percentage Composition
1.	Root bark	Camphor	60%
2.	Leaf	Eugenol	70 to 95%
		Cinnamaldehyde	1 to 5%
3.	Flowers	Trans-alpha Bergamotene	7.97%
4.	Buds	Oxygenated terpenoids	9%
		Terpene hydrocarbons	78%
		α-copaene	23.05%
		α-Bergamotene	27.38%
5.	Bark	Eugenol	5 to 10%
		Cinnamaldehyde	65 to 80%
6.	Fruit	β-Caryophyllene	9 to 14%
		Trans-cinnamyl acetate	42 to 54%

4.5 SECONDARY METABOLITE PROFILING AND FUNCTIONAL COMPONENTS

Cinnamaldehyde is found to be one of the major constituents of several Cinnamon species (Gopal Rao Mallavarapur and Rajeswara Rao, 2007). Eugenol is the second major constituent in the cinnamon species. Besides, 2'-hydroxycinnamaldehyde, and 2'-benzoloxycinnamaldehyde, cinnzeylanine, cinnzeylanol, arabinoxylan are also observed. Polyphenols are the major secondary metabolites of this plant and take part in the oxidative process. Bioactive constituents are validated to have possible ways of natural antioxidants (Dragland et al., 2003) displaying potent antioxidant activity in the selected *in vitro* models (Su et al., 2007, Mathew et al., 2006). Cinnamon polyphenols enhance the superoxide dismutase activity, and improve the scavenging ability of oxygen free radicals in the specific organisms, and protect tissues from oxidative stress injury (Figure 4.5) (Li et al., 2013).

4.6 PHARMACOLOGICAL EFFECTS OF CINNAMON

The spice also possesses several functional properties because of its natural components. *In vivo* and *In vitro* studies revealed that it holds major components exhibit wide pharmacological effects such as a potential defensive role in type II diabetes mellitus (Allen et al., 2013), antioxidant activity (Shen et al., 2010), cardiovascular diseases, cancer (Lu et al., 2010), neurological disease (Ulbricht et al., 2011), antiviral, antihypertensive, cholesterol and lipid-lowering (Ping et al., 2010), anti-inflammatory activity (Koteswara et al., 2007), and antimicrobial action (Winias et al., 2011). Table 4.3 highlights the important pharmacological activities of cinnamon and its related compounds.

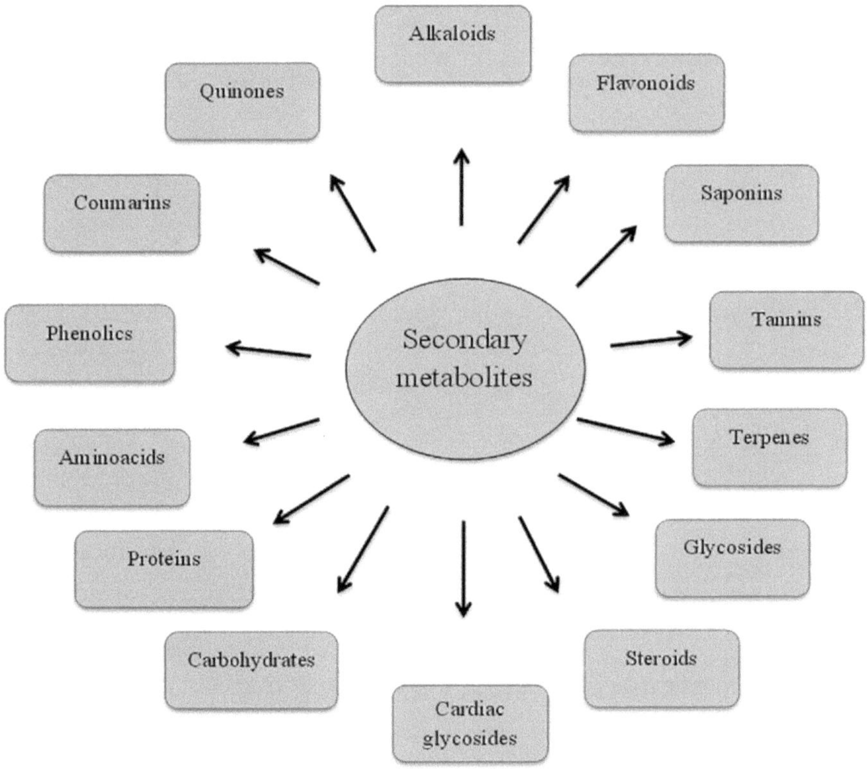

FIGURE 4.5 Major secondary metabolites of cinnamon.

Whole plant parts have wide applications, especially in the medicinal field. The major medicinal applications of *C. zeylanicum* are given in Figure 4.6.

TABLE 4.3 Pharmacological Effects of Components of Cinnamon Species

Cinnamon Species	Substance/ Type of Extract/ Components	Pharmacological Activity	References
C. philippinense	Cinnamophilin	Improvement of bronchoconstriction arrhythmia	Su et al. (1999)
C. cassia	Cinnamaldehyde	Anticancer, immunomodulatory, antiviral	Nishida et al. (2003); Hayashi et al. (2007)
C. zeylanicum *C. cassia*	Aqueous extract	Anti-ulcer	Keller et al. (1992)
C. cassia	Aqueous extract	Anti-inflammatory	Nagai et al. (1982a, b)
C. zeylanicum	Water, essential oil, and alcoholic extracts	Antioxidant	Chericoni et al. (2005)
C. cassia	Plant	Cholesterol and lipid-lowering	Khan et al. (2003)
C. zeylanicum *C. cassia*	Cinnamaldehyde aqueous extract	Antidiabetes	Wang et al. (2007)
C. burmannii	Acetic acid extract	Anti-hypertension	Preuss et al. (2006)
C. cassia	Aqueous extract	Depression, ability to improve the central nervous system	Iwasaki et al. (2008)
C. cassia	Methylene chloride extracts and ethanol	Gastroprotection	Tabak et al. (1999)

4.7 ANTI-DIABETIC ACTIVITY OF CINNAMON

The anti-diabetic property of *C. zeylanicum* was studied *in vitro* and *in vivo* conditions by many researchers. In an experimental study conducted by Verspohl et al. (2005) plant extracts from the species, *C. zeylanicum* was given to experimental rats to assess the plasma insulin and blood glucose level under diverse circumstances. The experimental results exhibit that *C. zeylanicum* extracts have superior activity on diabetes. Glucose tolerance test (GTT), specified that blood glucose levels dramatically reduces; however, it is not observed in rats that were not subjected to glucose treatments. In a study conducted by Verspohl et al. (2005), elevated levels of plasma insulin were observed. Besides, the insulin release from an insulin-secreting cell line (INS-1) was also detected as a result of the stimulatory *in vitro* effect. *C. zeylanicum* also brings down the activity of postprandial intestinal glucose absorption by enzyme inhibition mechanism which is associated with provoked cellular glucose uptake process by glucose transporter type 4 (GLUT4) membrane translocation mechanism, glycogen synthesis, carbohydrate assimilation (pancreatic α–amylase and α-glucosidase), and stimulated glucose metabolism, inhibition of gluconeogenesis by significant regulatory enzymes and also provoke in the insulin release and simultaneously increases insulin receptor activity (Ranasinghe et al., 2012, 2013). The active compound which is responsible for the beneficiary action is predicted as Cinnamtannin B1. Cao et al. (2007) analyzed the anti-diabetic effect of the cinnamon plant under *In vivo* conditions. Weight loss reduction related to diabetes, decreased blood sugar levels, reduced hemoglobin A1c, reduction in bad cholesterol level and increase in the circulating insulin levels. Positive results on nephropathy and diabetic neuropathy were also exhibited by this plant.

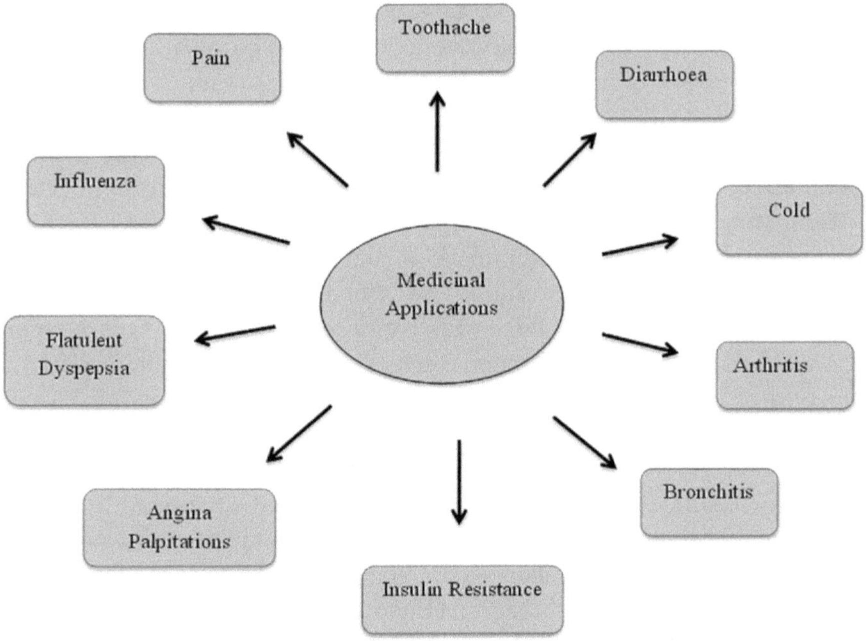

FIGURE 4.6 Applications of cinnamon.

Further investigation on the activity of cinnamon on the protein, GLUT4, and tristetraprolin (TTP/ZFP36) in animal model mouse 3T3-L1 adipocytes and mRNA levels of insulin receptor (IR) were monitored. Results indicated that cinnamon could increase the protein numbers involved in an anti-inflammatory/anti-angiogenesis response, glucose transport, and insulin signaling. Kim and coworkers studied the possible anti-diabetic effect of cinnamon bark extract in a type II diabetic animal model C57BIKsj db/db. Crude cinnamon extract was administrated at different dosages such as 50, 100, 150, and 200 mg/kg. When compared with the control, in treatment, the blood glucose level decreases significantly, especially in the 200 mg/kg group. Besides, it was observed that serum insulin levels and High-Density Lipoproteins – cholesterol levels were found to be considerably higher and the concentrations of intestinal α-glucosidase activity, triglyceride, and total cholesterol were observed to be less after the administration of 6 weeks of the extracts. The data indicated that extract from cinnamon has a controlling activity in lipids and blood glucose level and it exhibits a blood glucose-suppressing effect by reducing the absorption of carbohydrates in the small intestine and improving insulin sensitivity (Cao et al., 2007).

In another study by Kim et al. (2006) treatments were given at 200 mg/kg/day dosage for 12 weeks. Reduction in the blood glucose level and improved levels of glutathione peroxidase (GPX), glutathione-S-transferase (GST) activity, superoxide dismutase (SOD), reduced glutathione, and the glutathione reductase (GR) and catalase (CAT) in the liver was monitored. The cinnamon extract considerably lowers the lipid peroxidation state. These data proved that cinnamon extract could be efficient for preventing diabetic complications and hyperglycemia. Reduction in the plasma glucose concentration by cinnamaldehyde was observed in streptozotocin-induced diabetic rats by Subash et al. (2007) in a significant and dose-dependent manner. Shen et al. stated that cinnamon extracts encourage glucose transportation by glucose transporter 4 in muscles and brown adipose tissue. Optimistic effects on type 1 and 2 diabetes mellitus by cinnamon were observed in clinical research studies. In an experimental study by Anand et al. (2010) Cinnamon stimulated a compound, IRS-1 in rats, which further promotes glucose uptake in muscle tissue by stimulating insulin secretion. It also enhances the glucose uptake from the bloodstream towards muscle tissue by inducing Glucose transporter 4 (Qin et al., 2009). An experimental study was conducted by Khan et al. (2003) with nearly

60 patients around the age group of 50. The patients were separated into six groups. Each group has nearly 10 patients and the groups 1 to 3 were provided with 1 g, 3 g, or 6 grams of cinnamon for a period of 40 days every day. On the other hand, groups 4 to 6 were given the placebo. The total cholesterol, fasting glucose, triglycerides, and LDL cholesterol of all the patients were monitored. The recorded results revealed that no substantial changes were detected in the placebo group over the 40 days. However, in the treated groups, the reductions in triglycerides (23% to 30%), fasting glucose (18% to 29%), LDL cholesterol (7% to 27%) and total cholesterol (12% to 26%), were observed. Also, high doses and long-term ingestion of cinnamon were found to be safe. A related study was conducted in Germany, wherein 79 type 2 diabetes patients were examined. In this study, nearly half of the patients were treated as a placebo while the others received 3 grams of cinnamon every day for 4 months. In this experimental study, it was observed that between the two groups there is no significant difference was identified for HDL or LDL cholesterol, HgbA1c, or triglycerides, while the fasting glucose levels reduced about 7% high in the group receiving cinnamon (Mang et al., 2006). Several pieces of evidence propose that cinnamon extracts possess constituents to improve the insulin action; however, the effects of cinnamon in the case of non-insulin stimulated glucose uptake still need additional examination. An experimental study stated the effects of cinnamaldehyde under both basal and insulin-stimulated conditions on the glucose transport action of GLUT1 in L929 fibroblast cells where glucose uptake is triggered by glucose deficiency. It was observed that cinnamaldehyde has a twin action on the glucose transport activity of GLUT1. It enhances the glucose uptake under basal conditions and reaches a 3.5 fold extreme stimulation at 2.0 mM. On the other hand, cinnamaldehyde hinders glucose uptake activation by glucose deprivation significantly. In another experiment by Rind et al. (2010) the effect of *Cinnamonum zeylanicum* aqueous extract was analyzed for anti-diabetic activity. Cholesterol, serum glucose, body weight in normal male rabbits, and Alloxan-treated diabetes male rabbits and urea level, were monitored after oral administration of the aqueous extract. After 12 hours of fasting, male rabbits were induced diabetes mellitus using Alloxan monohydrate (120 mg/kg). It was observed that after 7 days of alloxan monohydrate induction, animals were reported with fasting serum glucose levels higher than 250 mg/dl and were found to have diabetes mellitus. The mortality rate was observed as 50%. In the

preliminary step of experimental work, 21 male rabbits were segregated into7 experimental groups and each one comprises nearly 3 male rabbits. The first group was considered as a normal one and the second group was treated as diabetic control and in the third to the seventh group, male rabbits treated with different percentage of aqueous cinnamon extract like 0.5%, 0.75%, 1.0%, 1.25%, 1.5% to evaluate the dose efficacy against diabetes. The further second step of experimental work was carried out by selecting 9 male rabbits. Each group possesses 3 male rabbits, wherein the first and second groups were considered as a normal one and diabetic controls whereas the third group is administered with a cinnamon extract at a high dosage level to evaluate its efficacy against serum glucose level along with serum cholesterol level, serum urea level, and body weight. The entire study was intended nearly for 2 months. Selected biochemical parameters were analyzed with blood serum which was collected every third day. It was noted that in the first stage of experimental work, 1% aqueous cinnamon extract-treated male rabbits exhibited a noteworthy hypoglycemic effect. A decrease in serum glucose level up to 52.38% was recorded, whereas, in the second step of experimental work, 1% cinnamon extract regulated 34.55% serum urea level, 35.25% serum cholesterol level, 54.11% serum glucose level, and body weight gradually decreases. In this study, it is determined that cinnamon aqueous extract administered orally brings down cholesterol, urea level, and serum glucose in diabetes-induced male rabbits (Rind et al., 2010).

4.8 CONCLUSION

Many studies pointed out the possible role of *Cinnamomum zeylanicum* in the prevention and controlling of diabetes mellitus. Studies carried out by the United States Department of Agriculture also recommended that components with biological activity relevant for insulin control may be present in the water-soluble, rather than the fat-soluble fractions. Oral administration of cinnamaldehyde has the potential to increase hepatic glycogen levels in animal models, probably because of the glycogen synthase reactivation system as an outcome of an increased amount of insulin secretion. It was concluded that *Cinnamomum zeylanicum* has potential applications in diabetes.

CONFLICT OF INTEREST

There is no conflict of interest.

KEYWORDS

- ***Cinnamomum zeylanicum***
- **cinnamon**
- **diabetes mellitus (DM)**
- **fasting blood glucose**
- **hyperglycemia**

REFERENCES

Allen, R. W., Schwartzman, E., Baker, W., L., Coleman, C., I., & Phung, O. J., (2013). Cinnamon use in type 2 diabetes: An updated systematic review and meta-analysis. *Annu. Fam. Med., 11*(5), 452–459.

Anand, P., Murali, K. Y., Tandon, V., Murthy, P., S., & Chandra, R., (2010). Insulinotropic effect of cinnamaldehyde on transcriptional regulation of pyruvate kinase, phosphoenolpyruvate carboxykinase, and GLUT4 translocation in experimental diabetic rats. *Chem. Biol. Interact., 186*(1), 72–81.

Anderson, R., A., Broadhurst, C., L., Polansky, M. M., Schmidt, W. F., Khan, A., & Flanagan, V. P., (2004). Isolation and characterization of polyphenol type-A polymers from cinnamon with insulin-like biological activity. *J. Agric. Food Chem., 52*(1), 65–70.

Chericoni, S., Prieto, J. M., Iacopini, P., Cioni, P., & Morelli, I., (2005). *In vitro* activity of the essential oil of *Cinnamomum zeylanicum* and eugenol in peroxynitrite-induced oxidative processes. *J. Agric. Food Chem., 53,* 4762–4765.

Dragland, S., Senoo, H., Wake, K., Holte, K., & Blomhoff, R., (2003). Several culinary and medicinal herbs are important sources of dietary antioxidants. *Hum. Nutr. Metab., 133,* 1286–1290.

Geng, S., Cui, Z., Huang, X., Chen, Y., Xu, D., & Xiong, P., (2011). Variations in essential oil yield and composition during *Cinnamomum cassia* bark growth. *Ind. Crops Prod., 33*(1), 248–252.

Gopal, R. M., & Rajeswara, R. B. R., (2007). *Aromatic Plants from Asia Their Chemistry and Application in Food and Therapy* (pp. 49–75). Har Krishan Bhalla & Sons.

Hayashi, K., Imanishi, N., Kashiwayama, Y., Kawano, A., Terasawa, K., Shimada, Y., & Ochiai, H., (2007). Inhibitory effect of cinnamaldehyde, derived from *Cinnamomi cortex,* on the growth of influenza A/PR/8 virus *in vitro* and *in vivo. Antiviral Res., 74,* 1–8.

He, Z., D., Qiao, C. F., Han, Q. B., Cheng, C. L., Xu, H. X., & Jiang, R. W., (2005). Authentication and quantitative analysis on the chemical profile of cassia bark (*Cortex cinnamomi*) by high-pressure liquid chromatography. *J. Agric. Food Chem., 53*(7), 2424–2428.

Heping, C., Marilyn, M. P., & Richard, A. A., (2007). Cinnamon extract and polyphenols affect the expression of tristetraprolin, insulin receptor, and glucose transporter 4 in mouse 3T3-L1 adipocytes. *Arch Biochem Biophys., 459*(2), 214–222.

Iwasaki, Y., Tanabe, M., Kobata, K., Watanabe, T., TRPA1 agonists—allyl isothiocyanate and cinnamaldehyde—induce adrenaline secretion. *Biosci Biotechnol Biochem., 72,* 2608–2614.

Keller, K., (1992). Cinnamomum species. In: Desmet. P. A. G. M., (ed.), *The Adverse Effect of Herbal Drugs* (Vol. I, pp. 105–114). Berlin: Springer-Verlag.

Khan, A., Safdar, M., Ali, K. M. M., Khattak, K. N., & Anderson, R. A., (2003). Cinnamon improves glucose and lipids of people with type 2 diabetes. *Diabetes Care., 26*(12), 3215–3218.

Kim, S. H., Hyun, S. H., & Choung, S. Y., (2006). Anti-diabetic effect of cinnamon extract on blood glucose in db/db mice. *J. Ethnopharmacol., 104,* 119–123.

Koteswara, Y., Fang, S., & Tzeng, Y., (2007). Evaluation of the anti-inflammatory and anti-proliferation tumoral cells. *J. Ethnopharmacol., 114,* 78–85.

Krishnamoorthy, B., Rema, J., Zachariah, T. J., Jose, A., & Gopalam, A., (1996). Navasree and nithyasree-two high yielding and high-quality cinnamon (*Cinnamomum Verum*-Bercht &Presl.). *J. Spices Aromat. Crops, 5*(1), 28–33.

Lee, S., Xu, W., Lin, L., Yang, J., & Liu, C., (2013). Chemical composition and hypoglycemic and pancreas-protective effect of leaf essential oil from indigenous cinnamon (*Cinnamomum osmophloeum* Kanehira). *J. Agric. Food Chem., 61,* 4905–4913.

Li, R., Liang, T., Xu, L., Li, Y., Zhang, S., & Duan, X., (2013). Protective effect of cinnamon polyphenols against STZ diabetic mice fed high-sugar, high-fat diet, and its underlying mechanism. *Food Chem. Toxicol., 51,* 419–425.

Lu, J., Zhang, K., Nam, S., Anderson, A., Jove, R., & Wen, W., (2010). Novel angiogenesis inhibitory activity in cinnamon extract blocks VEGFR2 kinase and downstream signaling. *Carcinogenesis, 31*(3), 481–488.

Mang, B., Wolters, M., Schmitt, B., Kelb, K., Lichtinghagen, R., & Stichtenoth, D., O., (2006). Effects of a Cinnamon extract on plasma glucose, HbA, and serum lipids in diabetes mellitus type 2. *Eur. J. Clin. Invest., 36*(5), 340–344.

Mathew, S., & Abraham, T. E., (2006). Studies on the antioxidant activities of cinnamon (*Cinnamomum verum*) bark extracts, through various *in vitro* models. *Food Chem., 94*(4), 520–528.

Mbaveng, A. T., & Kuete, V., (2017). *Medicinal Spices and Vegetables from Africa, Therapeutic Potential Against metabolic, Inflammatory, Infectious and Systemic Diseases*, 385–395.

Nagai, H., Shimazawa, T., Matsuura, N., & Koda, A., (1982a). Immunopharmacological studies of the aqueous extract of *Cinnamomum cassia* (CCAq). I. anti-allergic action. *Jpn. J. Pharmacol., 32,* 813–822.

Nagai, H., Takizawa, T., Nishiyori, T., & Koda, A., (1982b). Experimental glomerulonephritis in mice as a model for immunopharmacological studies. *Jpn. J. Pharmacol., 32,* 1117–1124.

Nishida, S., Kikuichi, S., Yoshioka, S., Tsubaki, M., Fujii, Y., Matsuda, H., Kubo, M., & Irimajiri, K., (2003). Induction of apoptosis in HL-60 cells treated with medicinal herbs. *Am. J. Chin. Med., 31,* 551–562.

Pasupuleti, V. R., & Siew, H. G., (2014). Cinnamon: A multifaceted medicinal plant. *Evid Based Complement Alternat Med.*

Ping, H., Zhang, G., & Ren, G., (2010). Anti-diabetic effects of cinnamon oil in diabetic KK-Ay mice. *Food Chem. Toxicol., 48,* 2344–2349.

Preuss, H. G., Echard, B., Polansky, M. M., & Anderson, R., (2006). Whole cinnamon and aqueous extracts ameliorate sucrose-induced blood pressure elevations in spontaneously hypertensive rats. *J. Am. Coll. Nutr., 25,* 144–150.

Qin, B., Polansky, M., M., Sato, Y., Adeli, K., & Anderson, R. A., (2009). Cinnamon extract inhibits the postprandial overproduction of apolipoprotein B48-containing lipoproteins in fructose-fed animals. *J. Nutr. Biochem., 20*(11), 901–908.

Ranasinghe, P., Jayawardana, R., Galappaththy, P., Constantine, G. R., De Vas, G. N., & Katulanda, P., (2012). Efficacy and safety of 'true' cinnamon (*Cinnamomum zeylanicum*) as a pharmaceutical agent in diabetes: A systematic review and meta-analysis. *Diab. Med., 29,* 1480–1492.

Ranasinghe, P., Pigera, S., Premakumara, G. A., Galappaththy, P., Constantine, G. R., & Katulanda, P., (2013). Medicinal properties of 'true' cinnamon (*Cinnamomum zeylanicum*): A systematic review. *BMC Complement Altern. Med., 13,* 275.

Rind, N. A., Umar, D. M., Malik, S. A., & Kumar, M., (2010). Anti-diabetic effect of cinnamon (*Cinnamonum Zeylanicum*) used to control diabetes mellitus in alloxan-induced diabetic Male rabbits. *Pak. J. Biotechnol., 7*(1), 81–90.

Senanayake, U. M., & Wijesekera, R. O. B., (1989). Volatiles of the Cinnamomum species. *Proc. 11th International Congress Essential Oils, Fragrances, and Flavors, 4,* 103–120.

Senanayake, U. M., (1997). *The Nature, Description, and Biosynthesis of Volatiles of Cinnamomum spp.* Ph.D. thesis, University of New South Wales, Kensington, Australia.

Shan, B., Cai, Y. Z., Brooks, J. D., & Corke, H., (2007). Antibacterial properties and major bioactive components of cinnamon stick (*Cinnamomum burmannii*): Activity against foodborne pathogenic bacteria. *J. Agric. Food Chem., 55*(14), 5484–5490.

Shen, Y., Fukushima, M., Ito, Y., Muraki, E., Hosono, T., & Seki, T., (2010). Verification of the anti-diabetic effects of cinnamon (*Cinnamomum zeylanicum*) using insulin-uncontrolled type 1 diabetic rats and cultured adipocytes. *Biosci. Biotechnol. Biochem., 74*(12), 2418–2425.

Shumaila, G., & Mahpara, S., (2009). Proximate composition and mineral analysis of cinnamon. *Pakistan J. Nutr., 8*(9), 1456–1460.

Singh, G., Maurya, S., DeLampasona, M. P., & Catalan, C. A. N., (2007). A comparison of chemical, antioxidant, and antimicrobial studies of cinnamon leaf and bark volatile oils, oleoresins, and their constituents. *Food Chem Toxicol., 45*(9), 1650–1661.

Su, L., Yin, J. J., Charles, D., Zhou, K., Moore, J., & Yu, L., (2007). Total phenolic contents, chelating capacities, and radical-scavenging properties of black peppercorn, nutmeg, rosehip, cinnamon and oregano leaf (Lucy). *Food Chem.,* 100(3), 990–997.

Su, M. J., Chen, W. P., Lo, T. Y., & Wu, T. S., (1999). Ionic mechanisms for the antiarrhythmic action of cinnamophilin in rat heart. *J. Biomed. Sci., 6,* 376–386.

Subash, B. P., Prabuseenivasan, S., & Ignacimuthu, S., (2007). Cinnamaldehyde-a potential anti-diabetic agent. *Phytomedicine., 14,* 15–22.

Tabak, M., Armon, R., & Neeman, I., (1999). Cinnamon extracts inhibitory effect on Helicobacter pylori. *J. Ethnopharmacol., 67,* 269–277.

Thantsin, K., Zhang, Q., Yang, J., & Wang, Q., (2008). Composition of semivolatile compounds of 10 *Cinnamomum* species from China and Myanmar. *Nat. Prod. Res.,* 22(7), 576–583.

Thomas, J., (2012). *Handbook of Herbs and Spices*. Woodhead Publishing Limited.

Ulbricht, C., Seamon, E., Windsor, R. C., Armbruester, N., Bryan, J. K., & Costa, D., (2011). An evidence-based systematic review of cinnamon (*Cinnamomum* spp.) by the natural standard research collaboration. *J. Diet. Suppl., 8*(4), 378–454.

Vangalapati, M., Sree, S. N., Surya, P. D. V., & Avanigadda, S., (2012). A review on pharmacological activities and clinical effects of Cinnamon species. *Res. J. Pharm. Biol. Chem. Sci., 3*(1), 653–663.

Verspohl, E. J., Bauer, K., & Neddermann, E., (2005). Anti-diabetic effect of *Cinnamomum cassia* and *Cinnamomum zeylanicum in vivo* and *in vitro*. *Phytother. Res., 19*(3), 203–206.

Wang, J. G., Anderson, R. A., Graham, G. M. III., Chu, M. C., Sauer, M. V., Guarnaccia, M. M., & Lobo, R. A., (2007). The effect of cinnamon extract on insulin resistance parameters in polycystic ovary syndrome: A pilot study. *Fertil. Steril., 88,* 240–243.

Wang, Y. C., Ahmad-Mudzaqqir, M. Y., & Wan-Nurdiyana, W. A., (2014). Extraction of Essential Oil from Cinnamon (*Cinnamomum zeylanicum*). *Orient J. Chem., 30*(1), 37–47.

Winias, S., Retno, A., Magfiroh, R., Ryan, M., Rahayu, R. P., (2011). Effect of cinnamaldehyde from cinnamon extract as a natural preservative alternative to the growth of *Staphylococcus aureus* bacteria. *Indonesia. J. Trop. Infect. Dis., 2*(1), 38–41.

Yang, J., Chen, L., H., Zhang, Q., Lai, M. X., & Wang, Q., (2007). Quality assessment of cortex cinnamomi by HPLC chemical fingerprint, principle component analysis, and cluster analysis. *J. Sep. Sci., 30*(9), 1276–1283.

CHAPTER 5

Antidiabetic Potential of the Insulin Plant (*Costus pictus*)

KAVITA RANI,[1] SWATI PRIYA,[1] NARENDER SINGH,[1] and RUCHI BANSAL[2]

[1]*Department of Botany, Kurukshetra University, Kurukshetra, Haryana – 136119, India, E-mails: kavitadahiya05@gmail.com (K. Rani); swatipriya9013@gmail.com (S. Priya); nsheorankuk@yahoo.com (N. Singh)*

[2]*Division of Germplasm Evaluation, ICAR-National Bureau of Plant Genetic Resources, New Delhi – 110012, India, E-mail: ruchibansal06@gmail.com*

ABSTRACT

Costus pictus D. Don (Syn. *Costus igneus* Nak) commonly known as "Spiral flag" or "The Insulin Plant" is a herbal cure for diabetes. Phytochemical composition consists of carbohydrates, ascorbic acid, α-tocopherol, flavonoids, terpenoids, proteins, steroids, and β-carotene. The plant extract possesses different therapeutic activities such as diuretic, hypolipidemic, antimicrobial, antioxidant, and anti-cancerous properties. Molecular signature molecules, i.e., bixin, geranial, and geraniol biosynthesis associated with its antidiabetic principles have been identified using transcriptomic approaches.

Antidiabetic Plants for Drug Discovery: Pharmacology, Secondary Metabolite Profiling, and Ingredients with Insulin Mimetic Activity. Deepu Pandita, MPhil, Anu Pandita, MSc, & Chander Bhanu, MBBS, PGDDCE (Eds.)
© 2023 Apple Academic Press, Inc. Co-published with CRC Press (Taylor & Francis)

5.1 INTRODUCTION

In the running era, most of the people often tend to be more health-conscious due to sedentary lifestyle, stress, improper diet, low immunity and hazardous climatic changes. Nowadays, due to varied lifestyles, the air, water and even the food we eat are all polluted which weakens our immune system making us prone to fatal diseases. Diabetes is one among them. Since long, this disease is increasing the morbidity and mortality rate across the world. The International Diabetes Federation (IDF) analyzed data from population-based epidemiological studies and approximated the global burden of diabetes to 110 million in 1994 and that it likely doubled up to 239 million by 2010. WHO predicted the global burden at 135 million in 1995, with the number touching 299 million mark by the year 2025. In 2019, 9.3% (463 million people) of global diabetes pervasiveness has been recorded in a study of International Diabetic Federation, which is predicted to rise up to 10.2%, i.e., 578 million by 2030 and 10.9%, i.e., 700 million by 2045 (Colagiuri et al., 2019). World Health Organization has declared it the 7th topmost cause for death at world level (W.H.O., 2013). The World Health Organization also declared 14 November as World Diabetes Day in 2006, on the birth date of Fredrick Baiting who co-discovered insulin in 1922 with Charles Beat. At first animal insulin was extracted from pig and was used to treat people, later on, synthetic insulin was manufactured around 1978. Therefore, human insulin is nowadays found in varied forms (Short-acting, Rapid-acting, etc.), according to a person's suitability (A.D.A report, 2018).

Meditation, regular exercise, a healthy diet, high intake of fiber as well as consuming low sugars, is helpful in controlling diabetes. There are different ancient medicinal systems, which are prevalent throughout the world, which include Traditional Chinese herbal medication, Ayurvedic Herbalism, Western Herbal treatments—which originated from Greece and Rome to Europe and then widened to North and South America also, Traditional Arabic and Islamic medicines in most of the Mediterranean countries and Ayurvedic, Yoga, Unani, Siddha, Homeopathy in India are common. In past centuries our ancestors used the plant parts and their extract for treating diseases. Medicinal plants are the principal source of drugs and herbal products due to their therapeutic effects (Lown, 1993) in general health problems (Anwannil and Atta, 2006). Different studies have focused on the identification of the specific drug targets at the molecular level for systematic evaluation of herbal therapeutics for the treatment of diabetes (Chawla et al., 2011; Ramasamy et al., 2012). The present chapter provides an overview of diabetes, botanical properties and distribution of insulin plants, its bioactive compounds and therapeutic potential.

5.2 DIABETES AND ITS TYPES

Diabetes is a chronic metabolic disorder with impaired functioning of insulin leading to malfunctioning of carbohydrate metabolism that results in elevation of blood glucose level (A.D.A report, 2009). Two German researchers named Oskar Minkowaski and Joseph Von Mering in 1899, when isolated the pancreas from a dog's body found it suffering from diabetes and they observed that, pancreas were responsible for producing a particular hormone which controls body sugar level. In 1910, Sir Edward Albert Sharpay named that particular hormone produced by pancreas as "insulin." Insulin is the one whose absence or malfunction causes diabetic symptoms in the body. The term "Insulin" came from the Latin word 'Insula' which means "island." Pancreas has β cells which generate insulin. In the healthy human body, carbohydrates are converted into glucose passing through the digestive tract and then it diffuses into the main blood stream (Taylor et al., 1994). The elevated glucose level gives a positive feedback signal to our pancreas, which starts producing more insulin to facilitate the entry of glucose inside adipose tissues. These cells use glucose in different metabolic reactions. In a diabetic person problems lie either in their insulin production or are found "insulin resistance" (Taylor et al., 1994).

American Diabetes Association (ADA) in 2018, has classified diabetes in different categories: Type 1 *Diabetes mellitus* (T1DM) involves β-cell autoimmune destruction, resulting into absolute insulin deficiency; Type 2 *Diabetes mellitus* (T2DM) has impaired insulin secretion leading to frequent development of resistance to insulin, Gestational *Diabetes mellitus* (GDM) is diagnosed during gestation period, particularly in 2nd or 3rd trimester of pregnancy. Specific forms of diabetes, such as exocrine pancreas disorder, monogenic diabetes syndromes, and drug-induced or chemical induced diabetes may be triggered by many other reasons. Out of these T2DM is the most prevalent form because of its deleterious effects.

5.2.1 TYPE 1 DIABETES MELLITUS (TIDM)

Type 1 *Diabetes mellitus* is commonly known as juvenile-onset diabetes or insulin-dependent diabetes (ADA, 2009). In T1DM, the body is reluctant to produce insulin at a much earlier stage (infants and children) of life, and insulin-producing β cells of the pancreas are destroyed by the auto-immune system (ADA, 2018). Autoimmune β cells destruction could be triggered by genetic as well as environmental factors (Notkins et al., 2002; Knip et al., 2012; Lacroix, M et al., 2013). Destructive β cells are unable to produce enough insulin and cause accumulation of sugar in the bloodstream rather than facilitating inside cells (Notkins et al., 2002). However, the body cannot use this extra glucose as a fuel for energy. The person suffering from T1DM may also be prone to some additional autoimmune diseases viz., Addison's disease, vitiligo, Graves' disease, Hashimoto's thyroiditis, autoimmune hepatitis, myasthenia gravis and pernicious anemia. It has harmful long-term effects such as cardiovascular disease, blindness, kidney failure and many if not treated on time (ADA report, 2009). Hence elevated levels of glucose in blood results in polyurea (excessive excretion of urine), polydipsia (excessive thirst), fatigue, vision changes, unexpected weight loss and damage tissues of the body. Its harmful long-term effects are: cardiovascular disease, blindness, kidney failure and many if not treated on time. Treatment of this type is seen by taking oral medication or insulin injections so that insulin deficiency can be mitigated.

5.2.2 TYPE 2 DIABETES MELLITUS (T2DM)

It is also known as adult-onset diabetes, encompasses insulin deficiency or insulin resistance. It can be hereditary or a genetic disorder depending upon family history as well (ADA, 2009). T2DM is the most commonly reported form of diabetes. According to CDC report 2017, 90–95% diabetic population, i.e., every one out of four persons suffering from T2DM do not know about the disease. T2DM can be treated, using various drugs (statin, metformin), surgeries, insulin therapy, etc. T2DM mostly affects when a person enters adulthood. This mostly happens due to excess body weight and physical inactivity. In Type 2 diabetes, insulin resistance occurs. Every nine people out of 10 with diabetes suffer with Type 2 form

(ADA Report, 2018). It is found mostly those who are around 40 years old and are heavyweight (ADA Report, 2009). It is of greater concern now a days as because in T2DM, the keyhole is rusty, that means the body produces insulin, but the body's cells get resistant to it. As it seems, we have keys, but there is trouble in recognition by the cells. Symptoms like increased hunger and thirst, dry mouth, unexplained weight loss, frequent urination, weakness, numbness, blurred vision, dry and itching skin, slow-healing sores, frequent yeast infection, etc., are common (IDF, 2006). Chronic renal failure, cardiovascular disease, nerve damage and retinal damage that can cause erectile dysfunction and poor healing are some major long-term complications related to diabetes. Type 2 diabetes crises have become a world spread severe health issue due to rapid cultural and social shifts, rising urbanization, nutritional transition, sedentary lifestyle and behavioral patterns (IDF, 2006). Some synthetic drugs (such as statin, metformin, insulin therapy, etc.), transplantation, surgery and weight loss, etc., treatments are used to control T2DM but, none of these found sufficient procurement of this disease. It is imperative to explore permanent solutions in order to cure this epidemic disease with less or no side effects of any compounds.

5.2.3 GESTATIONAL DIABETES (GDM)

It was firstly reported in around 20–24 weeks of gestation period because the hormones which placenta produces in 2nd and 3rd trimester of pregnancy could be "insulin resistance" (Lacroix et al., 2013). So it may cause complications during delivery. In the USA around 4% of total pregnancies are complicated by GDM (ADA Report, 2018).

5.2.3.1 ROLE OF INSULIN IN BODY METABOLISM

Insulin plays a fundamental role in stimulation of intermediary metabolisms and it decides whether the fuel gets stored in the form of fat or oxidized. Ultimately insulin controls not only the effect of carbohydrate and lipid metabolism but also significantly influences the mineral and protein metabolism. Consequently, diabetes is regarded as a primary disturbance in carbohydrates, lipids and protein metabolism due to impairment of insulin signaling which creates devastating effects on the whole body. Insulin stimulates three kinds of mechanisms in liver, muscle and fat tissues in order to uptake sugar from blood so that tissues can use glucose as a fuel and blood glucose drops. Role of insulin on carbohydrates, lipid and protein metabolisms are as follows.

5.2.3.1.1 Carbohydrate Metabolism

Uptake of blood glucose into cells (primarily myocytes and adipocytes with the exception of central nervous system cells) is regulated by the insulin secreted from the pancreas (Qaid and Abdelrahman, 2016). Insulin deficiency or the insensitivity of insulin receptor is the key cause of *Diabetes mellitus* (Czech, 1977). Much of the carbohydrate is converted to the monosaccharide glucose within a few hours, and released into the bloodstream. Insulin is released by Pancreatic β cells as a retaliation to increasing glucose level in blood (e.g., after a meal). Insulin stimulates the absorption of glucose from blood in most body cells, often used as fuel for energy production or for storage. Insulin primarily regulates the glucose to glycogen conversion cascade used for storage in the liver and myocytes (Czech, 1976). As the glucose level declines, simultaneously secretion of insulin from β cells falls down which results in reverse conversion of glycogen to glucose (process called Glycogenolysis) to provide immediate energy and re-enters the bloodstream to maintain blood glucose level during fasting (Chakravarthy et al.,1981).

An increased insulin level stimulates many anabolic processes viz. cell growth and replication, translation, and fat storage. Insulin acts as a principal signal for induction of several bi-directional metabolic processes (such as catabolic to anabolic direction and vice versa).

5.2.3.1.2 Lipid Metabolism

When the level of insulin declines, free fatty acid (FFA) levels of basal plasma and lipid oxidation elevates abnormally and fails to decline afterwards (DeFronzo, 1988). Production rate of fatty acid elevates up to three-fold. Insulin stimulates the degradation of food molecules into glucose for the production of energy and further accumulation of glucose in the form of glycogen in the liver and skeletal muscles (Qaid and Abdelrahman, 2016). Furthermore, insulin also stimulates liver cells (hepatocytes) to synthesize triglycerides and its storage in adipose tissue in the form of fat molecules (Suzuki, 1990). Insulin-mediated inhibition of lipolysis is to counter the increased storage of triglycerides in adipocytes (Suzuki, 1990). With impaired insulin signaling and increased catabolic hormone response in a diabetic individual, the rate of triglycerides hydrolysis enhanced, resulting in a threefold increase in plasma triglycerides, leading to a simultaneous increase in keto acids derived from oxidation of free fatty acid in hepatocytes and plasma free fatty acids level (Qaid and Abdelrahman, 2016). Numerous tissues (including liver but with exception of the brain) uptake the free fatty acid and metabolize those to gain energy (DeFronzo, 1988).

In the mitochondria, fatty acyl-CoA is oxidized using the enzyme carnitine palmitoyltransferase I for the production of energy (McGarry and Foster, 1980). The enzyme is impeded by the high levels of malonyl-CoA that is the normal outcome of the existence of insulin (McGarry et al., 1978). Thus the increased transport of fatty acyl-CoA into the mitochondria can conventionally be interpreted as the amount of malonyl-CoA decreased in the deficiency of insulin. Further mitochondrial oxidation of fatty acids results into acetyl-CoA, which could be oxidized in the Krebs cycle or could be substantially metabolized into ketone bodies, acetoacetate and β-hydroxybutyrate in hepatocytes, unlike the TCA cycle oxidation of it (Qaid and Abdelrahman, 2016). These ketone bodies were used by the heart, skeletal muscle and brain for energy production after leaving their site of production, i.e., liver (Suzuki, 1990).

While suffering from diabetes, the accessibility of free fatty acids and ketone bodies increase and consecutively aggravate by resulting hyperglycemia due to the reduced utilization of glucose. Smelling the breath of the one with increased production of ketone bodies can easily lead to diagnosis of ketoacidosis (Qaid and Abdelrahman, 2016). The plasma triglycerides are stored in 6 adipocytes by the activity of lipoprotein lipase (LPL) that require insulin and hypertriglyceridemia, which is the result of the absence of later (Kameshwara et al., 2003).

5.2.3.1.3 Protein Metabolism

Fluckey (1996) comprehended that the deficiency of insulin will lead to an increased catabolism of protein. As overall metabolism can be affected by the absence or presence of insulin, it can thus act as an inducer or repressor for regulation of the several genes and can be upheld accountable for increased protein synthesis and reduced protein degradation rates (Qaid and Abdelrahman, 2016). In non-insulin dependent diabetes, hyperglycemia can be an outcome of increased hepatic gluconeogenesis as the result of increased proteolysis rate that leads to an increase in amino acids levels in plasma, acting as precursors for renal and hepatic gluconeogenesis (Karuayaka et al., 1984).

5.3 TAXONOMICAL AND MOLECULAR STUDY OF *COSTUS PICTUS (SYN. IGNEUS)* AND ITS IMPLICATIONS IN DRUG DISCOVERY

Traditional medicinal systems use the hypoglycemic properties of plant derivatives around the world to cure different diseases (Laha et al., 2019). Traditionally, medicinal plants consumption, balanced, low sugar diet and physical exercises are used as an alternative form of treatment for *Diabetes mellitus*. Only 30% of 1,200 plants with antidiabetic activities that are used to treat *Diabetes mellitus* were pharmacologically and chemically investigated (Laha et al., 2019). As a result, there is an intense interest in the discovery of more effective antidiabetic drugs in the research and therapeutic communities. This interest focused on the plant kingdom and ancient medicinal methods which have shown that numerous modern drugs have been obtained from plants through the study of medicinal practices by local indigenous groups (Balick and Cox, 1996; Gilani and Rahman, 2005). In the native Indian System of Medication, antidiabetic properties of many medicinal plants were examined, separately or in combinations that are effective in the management of diabetes (Jung et al., 2006). For example, *Catharanthus roseus* (L.) G. Don leaves and *Clerodendrum viscosum* are homogenized in water and one teaspoon is consumed every day before meal, *Lannea coromandelica* (Houtt.) Merr. B barks and roots are soaked overnight in water and drinking this water on an empty stomach the next morning, *Alstonia scholaris* (L.)R.Br. Green leaves are chewed prior to meals, *Terminalia chebula* freshly collected fruits are taken every day. During unavailability of fresh fruits, dried fruits can be used, by overnight drenching in water and drinking it the next morning in empty stomach, half a cup of extract obtained from pressed leaves of *Momordica charantia* L. is taken in the morning while the stomach is empty. *Costus pictus* is one among these herbs that may be used as an alternative therapeutic agent

(Laha et al., 2019). With its abundance, less or nil side effects, and cost-effectiveness, this plant becomes more suitable for use (Laha et al., 2019). This chapter focuses here on the therapeutic potential of insulin plants in lowering T2DM effect.

5.3.1 THE PROFILE OF COSTUS PICTUS

5.3.1.1 SYSTEMATIC POSITION

Botanical name	***Costus pictus* D. Don**
Domain:	Eukaryota
Kingdom:	Plantae
Phylum:	Tracheophyta
Subphylum:	Euphyllophytina
Class:	Liliopsida
Subclass:	Commelinidae
Order:	Zingiberales
Family:	Costaceae
Subfamily:	Asteroidea
Genus:	*Costus*
Specific epithet:	*Pictus*

Costus pictus D. Don (syn. *Costus igneus* N.E.Br., *Costus mexicanus* Liebm ex Petersen or *Costus congenitus* Rowle), popularly known as Spiral flag/Insulin plant/ fiery costus, has its origin to South and Central America. The Costaceae family was mentioned first by Nakai in 1941 to the family rank (Pazhanichamy et al., 2010), to which it is found identical and easily recognizable members in order Zingiberales (Dahlgren et al., 1985). Various morphological and anatomical features, e.g., well-developed aerial shoots including hardy and sometimes branched stem, shows up its distinguished position from other families including Musaceae and Zingiberaceae. The most unique feature is its branched aerial shoots that have a unique spiral monistichous phyllotaxy, i.e., one-sided arrangement of leaves (Kirchoff and Rutishauser, 1990). The family Costaceae also have unique floral pattern within the order Zingiberales, by having a single fertile stamen along with five sterile stamens fused together to form a bulky petaloid labellum which highlights the floral orientation (Troll, 1928; Kirchoff., 1988). The Costaceae family has four genera with about 200 species (Airy Shaw, 1973). *Costus* is reported as a large genus within the family covering 150 species under it. Family Costaceae are mainly tropical in distribution (Hickey, 1981; Humphries, 1985). Its availability is marked often in the forest and savannah region. *Costus* plants are very much distinctive with their morphological pantropical distribution (Katoriya et al., 2016). Some evidence is in support of considering the Costaceae, as a separate family, based on several diagnostic anatomical characters of the vegetative parts (Tomlinson, 1956). These vegetative parts include the uniseriate filamentous hairs, isodiametric, irregular, or transversely extended thin-walled epidermal cells, tetracytic stomata (stomata with subsidiary cells), slightly asymmetrical nature of guard cells, colors and continuous hypodermis, clearly differentiated cortex, well developed vascular bundle at each nod restriction of silica to internal tissue close to vascular bundles, solitary bodies stellately spherical or drug-like vessels restricted for root, etc. (Edeoga, 1991). Recorded information on *Costus pictus/igneus* is mostly on leaf, rhizome and root anatomy other than anatomical findings of some Costaceae members (Figure 5.1).

The rhizome is circular and has smooth surfaces (Pazhanichamy et al., 2010). It has a clear distinct layer of epidermis, which comprises constricted oblong thin-walled cells (Pazhanichamy et al., 2010). Inner to the cortex is a broad cortical zone in the radial plane (Pazhanichamy et al., 2010). The cortical cells structure is circular yet wavy, thin-walled, compact and are widely spread. Small spherical vascular bundles are randomly distributed in the cortex (Pazhanichamy et al., 2010). The cortical bundles are collateral with an angular fairly wide and thick-walled xylem element having a tiny cluster of phloem elements. Distinct clusters of xylem elements, associated with the endo-dermoid layer are present (Pazhanichamy et al., 2010). The xylem elements are present in tangential multiples; they are wide, angular, thick-walled and measure 50 μm in diameter. Phloem elements also occur as a thin sheath placed between the xylem elements and endodermis (Pazhanichamy *et al.*, 2010). The stele has an outer boundary of the endo-dermoid layer. The stelar bundles are many and are scattered in the stele. They are also collateral and different in orientation. The central (stelar) vascular bundles have wide, angular, thin-walled meta-xylem elements and three to four protoxylem elements. The metaxylem elements are 70μm wide. Thick masses of phloem are present near the metaxylem elements. The phloem elements are also wide and distinctive.

FIGURE 5.1 *Costus pictus* plant.

Source: Photo by Sekh Sayantan, Burdwan Eco Garden, Bugwood.org. https://creativecommons.org/licenses/by/3.0/us/

The leaf is thin, smooth even surfaces, isobilateral and has similar adaxial and abaxial sides. The leaf consists of bilayers of thin epidermal cells and layers of broad, tangentially ovate, thin-walled mesophyll cells four in number (Pazhanichamy et al., 2010). Both epidermal layers are tapered, tangentially flat with thin-walled cells and are 10–20 μm thick (Katoriya et al., 2016). The mesophyll cells measure 100–140 μm thick.

Prominent vascular bundles are placed in the median of the lamina. Vascular bundles are collateral having a wide mass of xylem elements with a small cluster of phloem. At the phloem end of the bundle, there is a thick layer of lignified sclerenchymatous cells. The xylem elements are angular and thick walled. The vascular bundles have no distinct bundle sheath cells (Pazhanichamy et al., 2010).

Thin roots are present, having wide superficial sequenced periderm and narrow homogenous parenchyma cells in the cortex (Pazhanichamy et al., 2010). Older roots get thickened as the plant grows older; the periderm is a bit wider and consists of polyhedral, randomly arranged thin-walled cells. The cortex is broad and parenchymatous with thin-walled small compact cells (Katoriya et al., 2016). The periderm and the cortex are wider. The stele has approximately 10 exarch xylem and phloem alternating with each other's strands. The stele has separate endodermis with cylindrical narrow cells. Inner to the endodermis is a thin layer of hyaline cells, which are rectangular to circular in peripheral-line (Pazhanichamy et al., 2010). The xylem strands are numerous and each strand has circular, thick-walled metaxylem elements and thin protoxylem elements. A prominent mass of phloem is spread in between the xylem. No sclerenchymatous tissue found in the stele.

5.3.2 PHYTOCHEMISTRY

Medicinal plants are the source of drugs and natural products nearly for all kinds of health issues (Lown, 1993; Dahanuka et al., 2000; Anwannil and Atta, 2006). According to Dahanuka et al. (2000), most oxidative diseases like diabetes, cancer can be treated and prevented by the using potential bioactive compounds found by medicinal plants, which are very actively used in traditional medicinal systems (Wang et al., 2007; Okeke et al., 1999). Noticing the widespread usage of medicinal benefits of medicinal plants in infectious disease treatment and diabetes in several countries may be cultured due to the high cost of effective drugs. Scientific data have reported various herbal species as potential drugs for having antidiabetic activity that are being tested on model organisms. Researchers have described the various mechanisms used by medicinal plants for having antidiabetic effects including, recreation of Pancreatic β cells, glucose absorption inhibition, stimulation of insulin secretion, glucose tolerance improved, and lipid-lowering (Figure 5.3) (Anwannil and Atta, 2006;

Laha et al., 2019). Herbal medicine may therefore be an essential one with rich bioactive components. (Thomas et al., 2014).

On sequential phytochemical screening of *C. pictus (*syn *igneus)* plant, it is found to be rich in protein, iron, αtocopherol, ascorbic acid, steroids, βcarotene, flavonoids and terpenoids (Devi et al., 2010; Shankarappa et al., 2011). Preliminary studies revealed that methanolic extract of wild varieties had higher amounts of proteins, carbohydrates, terpenoids, tannins, alkaloids, and flavonoids (Jothivel et al., 2007). On further analysis of extract, steroids were found in all types of extracts but ethanol extract, also contained alkaloids along with it. The main components extracted from a fraction of ether were found bis(2'Ethylhexyl)1, 2benzenedicarboxylate; ergastanol and αtocopherol (George et al., 2007). Stem showed the occurrence of stigmasterol and a terpenoid, i.e., lupeol (Manjula et al., 2012). Kalailingam et al. (2011) found saponins and flavonoids in ethanol extract of rhizome of *C. pictus*. Rhizome and leaf of the insulin plant showed an appreciable amount of some trace elements as well (Jayasri et al., 2008). Steam distillation of various parts of the insulin plant has yielded many essential oils. The extracts of various parts contain the essential oils in proportion as given in Figure 5.2.

5.3.3 BIOACTIVE COMPOUNDS

Different clinical studies have been done on Insulin plant extract (obtained from its leaves, stem, rhizoids and roots) to study the effect of bioactive compounds that have hypoglycemic activities in different animal model systems (Zong et al., 2007). Quercetin (flavonoid) and diosgenin (steroidal sapogenin) were obtained as major bioactive compounds from *C. igneus/ pictus* leaves and rhizome, respectively, which controls gluconeogenesis by stimulating glycolytic enzymes in diabetic animals (Kalailingam P et al., 2011).

5.3.3.1 MODE OF ACTION FOR BIOACTIVE COMPOUNDS

5.3.3.1.1 Corosolic Acid

In humans, corosolic acid treatment has resulted in a decline in postprandial level of plasma glucose (Fukushima et al., 2006). In other experimental studies, corosolic acid inhibits the differentiation and simultaneously down-regulates the expression of PPAR-γ (peroxisome proliferators-activated receptor) as well as enhancer-binding protein (C/EBP-α) mRNA in 3T3-L1 adipocyte cells (Zong et al., 2007). It also elevates the glucose uptake (Zong et al., 2007). Moreover, corosolic acid treatment on KK-Ay mouse study revealed that it reduces blood and plasma glucose levels, suggesting an improved glucose metabolism by decreasing resistance to insulin (Miura et al., 2006). All of these reports state that corosolic acid elicits antidiabetic effects and could be beneficial for humans.

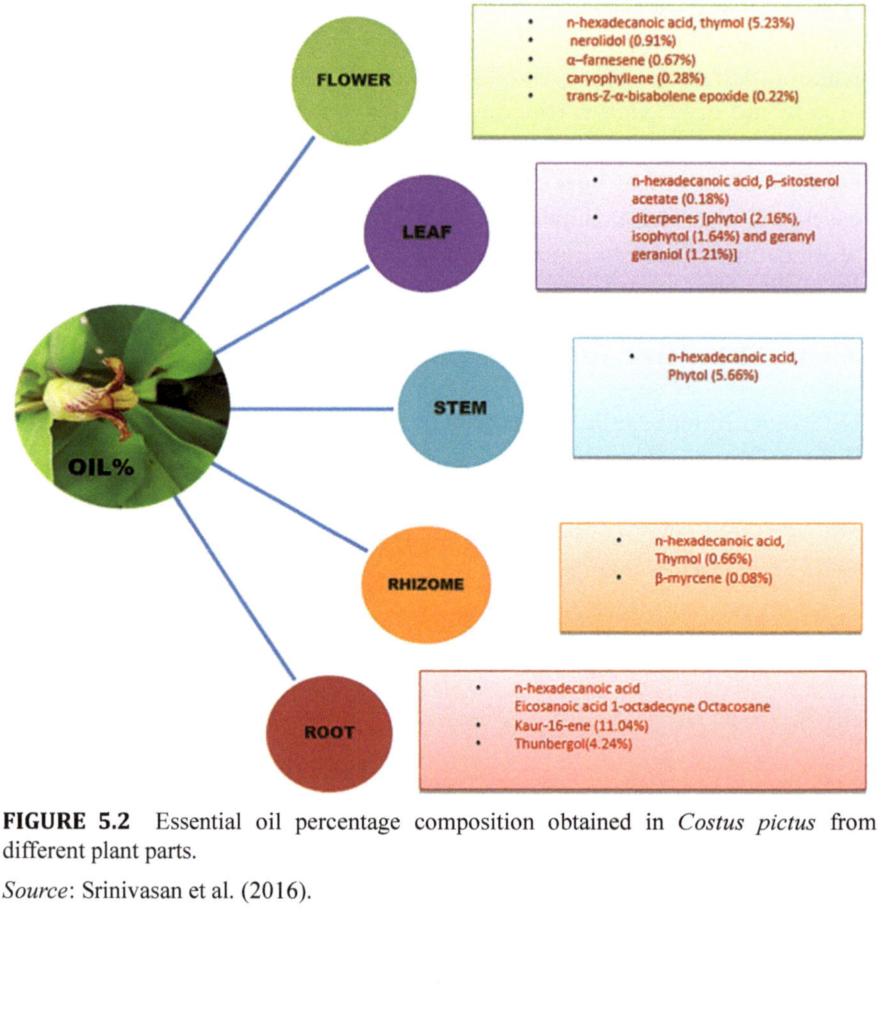

FIGURE 5.2 Essential oil percentage composition obtained in *Costus pictus* from different plant parts.

Source: Srinivasan et al. (2016).

5.3.3.1.2 Diosgenin

The amount of diosgenin present in the *C. pictus* root was found to be 1.32% w/w (dry weight basis) (Vijayan et al., 2016). Diosgenin reportedly lowers the plasma sugar level in STZ-stimulated diabetic rats by altering several hepatic rate-limiting enzymes (e.g., ATP-cytratelyase, pyruvate kinase and G-6-P dehydrogenase) associated in glucose metabolism signifying its antidiabetic capability (Mc Anuff et al., 2005). Diosgenin is effective against decreasing plasma sugar levels in diabetic rats. Uemura et al. (2010) reported that diosgenin is helpful to mitigate diabetes by stimulating 3T3-L1 adipocyte cell differentiation and by inhibiting inflammation of adipose tissues to increase insulin-dependent glucose uptake. Therefore, diosgenin enhances the insulin action in adipose tissues and skeletal muscles which could be helpful in glucose metabolic disorder related to obesity. For this perspective, another research model was observed that diosgenin is useful in total cholesterol levels reduction and led to enhancement in the plasma high-density lipoprotein (HDL) cholesterol level in diet-induced hypercholesterolemia rats (Son et al., 2007). Diosgenin treatment in diabetic rats led to a reduction in hyperglycemia, hypercholesterolemia and hyperglyceridemia (Son et al., 2007). In another report, diosgenin treated diabetic rats, the activity of glucokinase reduced, while an elevation was observed in the activities of enzymes glucose-6-phosphatase (G6Pase) and fructose-1, 6-bisphosphatase (FBPase) present in liver (Jesus et al., 2016). Supplementation with diosgenin in diabetic rats, blood glucose level decreased by the regeneration or recuperation of injured pancreatic β-cells and enhances the insulin secretion by pancreas which further resulted to rise in plasma insulin level hence; blood glucose is controlled (Laha and Paul, 2019). These reports suggested that unlike other antidiabetic drugs, Diosgenin has hypoglycemic properties which are profoundly found in *Costus igneus* plant (Pazhanichamy et al., 2012; Jesus et al., 2016).

5.3.3.1.3 β-Sitosterol (BS)

β-sitosterol is another antidiabetic phytosterol, structurally similar to the cholesterol, found in *Costus igneus* (Bouic et al., 1999; Benny et al., 2019). BS showed hypoglycemic effects by enhancing the circulating insulin level but the mechanism is not fully studied. BS elevates the uptake of glucose in adipocytes which further provoke adipogenesis in differentiating preadipocytes (Benny et al., 2019) (Table 5.1). Some studies revealed that BS down-regulates the gene expression of GLUT4, mimicking insulin, but clinical evidence is not manifested till date (Bouic et al., 1999).

TABLE 5.1 Mechanism Involved in Antidiabetic Effects of Some Major Isolated Compounds of *Costus pictus*

Constituents Nature	Major Bioactive Compound	Mechanism Involved for Antidiabetic Activity	Molecular Formula	References
Triterpenoid	Corosolic acid	Enhancing glucose uptake activity	$C_{30}H_{48}O_4$	Fukushima et al. (2006); Joshi et al. (2013); Laha and Paul (2019)
	Ursolic acid	Enhancing insulin receptor	$C_{30}H_{48}O_3$	Shi et al. (2008)
Steroid	Diosgenin	Hyperglycemic property	$C_{27}H_{42}O_3$	Lambert et al. (1988); Vijayan et al. (2016); Laha and Paul (2019)
	Stigmasterol	Increase plasma insulin level and glucose uptake activity	$C_{29}H_{48}O$	Manjula et al. (2013); Ward et al. (2017)
	β-sitosterol	Increase circulating insulin level	$C_{29}H_{50}O$	Lambert et al. (1988); Bouic et al. (1999); Laha and Paul (2019)
Flavonoid	Quercetin	Antihyperglycemic effect and hypolipidemic effect	$C_{15}H_{14}O_9$	Alam et al. (2014); Ramya and Chauhan (2015); Laha and Paul (2019)

TABLE 5.1 *(Continued)*

Constituents Nature	Major Bioactive Compound	Mechanism Involved for Antidiabetic Activity	Molecular Formula	References
Phenols	Catechin	Inhibit alpha-glucosidase and antioxidant activity	$C_{15}H_{14}O_6$	Kalailingam et al. (2013); Laha and Paul (2019)
Fatty acid	Hexadecanoic acid	Enhance insulin production in INS-1	$C_{16}H_{32}O_2$	Joshi et al. (2013); Laha and Paul (2019)
Others insulin-like compound	Oleic acid	Hypoglycemic property	$C_{18}H_{34}O_2$	Hoshino et al. (2017); Laha and Paul (2019)

5.3.3.1.4 Stigmasterol

Manjula et al. (2013) observed the presence of Stigmasterol in *Costus igneus*. Stigmasterol uplifts the cholesterol efflux in pancreatic cells and reduces LDL-induced pro-inflammatory cytokine release to stop β-cell dysfunctioning that are stimulated by glucolipotoxicity (Ward et al., 2017). Glucolipotoxicity accounts for a reduction in insulin release by induction of cholesterol deposition because high cholesterol level is a possible factor contributing to improper response by β cells. The direct effect of stigmasterol on β cells is not clearly known, but supposed to protect pancreatic β cells from glycolipids-toxicity by preventing free cholesterol deposition and reactive oxygen species improvising insulin secretion thus, increasing insulin content (Ward et al., 2017).

5.3.3.1.5 Quercetin

It is a member of the flavonoids class of polyphenols found in *Costus pictus*. Quercetin's antidiabetic function is a consequence of its insulin secretion effect. Quercetin is active in various metabolic processes like as glucose homeostasis; insulin secretion and stimulation, the inhibition of glucose absorption in the intestine, utilization of glucose in peripheral tissues, etc. (Figure 5.3) (Eid and Haddad, 2017). Quercetin increases the rate-limiting activity of enzymes in glycogen synthesis as well as enhances the activity of antioxidant enzymes (like GPX, SOD, CAT, etc.), and down-regulates glucose absorption in the small intestine manifested by inhibiting GLUT 2.

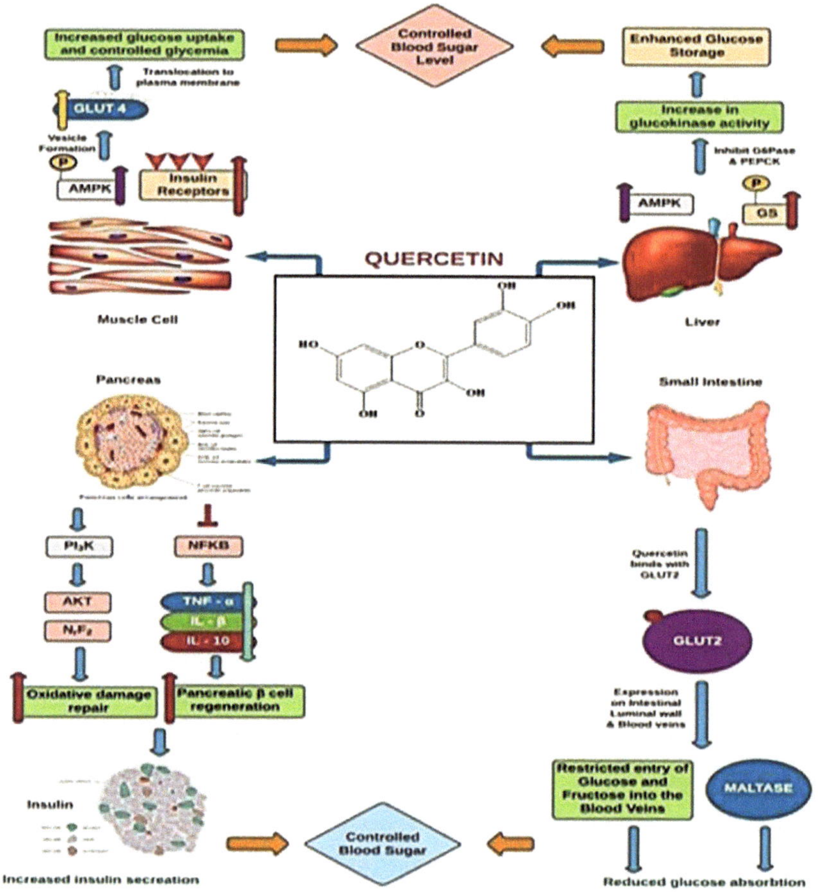

FIGURE 5.3 Role of quercetin in various metabolic pathways in different parts of the body for controlling blood sugar level.

Quercetin significantly prevents the lipid peroxidation of pancreatic cells and also downregulates the activities of carbohydrate metabolizing enzymes viz. α glucosidase and α amylase based on concentration (Oboh et al., 2014). The inhibition of both enzymes could slow down the blood glucose increase and prevent the pancreas damage from the pancreatic lipid peroxidation which could provide a repaired mechanism for the production and secretion of insulin. *In vitro* studies have elucidated that hypoglycemic properties of quercetin are due to GLUT4 mediated activation of AMPK (Alam et al., 2014). Diabetes is linked with a variety of other complications such as retinopathy, neuropathy, nephropathy and diabetic cataracts. Such complications often arises due to sorbitol formation, a sugar alcohol formed from glucose due to enzyme catalysis done by aldose reductase (Kelly, 2011). Quercetin has been reported for inhibiting aldose reductase activity and slowing down the sorbitol formation. Quercetin also improves glucose absorption by target tissue; improving the recovery of cell proliferation; inhibition of insulin-mediated activation of PI3K thus enhancing calcium uptake; increasing insulin secretion by renewal of β cells of pancreas (Coskun et al., 2005; Stewart et al., 2009). Hence, quercetin could be important for its therapeutic benefits in the management of diabetes (Kawahara et al., 2009).

5.3.4 MEDICINAL PROPERTIES OF COSTUS PICTUS

5.3.4.1 ANTIDIABETIC ACTIVITY

Different studies have been performed *in-vitro* to analyze the antidiabetic activity of insulin plants (Table 5.2). The results of all performed experiments showed that the most appropriate mechanism to rationalize the antidiabetic property of insulin plant is: the inhibitory activities of amylase and glucosidase enzymes. Significant blood sugar level reduction and gradual rise in insulin level was observed in streptozotocin or alloxan-induced rats that were orally supplemented with ethanolic and methanolic leaf extract of *Costus igneus* (Jayasri et al., 2008) (Table 5.2). These antidiabetic properties were also observed on STZ-induced diabetic rat models. Studies of insulin plant is not only for antidiabetic properties but has also led to spotting of some essential trace elements and carbohydrates hydrolyzing enzymes inhibition activities like glucosidase and amylase (Jayasri et al., 2009; Ferose Khan et al., 2016). On diabetic rats, the *in-vivo* study of *C. pictus* leaf was conducted to find out antihyperglycemic properties as well as secretion of insulin, where it demonstrated the lowered glucose concentration, increased insulin secretion, and enhanced glucose utilization (Gireesh et al., 2009). Al-Romaiyan et al. (2010), observed that costus extract stimulated secretion of insulin from the cells of islets of Langerhans.

Mathew and Vaeghese (2019) reduced fasting and post-meal levels of blood sugar, without knowing the mechanism measures of the antidiabetic activities. The insulin plant reduce the complications associated with diabetes; brings renal and hepatic parameters, reduces the amount of glycosylated hemoglobin, uplifts bodyweight, maintains the lipid profile, as well as level of insulin and indicates marked rectification in the histopathological examination in addition with its anti-diabetic properties. A wide research has been carried out to assess and explore the antidiabetic effect of insulin plants on animal models (Table 5.2). According to a study, to evaluate effectively producing glycemic control machinery in diabetic patients, 1 fresh leaf directly or a tablespoon per day of shade-dried leaf powder of *C. pictus* in combination with other herbs have shown the potential to cure the disease (Shetty et al., 2010). While, *in-vitro* ethanolic extract of leaves of *C. pictus* were studied for translocation of GLUT4 and glucose absorption activity, where no direct peripheral action was observed at doses of 300 gml^{-1} in comparison to metformin and insulin (Pareek et al., 2010). A study to evaluate the glucose homeostasis alteration in C57BLKS/J (KS) *db/db* mice (Hyperglycemic mice, showing progressive βcells depletion) via the use of leaves of *C. spicatus* showed no alteration in insulin sensitivity when leaves of the plant were given in form of tea preparation used dose of said tea of leaves in the result of intraperitoneal (IP) resistance to insulin testing done after the 10 week study period showing the inefficacy of plant in the treatment of hyperglycemia that are induced due to obesity (Keller et al., 2009). The methanolic and ethanolic leaf extract of *C. pictus*, when induced to alloxan and streptozotocin-induced rats is medicinally important in treatment of hyperglycemic activity as it demonstrated a significant reduction in blood sugar level (Jayashri et al., 2008; Kumar et al., 2010; Issac et al., 2011; Remya et al., 2013). Trace elements have been reported to have stimulatory antidiabetic properties (Narmadha et al., 2012). Oral administration of aqueous extract (250 $mgkg^{-1}$) of *C. pictus* on STZ-induced diabetic rats and *in-vitro* pancreatic islet cell culture was also reported to have hypoglycemic properties (Gireesh et al., 2009). Effect of carbohydrate hydrolyzing enzymes (α glucosidase and α amylase) in aqueous extract showed the highest inhibition at 100 $mgml^{-1}$ (Jayasri et al., 2009; Remu et al., 2009). *In-vitro* studies carried out on M1N6 membrane integrity and M1N6 monolayer showed a dosage-dependent rise in secretion of insulin and effects directly on islets of β cells to induce insulin secretion (Al Romaiyan et al., 2010). When leaves of *C. pictus* analyzed on a molecular basis, solid molecular transcripts for ABA, bixin, geraniol and geraniol biosynthesis is observed to be present by next-generation sequences base technique in contribution to therapeutic properties and presence of a bixin precursor a putative product of nor-bixin methyltransferases for antidiabetic property (Annadurai, 2012). Antiproliferative and cytotoxicity of ethanolic extract of *Costus* leaves when observed at lower concentration found to induce MOLT-4 cells death (Jyotrimoy et al., 2012), whereas at the concentration of 500 $mgkg^{-1}$ dose of the extract in comparison with compound β-L-arabinopyranose methyl glycoside as the reference exhibit the highest percentage reduction of glucose concentration in blood glucose level (Shiny et al., 2013).

TABLE 5.2 Antidiabetic Activity of *Costus pictus* Using Various Treatments on Animal Models

Plant Part Used	Extract Used	Results	References
Leaf	Ethanolic	Significant blood glucose Level lowering effect shown by hyperglycemic rat.	Dhanabal (2007); Bhat (2010)
		Synergetic hypoglycemic effect revealed by decreased serum lipid levels, restored insulin, hemoglobin in Wistar rats with induced STZ.	Jayasri et al. (2007); Mani et al. (2014)
		Loss of body weight prevented in Alloxan induced adult albino rats of Wistar strain.	Bhat et al. (2010); Nwakanma et al. (2020)
	Methanolic	Serum glucose level decreased while liver glycogen level and plasma insulin level increased significantly when Alloxan induced in Male Wistar Albino rat, however it found useful in hyperlipidemia manifested by improving lipid profile (cholesterol, phospholipids, HDL, LDL, VLDL, triglyceride).	Jothivel (2007); Tran et al. (2020)

TABLE 5.2 *(Continued)*

Plant Part Used	Extract Used	Results	References
		Fasting glucose level and serum lipid profile (triglyceride, total cholesterol), LDL, VLDL, decreases while HDL raised which ultimately indicated antidiabetic and hypolipidemic effects in male albino diabetic rats with induced STZ.	Pazhanichamy (2011); Tran et al. (2020)
		Useful in decreasing lipid peroxidation (LPO) and blood glucose level in Male albino Wistar rats with induced Alloxan.	Sethumathui and Nandakumar (2009); Tran et al. (2020)
		Reduction in level of blood glucose in diabetic albino Wistar rats (male).	Bhat and Asuti (2010)
	Crude aqueous extract	It stimulated glucose induced insulin secretion in STZ induced male albino Wistar rats; it regulates glucose secretion and homeostasis by muscarinic receptors.	Gireesh (2007)
		Showed antihyperglycemic effect on male albino diabetic rats with induced Alloxan.	Chimurkar et al. (2018); Tran et al. (2020)
	Powder	Helpful in lowering the blood glucose on STZ induction in male albino Wistar rat.	Devi and Urooj (2008)
		Decreased blood sugar level in Dexamethasone induced male Wistar rats preprandial and postprandial diabetic rat.	Shetty et al. (2010)
		Consistent and gradual decline was observed in blood glucose of STZ induced male albino rats of Wistar strain	Devi and Urooj (2011); Tran et al. (2020)
	Alcoholic	STZ induced Wistar rats Showed antidiabetic activity	Adiga et al. (2014); Tran et al. (2020)

TABLE 5.2 *(Continued)*

Plant Part Used	Extract Used	Results	References
Rhizome	Ethanolic	Blood glucose level reduced, carbohydrate metabolism and antioxidant enzymes activities are controlled significantly in STZ induced albino rats of Wistar strain.	Pazhanichamy (2011)
Whole plant	Ethanolic	Resulted in decreased blood glucose level; LDL; serum cholesterol and triglyceride level along with HDL increased gradually in Alloxan induced albino Wistar strain diabetic rat.	Palanivel et al. (2013)

5.3.4.2 ANTIBACTERIAL AND ANTIFUNGAL ACTIVITY

Reddy and Jose (2010) extracted essential oil from leaves and antibacterial activity was observed against both strains of gram-positive (*Bacillus cerenus, Staphylococcus aureus* and *S. faccalis*) and gram-negative bacteria (*Enterobacter facealis, Salmonella paratyphi, Escherichia coli, Protus vulgaris, Pseudomonas aeuroginosa, Serratia marcescens,* and *Klebsiella pneumonia*). Ethanolic extract showed moderate effect on bacterial (*S. aureus,* and *Escherichia coli*) and fungal (*C. albicans*) strains (Saraswathi et al., 2010; Gothandam et al., 2010) tested antibacterial activity in methanolic extract of *C. igneus* against gram-positive strains *(Micrococcus luteus, B. cereus, B. megaterium, S. aureus, S. lactis*) and gram-negative bacterial strains (*P. aeruginosa, Klebsiella pneumoniae, E. coli, E. aerogenes,* and *Salmonella typhimurium*) (Table 5.3). Different extracts were prepared from all parts of *C. pictus*, but the MeOH extracts from flower and stem showed maximum inhibition against the growth of tested microbes (*Klebsiella pneumonia, Shigella flexneri, B. subtilis, E. coli*) at 150 μgml^{-1} concentration (Sathuvan et al., 2012). Nagarajan et al. (2011) investigated *C. igneus* plant root powder derived from in vitro raised plants in Murashige and Skoog medium supplemented with IAA and IBA subjected to extraction using different solvents (methanol, acetone, and chloroform) against Gram-negative Bacteria cultures (*P. aeruginosa, K. pneumoniae, Salmonella sp,* and *P. vulgaris*) for antibacterial activities

(Table 5.3). Among all tested solvents and bacterial cultures, *Klebsiella pneumonia* established the susceptibility to both plant growth regulators (IAA and IBA) cultured roots.

5.3.4.3 ANTIOXIDANT ACTIVITY

Moderate antioxidant effects of alcoholic extract of *C. igneus* leaves have been studied (Dhanabal et al., 2017; Chacko et al., 2018). The antioxidant properties were investigated in leaves and rhizomes subjected to many of the solvents (aqueous, methanol, ethanol, and ethyl acetate, etc.), extract and were evaluated using several methods such as βcarotene, DPPH, deoxyribose, reducing power, superoxide anion and metal chelating assay at various concentrations (Table 5.3). Leaf and rhizome, observed to have about 89.5% and 90.0% antioxidant activities respectively, which is considerably good in reference to standard Butylated Hydroxytoluene (BTH 85%) at 400 μgml^{-1} concentration (Chacko et al., 2018). Out of all the tested extracts, MeOH extracts of both parts (leaf and rhizome) of the insulin plant encompasses greater antioxidant activity in comparison to other tested extracts (Jayasri et al., 2009). Others study result revealed that methanolic leaf extract induced considerable rise in glutathione peroxidase (GPx), catalase (CAT), superoxide dismutase (SOD), glutathione reductase (GR), vitamins (A,C, and E) and reduced glutathione(GSH), and due to this it is possibly effective in oxidative stress reduction and diseases induced by free radicals (Shivaprakash et al., 2014) (Table 5.3). Presence of phenolic compounds was observed to be responsible for antioxidant activities in the plant (Sethumathi et al., 2009). Quercetin and diosgenin encompasses significant antioxidant activities induced by reversal of oxidative stress in affected organs of diabetic rats (liver, pancreas, and kidneys) as well as provoking glycolytic enzymes, which ultimately controls gluconeogenesis process in diabetic rats (Kalailingam et al., 2011).

The extracts from rhizome and leaf of *C. pictus* shows about 89.5% and 90% antioxidant activity at 400 μgml^{-1} concentration assessed by beta carotene, Deoxyribose superoxide anion, DPPH, and metal chelating assay at varying concentration (Jayasari et al., 2008; Chacko et al., 2018). *C. pictus* leaf extract has shown significant inhibition of lipid peroxidation radical scavenging activity *in-vitro* for goat liver and RBC ghost (Kavitha, 2013). Also, the leaf extracts showed high percentage antioxidant and free radical scavenging activity when tested clinically for pre-oxidative damage in albino rats. In addition, the stem and root extracts also showed high antioxidant activity (Remya and Daniel, 2013). The methanolic root and stem extracts of the plant have high bioactive components and show active antioxidant properties (Maruthappan and Sree, 2010; Suchitra, 2017).

Methanol extract was investigated for antioxidant activity of plant extract against bacteria and fungi using different concentrations ranging between 100–500 μgml^{-1} (Ramya et al., 2015). Along with antioxidant activity, radical scavenging properties of roots and stem extracts of *Costus* were also investigated. Root extract reported better than stem, in terms of inhibition rate, high phenolic content, Vitamin E concentration. Polyphenols and antioxidants are also found to inhibit free radical inhibition properties (Ramya et al., 2015).

TABLE 5.3 Studies Carried Out to Investigate Various Other Biological Activities of *Costus pictus*

Biological Activities	Part Used	Extract Used	Results	References
Antioxidant	Leaves	–	The marked elevation in enzyme levels (like CAT, SOD and GSH) and an Inclusive recuperation of enzyme level reported at 600mg/kg dose along with huge reduction in level of MDA was observed.	Shivaprakash et al. (2014); Chacko et al. (2018)
		MeOH	Increase certain enzymes activities (SOD, CAT, GPase, GR, Vit. A, C, E, and reduced GHS	Sethumathi et al. (2009); Jayasri et al. (2009); Dhanabal et al. (2017)
	Roots	–	Highest phenol content responsible for maximum antioxidant property.	Ramya et al. (2015)

TABLE 5.3 *(Continued)*

Biological Activities	Part Used	Extract Used	Results	References
	Rhizome	EtOH	Quercetin and diosgenin both have a stimulatory effect on glycolytic enzymes and significantly control the gluconeogenesis process.	Jayasri et al. (2009); Kalailingnam et al. (2011)
Antimicrobial	–	MeOH	Significant antimicrobial effects on both Gram +ve and –ve bacteria.	Gothandam et al. (2010); Kala et al. (2014)
	Leaves	Essential oil	Significant antibacterial activity	Reddy and Josh (2010)
	Flower and stem	MeOH	Shown good antifungal and antibacterial activity	Sathuvan et al. (2012)
Hypolipidemic	–	–	Total LDL, VLDL, cholesterol, phospholipids and triglyceride level reduced significantly but a rise was found in HDL level.	Mani et al. (2014)
	Rhizome	MeOH	Major decline in lipid panel (total cholesterol, triglycerides, LDL cholesterol and VLDL) and reduction of HDL serum levels	Pazhanichamy et al. (2011)
	Whole plant	Alcoholic	Remarkable reduction was observed in lipid profile (cholesterol serum, triglycerides, and LDL) in hyperlipidemic rats.	Chacko et al. (2012)

TABLE 5.3 *(Continued)*

Biological Activities	Part Used	Extract Used	Results	References
Anti-inflammatory	Leaves	MEC	Compound β-amyrin was found inhibitory against paw edema	Krishnan et al. (2014)
Antiproliferative	Leaves	MeOH	Showed reduction in tumor size and 97.46% cytotoxicity.	Dhanasekaran et al. (2014)
		EtOH	Found anti-cancerous or antiproliferative effect against *in-vitro* mammalian fibrosarcoma cell lines (HT-1080).	Nadumane et al. (2011)
	Bark	Total extracts	Potentially active against cancerous cell lines (HT29 and A549 cells)	Sathuvan et al. (2012)
Diuretic and antiurolithiatic	Stem and rhizome	Aqueous	Found potent to help the construction of hydroxyapatite (HAP) crystals and CHPD crystals (a chief compound of Ca induced stone in urine) nucleation rate decreases.	Yuvarani et al. (2017)
	Whole plant	Aqueous	Clearance of Na and K was induced such as to be done by furosemide, representing its dieresis property.	Melendez-Camargo et al. (2006)
Putative	Stem	Aqueous	Elevated the formation of calcium oxalate dihydrate crystals, thus can be proved helpful in treatment of urinary stones	Manjula et al. (2012)

5.3.4.4 ANTI-PROLIFERATIVE AND ANTI-CANCEROUS ACTIVITY

The EtOH extract of *Costus* leaves showed anti-proliferative as well as anticancer potential in *in-vitro* mammalian fibrosarcoma (HT-1080) cells (Nadumane et al., 2011). Different bark extracts obtained from *C. pictus* displayed effective anti-cancerous activities against HT29 and A549 cells (Sathuvan et al., 2012) (Table 5.3). In addition, the methanolic extract of the plant is cytotoxic and can cause apoptosis of bone cancer on MCF7 Breast cells (Dhanasekaran et al., 2014).

5.3.5 MOLECULAR STUDIES (OMICS APPROACHES)

The emergence of techniques such as molecular markers has purely revolutionized the advancement in biological sciences. Applications of DNA based markers have been approached in various fields like DNA fingerprinting, taxonomy, analysis of genetic diversity, marker assisted selection, gene tagging, etc. Molecular marker is an approach to characterize plants for conservation and administration of plant genetic resources and in relation to understanding the full arrangement of hereditary lineage. Molecular marker technique is useful in characterization of cultivar, depiction of genetic relatedness, characterization and certification of purity or adulteration, security of intellectual property rights. After the emergence of DNA markers technique, other morphological characterization of species becomes less authentic. This technique is dominating because hardly any manipulation by the environment interferes with its result. Results of these techniques are quick, as once a marker gets associated with any desired character of need, the plantlet could be selected immediately for that desired character, there is no need to wait for its phenotypic expression (Sahu et al., 2015). Further research is required to determine the function of other regulators/signals participated in the antidiabetic impact of the *Costus* plant extracts.

5.3.5.1 METABOLOMICS

Metabolite profiling of *C. pictus* by various researchers provides adequate and full-proof technology for phytochemical fingerprinting dependent on plant parts collected. Chromatographic sophistication combined with mass spectrometry have facilitated a variety of new metabolite discoveries. Mutually complementary methodologies like TLC, GC-MS, LC-MS, HPLC and HPTLC were employed to obtain a comprehensive metabolomic profile (Ashwini et al., 2017; Nadumane et al., 2011; George et al., 2007). Near about eighteen major biochemical compounds are reported in saponified extract of *C. pictus* leaves using GC-MS (George et al., 2007; Devi, 2019) (Table 5.4). Besides these methanolic extracts of leaves are reported to contain flavonoids and phenols (Ashwini et al., 2015) as well as pentacyclic triterpenes α and β amyrins (Ashwini et al., 2017). Phytochemical studies of leaves have shown three major flavonoids namely Astragalin, kaempferol, quercetin and eight phenolic compounds that were Iso vitexin Galangin, Naringenin, Genistin, Onion, Licochalcone A, α and β amyrins (pentacyclic triterpene) (Table 5.4). Diosgenin, a steroid, was isolated and identified by using HPLC method (Nadumane et al., 2011). Veena et al. (2016) observed that it could also be characterized by using simple parameters such as odor, color, pH, solubility and physical appearance. Diosgenin has also been reported in petroleum ether extract of roots analyzed by GC-MS and HPLC methods (Srinivasan et al., 2016).

TABLE 5.4 Various Studies Carried Out for Compound Identification of *Costus pictus* Using Different Methodologies

Part Used	Extract Used	Compounds Identified	Methodology	References
Leaves	Saponified extracts	α-ergastanol, Bis (2'ethylhexyl)1-2-benzene di carboxylate, xanthene-3-one, tetradecanoic acid, pentadecanoic acid, hexadecanoic acid, heptadecanoic acid, heptadecene, carbonic acid, 4,8,12,16 tetramethyl-heptadecane-4olide, 7 tridecanone, docosanoic acid, 9-octadecenoic acid, octacosane, phytol, octadecanoic acid	GC-MS	George et al. (2007); Devi (2019)

TABLE 5.4 *(Continued)*

Part Used	Extract Used	Compounds Identified	Methodology	References
	Methanolic extract	B-arabinopyranose methyl glycoside	GC-MS, HPLC, TLC	Shiny et al. (2013)
	Methanolic extract	Flavonoids- Astragalin, kaempferol, quercetin and isoquercetin. Phenols- Genistin, Licochalcone A, Galangin, α and β amyrins (pentacyclic triterpene), Iso vitexin, Naringenin	LC-MS	Ashwini et al. (2015); Ashwini et al. (2017)
Stem	Ethanolic extract	Lupeol and stigmasterol	TLC, HPTLC	Manjula K et al. (2016)
Root	Petroleum ether extract	Steroid saponin diosgenin	GC-MS, HPLC	Srinivasan et al. (2016)

5.3.5.2 GENOMICS

A group of short endogenic, non-coding RNA molecules (~25 nucleotides) called microRNAs plays an important regulatory role in gene expression. The transcribed sequences of RNA of *C. pictus* were analyzed to specify conserved miRNA using homology approach, resulting in the identification of 42 microRNA's belonging to 13 families in *C. pictus* and an overall 109 genes have been identified as potential targets regulating different metabolic pathways. A total of 42 miRNA's homolog of 11 separate families, that is to say miR166; miR167; miR168; miR169; miR172; miR394; miR396; miR444; miR858 and miR5658 were observed which have been participated in development and growth, stress response, transduction of signals, and ubiquitin induced proteasome degradation pathways (Das et al., 2016). Genetic variation in vegetatively reproduced *C. pictus* species when studied on rhizomes on the basis of morphological, biochemistry, exchange of gas and water vapor and RAPD analysis, carried out in Kerala's three distinct agro-climatic regions namely Ernakulam, Kannur and Thiruvananthapuram. On evaluation, no substantial difference was observed in gas water vapor exchange parameters such as net photosynthesis rate, transcription rate, stomatal conductance, and did not even show any marked difference in PSII efficiency measured by chlorophyll fluorescence in the respective samples. The RAPD analyzes at three accessions with the aid of 40 primers and 10 decamers did not divulge any genetic discrepancies (Thomas and Palani, 2016).

Around 15 different geographical regions of India are selected as sites for collection of *C. pictus* plant, that generated 343 loci, 124 were polymorphic averaging 4.96 loci/primer using the RAPD marker with 25 primers, whereas using ISSR marker with 20 primer it generated 177 loci, 77 were polymorphic, average of 3.85 loci per primer with similarity coefficient range between 0.84–0.95, 0.86–0.96 and 0.86–0.99 (Naik et al., 2017). Graphically, The dendrogram did not reveal much variance with a low degree of divergence. The highest polymorphism levels in ISSR and RAPD were observed 42.47% and 35.57% respectively (Naik et al., 2017).

5.3.5.3 TRANSCRIPTOMIC STUDY FOR MEDICINAL PROPERTIES OF C. PICTUS

Wide transcriptome sequencing and further analysis of leaves of *C. pictus* done by using an Illumina GAIIx sequence have been reported to understand the molecular signatures associated with antidiabetic properties of the plant (Ramasamy et al., 2012). They identified about 55,000 transcripts and 69.2% of total identified transcripts are possible to be annotated. Transcriptome sequencing was performed on leaves of *C. pictus* to understand the molecular signatures (ABA, bixin and biosynthesis of geranial and geraniol) responsible for antidiabetic properties of the insulin plant. The occurrence of bixin was determined in MeOH extract of the *C. pictus* by using sequential extraction and HPLC techniques (Table 5.4). Another major transcript was acknowledged in order to anti-obesity, antidiabetic, and immuno- modulation is of Abscisic Acid (ABA) biosynthetic pathway. Some other significant transcripts were also identified in relation to biosynthesis of antioxidant, antimicrobial and antitumor metabolites of *C. pictus* leaves. They studied terpenoid pathways but also some other secondary metabolite biosynthesis pathways are noticed related to free radical scavengers, anti-microbial, antiglycation, antitumor and antidiabetic effects of *C.*

pictus leaves with the help of PlantCyc annotations. Terpenoid biosynthesis pathways were noticed in relation to bixin (10.49%), geranial and geraniol (8.95%) and abscisic acid (3.09%) biosynthesis transcripts, accounts to have antidiabetic properties (Takahashi et al., 2002; Takahashi et al., 2009).

Next-Generation Sequencing (NGS) technique is an important tool to classify molecular signatures in transcriptome belonging to physiological functions of plant tissues. The leaf transcriptome are sequenced using Illumina reversible dye terminator sequencing technology and used as combinations of bioinformatics tools in order to identify transcripts related with antidiabetic properties of *C. pictus*. Genomic research of *C. pictus*, a non-model traditional medicinal plant, is constrained by the limited amount of sequence data accessible to the public (Ramasamy et al., 2012).

Antioxidant properties reported from by-product of annotated pathways which include bixin biosynthesis (Silva et al., 2001), astaxanthin, canthaxanthin (Guerin et al., 2003), gossypol (Dodou et al., 2005), crocetin (Magesh et al., 2006), saponins (Meesa Pyod Suk et al., 2007), oleoresin (Singh et al., 2008) and these are strongly associated with *C. pictus* antioxidant properties. The end product of Menthol biosynthetic pathway is menthol (found predominantly 8.02%) which might have antitumor properties (Li et al., 2009).

Some other research has recorded by-products, obtained from interpreted pathways, potentially have the antitumor properties comprising vincristine and vinblastine (Jordan et al., 1998), geraniol (Carnesecchi et al., 2001), bixin (Takahashi et al., 2002), astaxanthin (Guerin et al., 2003), taxol (Rowinsky et al., 2004), all-trans-lycopene (Etminan et al., 2004), gossypol (Dodou et al., 2005), crocetin (Magesh et al., 2006), and perillyl alcohol (Wanger et al., 2002). In addition, 4.94% of the total terpenoid pathway transcript annotations linked to mevalonate path-I were observed. Isopentyl diphosphate (IPP) and its isomer dimethylallyl diphosphate (DMAPP), universal precursors in terpenoid category, are the end-products of mevalonate pathway (Goto et al., 2010). Transcripts corresponding to artemisinin biosynthetic pathway were also analyzed in pathway annotations. It has an end product artemisinin, reported for antimalarial activity (Meshnick et al., 2002). Transcripts annotations corresponding to biosynthetic pathways of gossypol, linalool, bergamotene, farnesene, oleoresin, capsidiol, saponins, phytoalexins and sesquiterpenoid phytoalexins, isopimaric acid recommend that they provide insect/herbivore/microbes defense mechanism to the plant.

5.4 CONCLUSION

Recent *in vitro* as well as *in vivo* experimental trials have facilitated the antiglycemic potential of phytochemicals derived from *C. pictus* (syn- *igneus*) plant parts extract. The antidiabetic actions of phytochemicals are associated to its bioactive components such as triterpenoids, alkaloids, flavonoids, steroids, proteins, fatty acids, carbohydrates, etc. Production of phytochemicals shows how they perform the modulatory effects on glucose levels in the blood via. stimulating insulin release, promoting pancreatic cells proliferation and insulin resistance reduction. In this chapter, an attempt has been made to infer the various biochemical compounds and their activities along with their chemical nature and solvents used to extract compounds from *Costus pictus*. In this compilation, the organic activities and chemical constituents of *Costus pictus* has been enlightened. For a Worldwide implementation of these natural plant resources, specific and rigorous proof-based approaches must be conducted to prove their quality, safety and effectiveness. Advances in biotechnology such as *in vitro* culture techniques, molecular biology, and metabolite profiling of metabolites other than terpenoids should be carried out to provide further new insights for knowing the detailed benefits of the plant against diabetes. Conservation as well as management of plant genetic resources must be done, and improved extraction procedures of drugs from *C. pictus* plants must be simplified. The implementation of emerging omics tools in combination with biotechnological techniques have revealed the potential role and significance of conservation of medicinal plant species that are now endangered or are vulnerable in nature.

KEYWORDS

- **blood glucose**
- **diabetes**
- **insulin**
- **spiral flag**
- **sugar**

REFERENCES

Aa, N., Idaguko, C., Co, E., Ji, M., Kc, O., & Sci, J., (2020). Histological studies of ethanolic extract of *Costus aferkergawl* leaves on the pancreas of alloxan-induced diabetes in Wistar rats. *J. Anat. Sci., 11*(2), 102–106.

Alam, M. M., Meerza, D., & Naseem, I., (2014). Protective effect of quercetin on hyperglycemia, oxidative stress and DNA damage in alloxan-induced type 2 diabetic mice. *Life Sci. J., 109*, 8–14.

AL-Ishaq, R. K., Abotaleb, M., Kubatka, P., Kajo, K., & Büsselberg, D., (2019). Flavonoids and their Antidiabetic effects: Cellular mechanisms and effects to improve blood sugar levels. *Biomolecules, 9*(9), 430.

Al-Romaiyan, A., Jayasri, M. A., Mathew, T. L., Huang, G. C., Amiel, S., Jones, P. M., & Persaud, S. J., (2010). *Costus pictus* extracts stimulate insulin secretion from mouse and human Islets of Langerhans *in vitro. Cell. Physiol. Biochem., 26*(6), 1051–1058.

American Diabetes Association, (2018). *Classification and Diagnosis of Diabetes: Standards of Medical Care in Diabetes Care., 41,* 13–27.

American Diabetes Association., (2009). Diagnosis and classification of *Diabetes mellitus. Diabetes Care, 32*(1), S62–S67.

Annadurai, R. S., Jayakumar, V., Mugasimangalam, R. C., Katta, M. A., Anand, S., Gopinathan, S., Sarma, S. P., et al., (2012). Next-generation sequencing and de novo transcriptome analysis of *Costus pictus* D. Don, a non-model plant with potent anti-diabetic properties. *BMC Genomics, 13,* 1–15.

Anwannil, H. G., & Atta, R., (2006). Trends in ethnopharmacology. *J. Ethnopharmacol., 100*, 43–49.

Ashwini, S., Bobby, Z., Joseph, M., Jacob, S. E., & Priya, R. P., (2015). Insulin plant (*Costus pictus*) extract improves insulin sensitivity and ameliorate atherogenic dyslipidemia in fructose-induced insulin-resistant rats: Molecular mechanism. *J. Function. Foods., 17,* 745–760.

Ashwini, S., Sridhar, B. Z., & Clectus, (2017). Insulin plant (*Costus pictus*) extracts restores thyroid hormone levels in experimental hypothyroidism. *Pharmacogn. Res., 9*(1), 51–59.

Bhat, V., Asuti, N., Kamat, A., Sikarwar, M. S., & Patil, M. B., (2010). Antidiabetic activity of insulin plant (*Costus igneus*) leaf extract in diabetic rats. *J. Pharma. Res., 3*(3), 608–611.

Bouic, P. J. D., Clark, A., Lamprecht, J., Freestone, M., Pool, E. J., Liebenberg, R. W., Kotze, D., & Jaarsveld, P. P. V., (1999). The effects of B-sitosterol (BSS) and B-sitosterol glucoside (BSSG) mixture on selected immune parameters of marathon runners: Inhibition of post-marathon immune suppression and inflammation. *Int. J. Sports Med., 20*(04), 258–262.

Carnesecchi, S., Schneider, Y., Ceraline, J., Duranton, B., Gosse, F., Seiler, N., & Raul, F., (2001). Geraniol, a component of plant essential oils, inhibits growth and polyamine biosynthesis in human colon cancer cells. *J. Pharmacol. Exp. Ther., 298,* 197–200.

Chacko, N., Shastry, C. S., & Shetty, P., (2018). Studies on the antioxidant activity of *Costus igneus* leaf extract. *H. J. D. Med., 10,* 9–15.

Chacko, N., Shastry, C. S., Shetty, P., Shyamma, P., D'souza, U., & Maulika, P., (2012). Anti-hyperlipidemic activity of *Costus igneus* in triton X-100 induced hyperlipidemic rats. *Int. J. Pharm. Chem. Sci., 1,* 813–818.

Chakravarthy, B. K., Gupta, S., Gambhir, S. S., & Gode, K. D., (1981). Pancreatic beta-cell regeneration in rats by (-)-epicatechin. *The Lancet, 2,* 759–760.

Chaudhary, J. D., Maurya, A. K., Tripathi, S., Ranga R., Kahrana, M., & Samyal, M. L., (2012). Estimation of the anti-proliferative potential of *pictus* on molt-4 human cancer cell lines. *Int. J. Bio Pharm. and Toxicol. Res., 2,* 244–250.

Chetty, S., Adiga, S., & Reddy, S., (2014). Evaluation of the effect of *Costus igneus* on learning and memory in normal and diabetic rats using passive avoidance task. *Int. J. Pharm. Pharm. Sci., 6*(2), 835–838.

Chimurkar, L., Kale, R., & Varma, S., (2018). Evaluation of *Costus igneus* on lipid profile status and anti-hyperglycemic activity in alloxan-induced diabetic rats. *Int. J. Res. Rev., 5,* 88–93.

Colagiuri, S., Guariguata, L., Motala, A. A., & Ogurtsova, K., (2019). Global and regional diabetes prevalence estimates for, 2019 and projections for 2030 and 2045: Results from the international diabetes federation diabetes Atlas (9th edn.). *Diabetes Res. Clin. Pract., 157,* 107–843.

Coskun, O., Kanter, M., Korkmaz, A., & Oter, S., (2005). Quercetin, a flavonoid antioxidant, prevents and protects streptozotocin-induced oxidative stress and beta-cell damage in rat pancreas. *Pharmacol. Res., 51*(2), 117–123.

Czech, M. P., (1976). Differential effects of sulphahydryl reagents on activation and deactivation of the fat cell hexose transport system, *J. Biol. Chem., 251,* 1164–1170.

Czech, M. P., (1977). Molecular basis of insulin action, *Ann. Rev. Biochem., 46,* 359–384.

Dahanuka, S. A., Kulkarni, R. A., & Rege, N. N., (2000). Pharmacology of medicinal plants and natural products. *Indian J. Pharmacol., 32,* 81–118.

Dahlgren, R. M. T., Clifford, H. T., & Yeo, P. F., (1985). *The Families of the Monocotyledons, Structure, Evolution and Taxonomy*. Springer Verlag, Berlin.

Das, A., Das, P., & Kalita, M. C., (2016). Computational identification target prediction and validation of conserved miRNAs in Insulin plant (*Costus pictus* D. Don). *Appl. Biochem. Biotechnol., 178,* 513–526.

DeFronzo, R. A., (1988). The triumvirate: β-cell, muscle, liver: A collusion responsible for NIDDM. *Diabetes, 37*(6), 667–687.

Devi, G. M. R., (2019). A comprehensive review on *Costus* pictus D. Don. *Int. J. Pharma. Sci. Res., 10*(7), 3187–3195.

Devi, V. D., & Urooj, A., (2010). Nutrient profile and antioxidant components of *Costus speciosus* Sm. and *Costus igneus* nak. *Indian J. Nat. Prod. Resour., 1,* 116–118.

Dhanabal, S. P., Kumar, A., Chandrasekar, R., John, S., Joseph, S., James, M., Joseph, L., et al., (2007). Hypoglycemic and antioxidant activities of *Costus Mexicans* (Costaceae). *Aryavaidyan., 21,* 5358.

Dhanasekaran, S., Akshaya, M., & Preethi, S., (2014). *In Vitro* anti-proliferative potential of leaves of *Costus igneus*. *Int. J. Innov. Eng. and Technol., 4*(4), 277–283.

Dodou, K., Anderson, R. J., & Small, D. A., & Groundwater, P. W., (2005). Investigations on gossypol: Past and present developments. *Expert. Opin. Investig. Drugs, 14*(11), 1419–1434.

Eid, H. M., & Haddad, P. S., (2017). The antidiabetic potential of Quercetin: Underlying mechanisms. *Curr. Med. Chem., 24*(4), 355–364.
Etminan, M., Takkouche, B., & Caamaño-Isorna, F., (2004). The role of tomato products and lycopene in the prevention of prostate cancer: A meta-analysis of observational studies. *Cancer Epidemiol. Biomarkers Prev., 13,* 340–345.
Ferosekhan, M., Ramu, A., & Ravikumar, S., (2016). Scientific evaluation of traditionally known insulin plant *Costus* species for the treatment of diabetes in humans. *Int. J. Curr. Res. Biosci. Plant Biol., 3*(6), 87–91.
Fluckey, J. D., Vary, T. C., Jepfferson, L. S., & Farrell, P. A., (1996). Augmented insulin action on rates of protein synthesis following resistance exercise in rats. *Am. J. Physiol., 33,* 331–319.
Fukushima, M., Matsuyama, F., Ueda, N., Egawa, K., Takemoto, J., Kajimoto, Y., & Mitsui, R., (2006). Effect of corosolic acid on post-challenge plasma glucose levels. *Diabetes Res. Clin. Pract., 73*(2), 174–177.
George, A., Thankamma, A., Devi, V. K. R., & Fernandez, A., (2007). Phytochemical investigation of Insulin plant (*Costus pictus*). *Asian J. Chem., 19,* 3427–3430.
Gireesh, G., & Paulose, C. S., (2007). *Muscarinic M1 Receptor Gene Expression in Streptozotocin-Induced Diabetic Rats: Regulation of Insulin Secretion by Aegle Marmelos and Costus Pictus Leaf Extracts* (pp. 470–474). Faculty of Science, Cochin University of Science and Technology, Kerala, India.
Gireesh, G., Thomas, S. K., Joseph, B., & Paulose, C. S., (2009). Antihyperglycemic and insulin secretory activity of *Costus pictus* leaf extract in streptozotocin-induced diabetic rats and in-vitro pancreatic islet of Langerhans culture. *J. Ethno. Pharmacol., 123*(3), 470–474.
Gothandam, K. M., Aishwarya, R., & Karthikeyan, S., (2010). Preliminary screening of antimicrobial properties of few medicinal plants. *J. Phytol.*, 1–6.
Goto, T., Takahashi, N., Hirai, S., & Kawada, T., (2010). Various terpenoids derived from herbal and dietary plants function as PPAR modulators and regulate carbohydrate and lipid metabolism. *PPAR Res.*, 483–958.
Guerin, M., Huntley, M. E., & Olaizola, M., (2003). Haematococcus astaxanthin: Applications for human health and nutrition. *Trends Biotechnol., 21*(5), 210–216.
Hickey, M., & King, C. J., (1981). *100 Families of Flowering Plants*. Cambridge University Press, Cambridge.
Hoshino, T., (2017). β-Amyrin biosynthesis: Catalytic mechanism and substrate recognition. *Org. Biomol. Chem., 15*(14), 2689–2691.
Humphries, C. J., (1985). Zingiberaceae, ginger, cardamom and turmeric. In: Heywood, V. H., (ed.), *Flowering Plants of the World* (pp. 296–298). Sci. Peche, Oxford.
Issac, S. T., & Alphonse, J. K., (2011). Comparative study of hypoglycemic activity of *pictus* and *C. igneus* in streptozotocin induced diabetic rats. *J. Pharm Res., 4,* 3628–3629.
Jayasri, M. A., Gunasekaran, S., Radha, A., & Mathew, T. L., (2008). Antidiabetic effect of *Costus pictus* leaves in normal and streptozotocin-induced diabetic rats. *Int. J. Diabetes Metab., 16*(3), 117–122.
Jayasri, M. A., Lazar, M., & Radha, A., (2009). A report on the antioxidant activity of leaves and rhizomes of *Costus pictus* D. Don. *Int. J. Integr. Biol., 5*(1), 20–26.

Jayasri, M. A., Radha, A., & Mathew, T. L., (2009). α-amylase and α-glucosidase inhibitory activity of *Costus pictus* don in the management of diabetes. *J. Herb. Med. Toxicol., 3*(1), 91–94.

Jesus, M., Martins, A. P., Gallardo, E., & Silvestre, S., (2016). Diosgenin: Recent highlights on pharmacology and analytical methodology. *J. Anal. Methods Chem.*

Jordan, A., Hadfield, J. A., Lawrence, N. J., & Mc Gown, A. T., (1998). Tubulin as a target for anti-cancer drugs: Agents which interact with the mitotic spindle. *Med. Res. Rev., 18*(4), 259–296.

Joshi, B. N., Hitendra, M., Hardikar, M., & Anjali, A. K., (2013). Orally active hypoglycaemic protein from *Costus igneus*: An *in-vitro* and *in-vivo* study. *Biochem. Biophy. Res. Commu., 436,* 278–282.

Joshi, N. P., Vikas, V., Sachin, V. S., Parwar, & Gadgil, J. N., (2013). Development and validation of HPLC method for simultaneous determination of bio-active markers corosolic acid, Asiatic acid and β-sitosterol from leaves of *Lagerstroemia speciosa* and from the marketed formulation. *Int. J. Pharm. Sci., 5*(4), 223–226.

Jothivel, N., Ponnusamy, S. P., Appachi, M., Singaravel, S., Rasilingam, D., Deivasigamani, K., & Thangavel, S., (2007). Antidiabetic activity of methanol leaf extract of *Costus pictus* D. Don in alloxan-induced diabetic rats. *J. Health Sci., 53*(6), 655–663.

Kala, S., (2014). Antimicrobial activity of coleus for skohlii (Wild) Briq and *Costus igneus* NE Br. *J. Pharm. Biol. Sci., 9,* 1–6.

Kalailingam, P., Balasubramanian, K., Kannaian, B., Mohammed, A. K. N., Sundaram, K. M., Tamilmani, E., & Kaliaperumal, R., (2013). Isolation and quantification of flavonoids from ethanol extract of *Costus igneus* rhizome (CiREE) and impact of CiREE on hypoglycaemic, electron microscopic studies of the pancreas in streptozotocin (STZ)-induced diabetic rats. *Biomed. Prev. Nutr., 3*(3), 285–297.

Kalailingam, P., Kaliaperumal, R., Shanmugam, K., & Tamilmani, E., (2011). Efficacy of Methanolic extract of *Costus igneus* rhizome on hypoglycemic, hypolipidemic activity in streptozotocin (STZ) diabetic rats and HPTLC analysis of its active constituents. *Int. Conf. Biosci., Biochem. Bioinfo., 26*(5), 318–321.

Kalailingam, P., Sekar, A. D., Samuel, J. S., Gandhirajan, P., Govindaraju, Y., Kesavan, M., Kaliaperumal, R., & Tamilmani, E., (2011). The efficacy of *Costus igneus* rhizome on carbohydrate metabolic, hepatoprotective and antioxidative enzymes in streptozotocin-induced diabetic rats. *J. Health Sci., 57*(1), 37–46.

Kameshwara, R. B., Keshavalu, M. M., & Appa, R. C., (2003). Evaluation of antidiabetic effect of *Momordica cymbalaria* fruit in alloxan-diabetic rats, *Fitoterapia., 74,* 7–13.

Karuayaka, E. H., Welihinda, J., Sirimanne, S. R., & Gowri, S., (1984). Inoral hypoglycemic activity of some medicinal plants of Sri Lanka. *J. Ethnopharmacol., 75,* 2717–2731.

Katoriya, V. S., Deokar, G., Kshirsagar, S., & Rupvate, S., (2016). A review on medicinal boon: *Costus* species. *Int. J. Inst. Pharm. Life Sci., 6*(4), 63.

Kavitha, R., (2013). In-vitro anti-oxidant studies of *Costus pictus*. *J. Res. Pharm. Biomed. Sci., 4*(1), 159–164.

Kawahara, T., Kawaguchi-Ihara, N., Okuhashi, Y., Itoh, M., Nara, N., & Tohda, S., (2009). Cyclopamine and qu*ercetin suppress* the growth of leukemia and lymphoma cells. *Anticancer Res., 29*(11), 4629–4632.

Keller, A. C., Vandebroek, I., Liu, Y., Balick, M. J., Kronenberg, F., Kennelly, E. J., & Brillantes, A. M. B., (2009). *Costus spicatus* tea failed to improve diabetic progression in C57BLKS/J db/db mice, a model of type 2 *Diabetes mellitus*. *J. Ethnopharmacol., 121*(2), 248–254.

Kelly, G. S., (2011). Quercetin. *Altern Med Rev., 16*(2), 172–194.

Kirchoff, B. K., & Rutishauser, R., (1990). The phyllotaxy of *Costus* (Costaceae). *Bot. Gaz. (Chicago), 151*, 88–105.

Kirchoff, B. K., (1988). Floral ontogeny and evolution in the ginger group of the *Zingiberales*. *Aspects of Floral Development, 45–56.*

Knip, M., & Simell, O., (2012). Environmental triggers of type 1 diabetes. *Cold Spring Harb. Perspect. Med., 2,* a007–690.

Krishnan, K., Mathew, L. E., Vijayalakshmi, N. R., & Helen, A., (2014). Anti-inflammatory potential of β-amyrin, a triterpenoid isolated from *Costus igneus*. *Inflammopharmacology, 22,* 373–385.

Kumar, A. R., Bastin, T. M., Jenifer, S., & Arumugam, M., (2010). Comparative evaluation of extracts of *igneus* (*C. pictus*) for hypoglycemic and hypolipidemic activity in alloxan diabetes rats. *Int. J. Pharm. Technol., 2,* 182–195.

Lacroix, M., Kina, E., & Hivert, M. F., (2013). Maternal/fetal determinants of insulin resistance in women during pregnancy and in offspring over life. *Curr. Diabetes Rep., 13*(2), 238–244.

Laha, S., & Paul, S., (2019). *Costus igneus*–A Therapeutic antidiabetic herb with active phytoconstituents. *Int. J. Pharm. Sci. Res., 10*(8), 1000–1009.

Lambert, N., Baccou, J., & Sauvaire, Y., (1988). Screening for diosgenin in rhizomes from 3 *Costus species* (*C.deistellii, C. igneus* and *C. lucanusianus*). *Planta. Medica., 54*(4), 366.

Li, Q., Wang, X., Yang, Z., Wang, B., & Li, S., (2009). Menthol induces cell death via the TRPM8 channel in the human bladder cancer cell line T24. *Oncology., 77*(6), 335–341.

Lown, J. W., (1993). Discovery and development of anthracycline antitumor antibiotic. *R. Soc. Chem., 165–176.*

Magesh, V., Singh, J. P., Selvendiran, K., Ekambaram, G., & Sakthisekaran, D., (2006). Antitumor activity of crocetin in accordance to tumor incidence, anti-oxidant status, drug-metabolizing enzymes and histopathological studies. *Mol. Cell. Biochem., 287,* 127–135.

Mani, P., Kanivalan, N., & Rajakumar, R., (2014). Antidiabetic and hypolipidemic effects of *Costus igneus* leaves extracts against Streptozotocin-induced diabetic albino rats. *Acta Biomedica Scientia, 1,* 74–79.

Manjula, K., Pazhanichamy, K., Kumaran, S., Eevera, T., Dale, K. C., & Rajendran, K., (2012). Growth characterization of calcium oxalate monohydrate crystals influenced by *Costus igneus* aqueous stem extract. *Int. J. Pharm. Pharm. Sci., 4*(1), 261–270.

Manjula, K., Rajendran, K., Eevera, T., & Kumaran, S., (2013). Quantitative estimation of lupeol and stigmasterol in *Costus igneus* by high-performance thin-layer chromatography. *J. Liq. Chromatogr. Relat. Technol., 36*(2), 197–212.

Mathew, F., & Varghese, B., (2019). A review on medicinal exploration of *Costus igneus*: The insulin plant. *Int. J. Pharm. Sci. Rev. Res., 54*(2), 51–57.

Mc Garry, J. D., & Foster, D. W., (1980). Regulation of hepatic fatty acid oxidation and ketone body production. *Annu. Rev. Biochem., 49,* 395–420.

Mc Garry, J. D., Takabayashi, Y., & Foster, D. W., (1978). The role of malonyl-CoA in the coordination of fatty acid synthesis and oxidation in isolated rat hepatocytes. *J. Biol. Chem., 253,* 8294–8300.

Meesapyodsuk, D., Balsevich, J., Reed, D. W., & Covello, P. S., (2007). Saponin biosynthesis in *Saponaria vaccaria*. cDNAs encoding β-amyrin synthase and a triterpene carboxylic acid glucosyltransferase. *Plant Physiol., 143*(2), 959–969.

Melendez-Camargo, M. E., Castillo-Nájera, R., Silva-Torres, R., & Campos-Aldrete, M. E., (2006). Evaluation of the diuretic effect of the aqueous extract of *Costus pictus* D. don in rat. *West Pharmacol. Soc., Proc., 72–74*.

Mering, J. V., & Minkowski, O., (1889). Diabetes mellitus nach pancreas extirpation. *Central Blatt Fürklinische Medicin, Leipzi., 10,* 383.

Meshnick, S. R., (2002). Artemisinin: Mechanisms of action, resistance and toxicity. *Int. J. Parasitol., 32*(13), 1655–1660.

Miura, T., Ueda, N., Yamada, K., Fukushima, M., Ishida, T., Kaneko, T., Matsuyama, F., & Seino, Y., (2006). Antidiabetic effects of corosolic acid in KK-Ay diabetic mice. *Biol. Pharm. Bull., 29*(3), 585–587.

Nadumane, V. K., Rajashekar, S., Narayana, P., Adinarayana, S., Vijayan, S., Prakash, S., & Sharma, S., (2011). Evaluation of the anticancer potential of *Costus pictus* on fibrosarcoma (HT-, 1080) cell line. *J. Nat. Pharm., 2*(2), 72–76.

Nagarajan, A., Arivalagan, U., & Rajagurua, P., (2011). *In vitro* root induction and studies on antibacterial activity of root extract of *Costus igneus* on clinically important human pathogens. *J. Microbiol. Biotechnol. Res., 1*(4), 67–76.

Naik, A., Parjapat, P., Krishnamurthy, R., & Pathak, J. M., (2017). Assessment of genetic diversity in *Costus pictus* accession based on RAPD and ISSR markers. *Biotech., 3*(7), 70.

Narmadha, R., Gopalakrishna, V. K., & Devaki, K., (2012). Hypoglycemic effect of *Costus pictus* don on alloxan-induced type 2 *Diabetes mellitus* in albino rats. *Asian Pac. J. Trop. Diseases., 2,* 117–123.

Notkins, A. L., (2002). Immunologic and genetic factors in type 1 diabetes. *J. Biol. Chem., 277*(46), 43545–43548.

Oboh, G., Ademosun, A. O., Ayeni, P. O., Omojokun, O. S., & Bello, F., (2014). Comparative effect of quercetin and rutin on α-amylase, α-glucosidase, and some pro-oxid*ant-induc*ed lipid peroxidation in rat pancreas. *Comp. Clin. Pathol., 24,* 1–8.

Okeke, I. N., Lamikanra, A., & Edelma, R., (1999). Socioeconomic and behavioral factors leading to acquired bacterial resistance to antibiotics in developing countries. *Emerg. Infect. Disease., 5,* 18–27.

Palanivel, V., Jihad, E. V. M., & Kumar, K. L. S., (2013). Evaluation of hypoglycemic activity of *Costus igneus* extract (whole plant) on alloxan induced diabetic rats. *Int. J. Adv. Pharm. Res., 1*(2), 9–19.

Pareek, A., Suthar, M., Godavarthi, A., Goyal, M., & Bansal, V., (2010). Negative regulation of glucose uptake by *Costus pictus* in L6 myotube cell line. *J. Pharm. Negat. Results, 1*(1), 24.

Pazhanichamy, K., Bhuvaneswari, K., Kunthavai, B., Eevera, T., & Rajendran, K., (2012). Isolation, characterization and quantification of diosgenin from *Costus igneus*. *J. Planar. Chromatogr., 25*(6), 566–570.

Qaid, M. M., & Mutassim, M., (2016). Abdelrahman. role of insulin and other related hormones in energy metabolism—A review. *Cogent Food & Agriculture, 2,* 1–18.

Ramya, U. S. K., & Chauhan, J. B., (2015). Phytochemical screening, antimicrobial activity and antioxidant activity of *Costus igneus*. *Eur. J. Mol. Bio and Biochem., 2*(2), 93–96.

Reddy, L. J., & Jose, B., (2010). Evaluation of antibacterial activity of the leaf essential oil *Costus pictus* don from South India. *Int. J. Curr. Pharm. Res., 2*(3), 68–70.

Remya, R., & Daniel, M., (2013). Phytochemical and pharmacognostic investigation of antidiabetic *Costus pictus* don. *Int. J. Pharm. Bio-Med. Res., 3*(1), 30–39.

Rowinsky, E. K., & Donehower, R. C., (1995). Paclitaxel (Taxol). *N. Engl. J. Med., 332,* 1004–1014.

Saraswathi, R., Lokesh, U., Venkatakrishnan, R., Meera, R., & Devi, P., (2010). Isolation and biological evaluation of steroid from stem of *Costus igneus*. *J. Chem. Pharma. Res., 2*(5), 444–448.

Sathuvan, M., Vignesh, A., Thangam, R., Palani, P., Rengasamy, R., & Murugesan, K., (2012). *In vitro* antioxidant and anticancer potential of bark of *costus pictus* D. don. *Asian Pac. J. Trop. Biomed., 2*(2), S741–S749.

Sayantan, S., (2012). *Forestry- Understory and Rangeland Plants*. Burdwan Eco Garden, West Bengal, Bugwood-UGA (Image No. 5476153).

Sethumathi, P. P., Nandhakumar, J., Sengottuvelu, S., Duraisam, R., Karthikeyan, D., Ravikumar, V. R., Malini, A., & Sivakumar, T., (2009). Antidiabetic and antioxidant activity of methanolic leaf extracts of *Costus pictus* D. Don in alloxan-induced diabetic rats. *Pharmacology Online, 1,* 1200–1213.

Shankarappa, L., Gopalakrishna, B., Jagadish, N. R., & Siddalingappa, G. S., (2011). Pharmacognostic and phytochemical analysis of *Costus ignitius*. *Int. Pharm. Sci., 1,* 36–41.

Shaw, H. K. A., & Wilis, J. C., (1973). *A Dictionary of the Flowering Plants and Ferns* (8th edn.). Cambridge University Press, Cambridge.

Shetty, A. J., Choudhury, D., Rejeesh, V. N., Kuruvilla, M., & Kotian, S., (2010). Effect of the insulin plant (*Costus igneus*) leaves on dexamethasone-induced hyperglycemia. *Int. J. Ayur. Res., 1*(2), 100–110.

Shetty, A. J., Parampalli, S. M., Bhandarkar, R., & Kotian, S., (2010). Effect of the insulin plant (*Costus igneus*) leaves on blood glucose levels in diabetic patients: A cross sectional study. *J. Clin. Diagn. Res., 4*(3), 2617–2621.

Shi, L., Zhang, W., Zhou, Y. Y., Zhang, Y. N., Li, J. Y., Hu, L. H., & Li, J., (2008). Corosolic acid stimulates glucose uptake via enhancing insulin receptor phosphorylation. *Eur. J. Pharmacol., 584*(1), 21–29.

Shiny, C. T., Yadav, K. S., Yadav, N. P., Lugman, S., & Palni, L. M. S., (2013). Comparative evaluation of *Costus pictus* Don leaf extracts against glucose challenged mice. *Ann. Phytomed., 2*(2), 57–62.

Shivaprakash, G., Thottacherry, E. D., Rai, S., Nandini, M., Kumarachandra, R., & Pallavi, L. C., (2014). Evaluation of antioxidant potential of *Costus igneus* in ethanol-induced peroxidative damage in albino rats. *J. Appl. Pharm. Sci., 4*(8), 52.

Silva, C. R., Antunes, L. M., & Bianchi, M. L., (2001). Antioxidant action of bixin against cisplatin-induced chromosome aberrations and lipid peroxidation in rats. *Pharmacol. Res., 43*(6), 561–566.

Singh, G., Kapoor, I. P., Singh, P., De Heluani, C. S., De Lampasona, M. P., & Catalan, C. A., (2008). Chemistry, antioxidant and antimicrobial investigations on essential oil and oleoresins of Zingiber officinale. *Food Chem. Toxicol., 46*(10), 3295–3302.

Son, I. S., Kim, J. H., Sohn, H. Y., Son, K. H., Kim, J. S., & Kwon, C. S., (2007). Antioxidative and hypolipidemic effects of diosgenin, a steroidal saponin of yam (*Dioscorea* spp.), on high-cholesterol fed rats. *Biosci., Biotechnol. Biochem., 71*(12), 3063–3071.

Srinivasan, G. V., Vijayan, K. K., Sharanappa, P., & Jagadeesh, K. S., (2016). Identification of chemical compounds in the essential oil from *Costus pictus* D. don plant parts antimicrobial studies isolation and quantification of diosgenin from its root. *J. Chem. Pharm. Res., 8*(7), 594–604.

Stewart, L. K., Wang, Z., Ribnicky, D., Soileau, J. L., Cefalu, W. T., & Gettys, T. W., (2009). Failure of dietary quercetin to alter the temporal progression of insulin resistance among tissues of C57BL/6J mice during the development of diet-induced obesity. *Diabetologia, 52*(3), 514–523.

Suzuki, H., Kawarabayasi, Y., Kondo, J., Abe, T., Nishikawa, K., Kimura, S., Hashimoto, T., & Yamamoto, T., (1990). Structure and regulation of rat long-chain acyl-CoA synthetase, *J. Biol. Chem., 265,* 8681–8685.

Takahashi, N., Goto, T., Taimatsu, A., Egawa, K., Katoh, S., Kusudo, T., Sakamoto, T., et al., (2009). Bixin regulates mRNA expression involved in adipogenesis and enhances insulin sensitivity in 3T3-L1 adipocytes through PPAR α-activation. *Biochem. Biophys. Res. Commun., 390,* 1372–1376.

Takahashi, N., Kawada, T., Goto, T., Yamamoto, T., Taimatsu, A., Matsui, N., Kimura, K., et al., (2002). Dual-action of isoprenols from herbal medicines on both PPARα and PPARα in 3T3-L1 adipocytes and HepG2 hepatocytes. *FEBS Lett., 514,* 315–322.

Taylor, S. I., Accili, D., & Imai, Y., (1994). Insulin resistance or insulin deficiency: Which is the primary cause of NIDDM. *Am. Diabetes Assoc., 43*(6), 735–740.

Thomas, S. C., & Palani, L. M. S., (2016). Taxonomic, molecular and physiological evaluation of *Costus pictus* don plants originally obtained from different parts of Kerala. *J. Appl. Fundamental Sci., 2*(2), 141–148.

Tomlinson, P. B., (1956). Studies in the systematic anatomy of the *Zingiberaceae. J. Linn. Soc. London, Bot., 55,* 547–592.

Tran, N., Pham, B., & Le, L., (2020). Bioactive compounds in antidiabetic plants: From herbal medicine to modern drug discovery. *Biology, 9,* 252.

Troll, W., (1928). *Organization und Gestalt im Bereich der Blute.* J. Springer. Berlin.

Uemura, T., Hirai, S., Mizoguchi, N., Goto, T., Lee, J. Y., Taketani, K., Uemura, T., et al., (2010). Diosgenin present in fenugreek improves glucose metabolism by promoting adipocyte differentiation and inhibiting inflammation in adipose tissues. *Mol. Nutr. Food Res., 54*(11), 1596–1608.

Urooj, A., (2008). Hypoglycemic potential of *Morus indica.* L and *Costus igneus.* Nak.—A preliminary study. *CSIR,* 614–616.

Veena, S., Deokar, G., & Kshirsaga, S., (2016). Pharmacognostic investigation, isolation and evaluation of diosgenin from *Costus pictus* don. *Int. J. Inst. Pharm. Life Sci., 6*(3), 389–404.

Wang, H., Liu, T., Chen, Y., & Shiuan, D., (2007). Protective effect of freeze-dried extract of vegetables and fruits on the hydroxyl radical-mediated oxidative damage of DNA and decrease of erythrocytes deformability. *Appl. Biochem. Biotechnol., 141*(2, 3), 241–249.

Wangner, J. E., Huff, J. L., Rust, W. L., Kingsley, K., & Plopper, G. E., (2002). Perillyl alcohol inhibits breast cell migration without affecting cell adhesion. *J. Biomed. Biotechnol., 2*(3), 136–140.

Ward, M. G., Li, G., Barbosa-Lorenzi, V. C., & Hao, M., (2017). Stigmasterol prevents glucolipotoxicity induced defects in glucose-stimulated insulin secretion. *Sci. Rep., 7*(1), 1–13.

World Health Organization (WHO), (2013). *Traditional Medicine Strategy, 2014–2023*. WHO: Geneva, Switzerland.

Yuvarani, T., Manjula, K., & Perumal, A. G., (2017). Growth characterization of calcium hydrogen phosphate dihydrate crystals influenced by *Costus igneus* aqueous extract. *Int. J. Pharm. Pharm. Sci., 9,* 173–178.

Zong, W., & Zhao, G. Y., (2007). Corosolic acid isolation from the leaves of *Eriobotrya japonica* showing the effects on carbohydrate metabolism and differentiation of 3T3-L1 adipocytes. *Asia Pac. J. Clin. Nutr., 16*(S1), 346–352.

CHAPTER 6

Health Benefits of *Trigonella foenum-graecum* with Regard to Metabolomics of Antidiabetic Efficiency

NEHA SHARMA,[1] RAJESH SINGH TOMAR,[1] and SHUCHI KAUSHIK[2]

[1]*Amity Institute of Biotechnology, Amity University, Gwalior, Madhya Pradesh, India, E-mail: nsharma2@gwa.amity.edu (N. Sharma)*

[2]*M. P. Forensic Science Laboratory, Madhya Pradesh, India, E-mail: shuchi.kaushik2@gmail.com*

ABSTRACT

On the worldwide level, different plants and their parts (leaves, seeds, and stem) have been used in conventional medicines and believed to be helpful to treat several diseases. *Trigonella foenum-graecum* (Fenugreek) have a collection of therapeutic activities like antidiabetic, hypo-cholesterolemic, anti-cancer, anti-inflammatory, anti-toxic, and anti-microbial. It also works as a lactation stimulant. Its antioxidant property is also well known. Fenugreek is enriched in useful bioactive compounds and used as antidiabetic nutraceutical. Diabetes mellitus is the most common metabolic disorders in the world associated with high sugar levels in the blood. The disorder is grouped into type I and type II in which diabetes II is a common type. Diabetes II is insulin dependent diabetes. Active compound of *Trigonella foenum-graecum* (Fenugreek) can carry out a positive role against high glucose level. The present review focused on conventional, therapeutic, and nutritional applications of *Trigonella foenum-graecum* as well focused on actions of bioactive compounds in the

Antidiabetic Plants for Drug Discovery: Pharmacology, Secondary Metabolite Profiling, and Ingredients with Insulin Mimetic Activity. Deepu Pandita, MPhil, Anu Pandita, MSc, & Chander Bhanu, MBBS, PGDDCE (Eds.)
© 2023 Apple Academic Press, Inc. Co-published with CRC Press (Taylor & Francis)

improvement of healthy life of human beings. This chapter will also focus on systems biology analysis including statistical approaches, analytical techniques, and bioinformatics tools for analyzing and maintenance of metabolomics data which will be helpful in understanding comprehensive system. Applications of systems medicine and metabolomics are based on *Trigonella foenum-graecum* (Fenugreek), will also be discussed along with other omics technologies, such as genomics, proteomics, and transcriptomics. Thus, this review will be helpful in the enriching the phytomedicine database on *Trigonella foenum-graecum* (Fenugreek) and update the translation of scientific understanding of researcher for high impact diabetes disease.

6.1 INTRODUCTION

Modern establishments and advanced technologies have changed the lifestyle of humans. The dependency on automatic tools and techniques makes humans sedentary. The fast pace required to keep oneself in the competitive world shifts human focus from health to monetary benefits and thus causes human negligence towards health. This is the foremost reason of developing everyday life associated health problems like diabetes (sugar dysregulation) and obesity (overweight and fatness). Diabetes or Diabetes Mellitus (DM) is a heterogeneous metabolic disorder in which our physiological system cannot regulate blood sugar level and thus results in increased glucose levels and deranged glucose metabolism. Glucose is the primary source of energy for our cells. Due to deficiency of insulin cells have insufficient supply of glucose or its resistance result in work efficiency loss and malfunctioning of insulin-dependent tissues like liver and muscles and thus weakness. The drastic effects of imbalance in sugar level in body does not restrict to only tissue weakness but also leads to a number of other secondary health problems including organs dysfunctioning in tissues which are not dependent on insulin-like heart tissues, blood vessels, peripheral nerves, renal medulla, and lens. It is thus considered to be a slow death due to tissue damage complications. Amending blood glucose is the paramount preventive measure but till date there is no known effective medicine accessible to attain this goal (Figure 6.1).

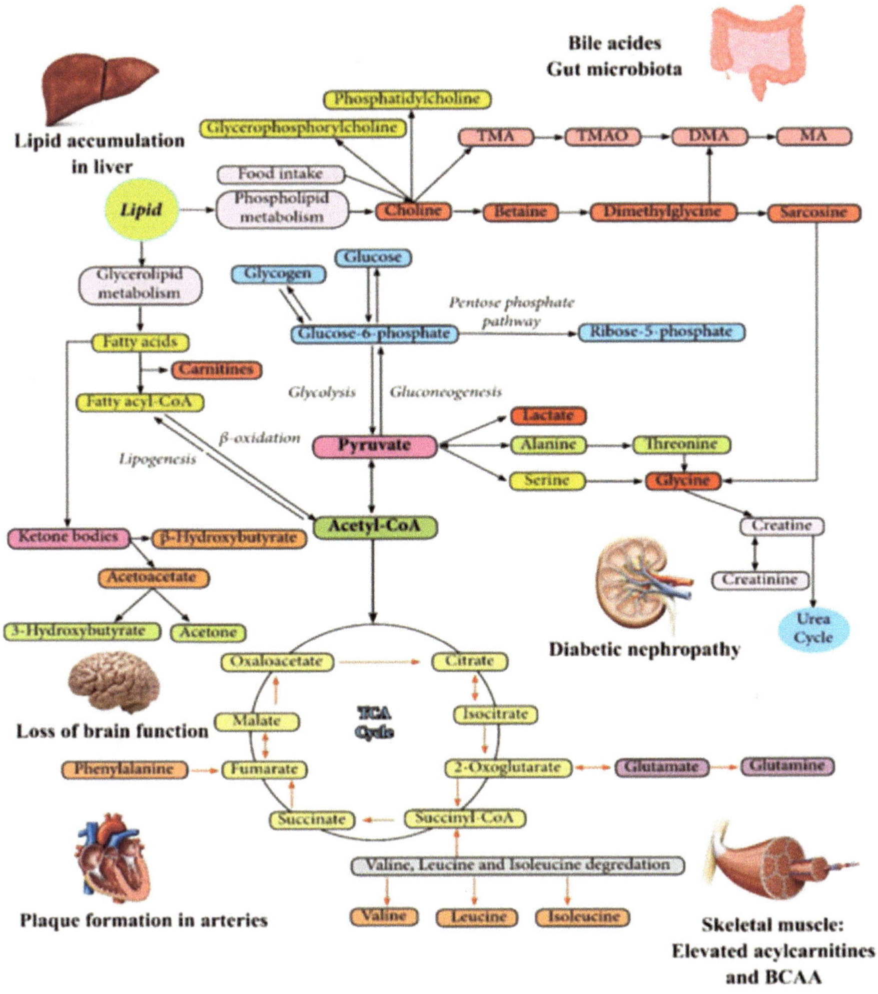

FIGURE 6.1 Schematic illustration of the metabolic pathways based on obesity and type 2 diabetes mellitus.

Source: *Reprinted with permission from* Hasanpour et al. (2020). © Elsevier.

The usefulness of plants and plant derived products have been known to mankind since time immemorial. The traditional knowledge of some indigenous plants having medicinal properties leads away to cost-effective and easily accessible treatments to some common diseases including cold, fever, pain, diarrhea, and also to some extent proven effective for serious complications such as diabetes, hypertension, and even cancer (Arumugam et al., 2013; Kooti et al., 2015). The therapeutic potential of these medicinal plants lies in their phytocomponents which are the secondary metabolic products like polyphenols and flavanones (Afrisham et al., 2015). The understanding of the biological role of these phytoconstituents paves the path to reveal the beneficial properties of medicinal plants possessing these active components (Durazzo et al., 2018).

The herb fenugreek aka methi, botanically known as *Trigonella foenum-graecum*, is indigenous to Africa and Asia continents and is found throughout India (Basch et al., 2003). The plant belongs to the family Fabaceae and is considered to have several medicinal properties. The plant parts especially the bitter seeds, are being used by the local people mainly as a part of spices in not only treating metabolic disorders such as diabetes and obesity (Dangi et al., 2004; Prasanna, 2000) but also found to possess antifungal and antioxidant properties (Khan et al., 2012; Gauttam and Kalia, 2013; Nagulapalli, 2017). The seeds have a variety of medicinal uses from bust enhancement to increasing the lactation in weaning mothers and from healing the wounds to sex stimulant (Gupta et al., 2001; Campbell-Tofte et al., 2012).

6.2 HEALTH BENEFITS OF *TRIGONELLA FOENUM-GRAECUM* (FENUGREEK)

Several human diseases mainly characterized by abnormal metabolism of carbohydrates, proteins, and fats have been reported to be caused by change in the food habits and lifestyle of modern man. The environmental factors, stress, and replacement of natural products with synthetic chemicals in agricultural and food processing practices have caused a drastic impact on humans and animals health. The effective management of these lifestyle related disorders is the need of the hour and thus shifts the focus of scientific society to natural products in order to find suitable therapeutic substances for these malfunctions. The plants are affluent in bioactive compounds, which are being progressively and intensively studied by researchers all over the globe to find solutions to these metabolic problems.

Fenugreek is an important herb which most suitably fits in the criteria of beneficial therapeutic plant against diabetes due to the presence of several phytocomponents in it such as alkaloids, carbohydrates, steroidal saponins, amino acids and minerals as given in Table 6.1. Its imperative medicinal properties cause its utilization in food and pharmaceutical industries as a treatment measure against various ailments like cancer, hypercholesterolemia, diabetes, and inflammation (Qamar et al., 2020; Wani et al., 2018).

TABLE 6.1 Various Phytocompounds and Their Content Present in Fenugreek Plant

SL. No.	Nutrient	Nutrient Content
1.	Carbohydrates	58% in seeds and 6% in leaves
2.	Proteins	23–26% in seeds and 4.4% in leaves
3.	Fats	0.9% in seeds
4.	Fiber	25% in seeds and 1.1% in leaves
5.	Potassium	603 mg/100 g
6.	Magnesium	42 mg/100 g
7.	Calcium	75 mg/100 g
8.	Zinc	2.4 mg/100 g
9.	Manganese	0.9 mg/100 g
10.	Copper	0.9 mg/100 g
11.	Iron	25.8 mg/100 g
12.	Vitamin C	220 mg/100 g
13.	β carotene	19 mg/100 g

6.2.1 CONVENTIONAL VALUES

Fenugreek is one of the popular cooking spices commonly used in kitchen. The rich dietary fiber, gum, protein, and volatile compounds content make it a suitable medicinal agent (Meghwal and Goswami, 2012). The use of its seeds and green leaves in traditional cuisines as vegetables or sprouts or to enhance the flavor, color, and modifying the texture of food, is an old practice of society (Srinivasan, 2006; Passano, 1995). Its extracts and powders are used as adhesive, emulsifier, and stabilizer in various food items across the globe like spice blends for soups and stews in the USA, baking bread in Egypt and Ethiopia and flavoring cheese in Switzerland (Reddy et al., 2019; Wani et al., 2018).

6.2.2 NUTRITIONAL VALUES

The nutrient rich food which is also having pharmaceutical potential is considered as nutraceutical. Fenugreek fulfills the criteria of being nutraceutical in terms of its functional and physiological properties which affirms its multipurpose roles in nutrition and medicines (Srivastava et al., 2020). A large number of different active principles and nutritious compounds are present in fenugreek plant parts as mentioned in Table 6.2. These phytocompounds are essential for the normal functioning of a healthy biological system, thus use of fenugreek in daily diet is beneficial to human health (Wani et al., 2018; Al-Jasass et al., 2012).

TABLE 6.2 The Types and Examples of Various Nutrients Present in Fenugreek Plant

SL. No.	Nutrients	Examples	References
1.	Alkaloids	Trigonella, trigocoumarin, nicotinic acid, trimethyl coumarin, choline	Nagulapalli (2017); Snehlata and Payal (2012)
2.	Protein amino acids	Aspartic acid, glutamic acid, leucine, tyrosine, phenylalanine, and minute quantity of cysteine and methionine	Feyzi et al. (2015)
3.	Non-protein amino acid (80% of dry seeds)	(2S, 3 R, 4S) – 4-hydroxyisoleucine	Feyzi et al. (2015)
4.	Minerals	Potassium, Magnesium, Calcium, Zinc, Manganese, Iron, Phosphorous, and Copper	Wani et al. (2018); Al-Jasass et al. (2012); Rao (2003)
5.	Vitamin	Vitamin C and β-carotene, riboflavin, thiamine, niacin	Wani et al. (2018); Al-Jasass et al. (2012); Rao (2003)
6.	Lipids	Triacylglycerols, diacylglycerols, monoacylglycerols, phosphatidylcholine phosphatidylethanolamine, phosphatidylinositol, free fatty acids	Chatterjee et al. (2010)
7.	Polyphenol compounds	Rhaponticin and Isovitexin	Wani et al. (2018); He et al. (2015)

TABLE 6.2 *(Continued)*

SL. No.	Nutrients	Examples	References
8.	Volatile oil	(Z)-1,5-Octadiene-3-one, linalool, 3-Amino-4,5-dimethyl-3, diacetyl, 1-Octene-3-one, sotolon, acetic acid; eugenol, butanoic acid, caproic acid, 3-Isobutyl-2-methoxypyrazine, isovaleric acid, 3-isopropyl-2-methoxypyrazine,	Wani et al. (2018)
9.	Steroids	Cholesterol and sitosterol	Snehlata and Payal (2012)
10.	Flavonoids	Vitexin, quercetin, luteolin, apigenin	Snehlata and Payal (2012)
11.	Saponins	Tigogenin, gitogenin, diosgenin	Snehlata and Payal (2012)

6.2.3 THERAPEUTIC VALUES

The unhealthy and sedentary lifestyle of modern man leads to health complications and furthermore gets deteriorated by the side effects of chemical drugs. Thus, to control the pathologies caused by inefficient therapies and to develop cost effective, easily available, and accessible drug modern medical world turns to natural sources. The fenugreek plant has the affluent amount of dietary fiber, phytochemicals, minerals, and amino acids in its seeds and leaves which make this plant a potential candidate for developing functional food and therapeutic products (Aasim et al., 2018). The therapeutic potential of this plant ranges from antidiabetic, anti-lipidemic, antioxidant, anticancer to anti-inflammatory. Many components like diosgenin, saponins, trigonelline, alkaloids, high fiber, 4-hydroxyleucine and 4-hydroxyisoleucine are responsible for the therapeutic potential of fenugreek plant (Saravanan et al., 2014; Patel et al., 2014; Rizvi and Mishra, 2013).

6.2.3.1 ANTIDIABETIC POTENTIAL

The antidiabetic effects of fenugreek plant extracts and its crude preparations have been successfully experimented by a number of scientists on animal model and human patients (Bordia et al., 1997; Khan et al., 2012; Sharma et al., 1990). Ai-Habori et al. in 2001 explained the role of Disogenin, a type of saponin and trigonelline an alkaloid, found in fenugreek plant in inhibiting glucose uptake *in-vitro*. SDF (Soluble Dietary Fiber) of *Trigonella foenum-graecum* also work in diabetes as assessed in type 1 and type 2 diabetic rats as well as better oral glucose tolerance was reported significantly (Hannan et al., 2006).

Various effects of Trigonella *foenum-graecum* based SDF is reported on (Hannan et al., 2006):

- Absorption of sucrose and glucose from the gut;
- Affect intestinal disaccharidase activity;
- Improved motility in gastrointestinal track;
- Effect on insulin secretion or enhance the peripheral insulin action;
- Effect on carbohydrate digestion and absorption.

In another study, the effect of complete seed powder (5% in the diet) is reported. Dry seed powder of fenugreek is used for oral administration for 21 days. This study is carried out on Wistar rats in which alloxan are used to induce diabetes. Kidney and liver tissues were used to determine the activity of NADP linked lipogenic enzymes, glycolytic enzymes and gluconeogenic associated enzymes. (Raju et al., 2001). After 21 days, it is reported that high glucose level is down as compared to control. The effects on biochemical pathways shown that fenugreek seed powder is a possible new therapeutic effect in diabetes type-1. The antidiabetic potential of fenugreek seeds powder is because of the presence of fiber content, steroid, and saponins (Sauvaire et al., 1996; Zia et al., 2001).

Abdel-Barry et al. in 1997 reported the hypoglycemic effect of leaves of *Trigonella foenum-graecum* extract prepared in aqueous solution when administered through peritoneum or through oral route. *Trigonella foenum-graecum* is considered as nutraceutical efficient food except glycemia and lipidemia role, which regulate the blood glucose level not only in diabetic persons but also in other patients whose diet is based on calorie-rich food. When powder of fenugreek seed is taken after meal, it could normalize cholesterol production in the liver. It is analyzed that fenugreek seeds contain total of 45.4% dietary fiber in which insoluble fiber is 32% and soluble fibers are 13.3%. Mannose and galactose are epimer of glucose and also found in seeds of fenugreek. Galactomannan is a soluble fiber and related to and anticholesterolemic and antiglycemic properties. Because of the presence of galactomannan, the usage of fenugreek based dilatory fiber is common in the food industry as a stabilizing and emulsifying agent. Flour of fenugreek (8–10%) is used to make western diet more nutritious like bread, muffins, cakes, and pizza (Roberts, 2011).

Hooda and Jood (2004) have shown in their studies that mixing of fenugreek powder with wheat flour increased 5–20% the lysine amino acid, whole protein concentration, fat, and minerals contents. It is found that germinated fenugreek flour is enhanced nutritional value in food supplements (Hooda and Jood, 2004).

The application of fenugreek-based galactomannan to manage blood glucose has been reported by Srichamroen et al. in their research article published in 2009. In one *in vitro* study, different concentration of galactomannan is used and it is observed that it significantly involves in reducing glucose level.

6.2.3.2 ANTILIPIDEMIC EFFECT OF TRIGONELLA

Fenugreek seed preparations showed effects on hepatic lipid concentrations and significantly lower serum triglycerides, cholesterol, and LDL-cholesterol while enhancing HDL-cholesterol (Annida et al., 2004; Hannan et al., 2003; Raju and Bird, 2006). This modulating activity of fenugreek seeds indicates its potential role in metabolic reactions involving lipids and glucose (Raju et al., 2001; Yadav et al., 2004). The hypocholesterolemia effect of fenugreek is attributed to the gum present in its seeds, which makes it a suitable functional food with glycemia and lipidemia effects (Roberts, 2011).

6.2.3.3 ANTIOXIDANT EFFECTS OF TRIGONELLA

Humans are obligate aerobes and thus aerobic metabolism is a mandatory part of human survival. The consumed oxygen molecules participate in the tightly regulated oxidation process of food particles. However, some factors negatively influence this process and produces reactive oxygen species which cause oxidative stress in the cells resulting in inflammation, tissue damage and disease. To combat these side effects, the human system has developed native antioxidant mechanism which involves the activity of several enzymes like superoxide dismutase (SOD), glutathione reductase (GR), catalase, and glutathione peroxidase (GPx). Several synthetic compounds having the potential to combat oxidative stress have been tested and verified through clinical trials and proved to be effective antioxidant supplements to patients suffering from pathogenesis induced by oxidative stress. However, to be effective, they are used in higher dose at which these often-become pro-oxidant and thus cause more suffering than healing. Therefore, anti-oxidants from natural sources like flavonoids and others from plants are successfully tested (Table 6.3).

TABLE 6.3 Although the Accurate Role of the *Trigonella* Seeds as Antioxidant Constituent is Not Known*

SL. No.	Name of Plant Part	Significant Role	References
1.	Seeds in aqueous solution	Influence lipid peroxidation and levels of antioxidants when ethanol toxicity was induced in rats. Antioxidant enzymes like SOD, catalase, GPx, glutathione-*S*-transferase (GST) and GR showed decreased activity in liver and brain	Thirunavukkarasu et al. (2003)
2.	Seeds in aqueous solution mixed with ethanol	Enzymatic leakage was inhibited, lipid peroxidation and antioxidant potential improved appreciably when compared with the activity of GSH and α-tocopherol.	Thirunavukkarasu et al. (2003)
3.	Fenugreek seeds extract rich with polyphenol compound	Significantly decreased hydrogen peroxide-induced oxidative stress reported in diabetic and normal human erythrocytes	Kaviarasan et al. (2004)
4.	Fenugreek leaves powder	Measuring the levels of catalase, superoxide dismutase, reduced glutathione (GSH) and thiobarbituric acid-reactive substances (TBARS). The fenugreek leaf powder supplement significantly increased the anti-oxidant system while lowering the lipid peroxidation simultaneously.	Annida et al. (2005)

TABLE 6.3 *(Continued)*

SL. No.	Name of Plant Part	Significant Role	References
5.	Germinating seeds extract	Reduced lipid peroxidation, Increases oxygen radical scavenging or absorbance capacity.	Dixit et al. (2005)
6.	Seeds powder	Active principle in the seeds have the potential to correct the malfunctional enzymatic activity of antioxidant enzymes.	Baquer et al. (2011)
7.	Aqueous extract of fenugreek with polyphenols and flavonoids	Reported maximum antioxidant activity	Xue et al. (2011)
8.	Aqueous extract of *Trigonella* seeds	Examined restoring kidney function during progression of diabetic nephropathy.	Xue et al. (2011); Tripathi and Chandra (2009)
		Alteration in the activities of catalase and SOD and catalase.	
		Decrease the concentrations of malondialdehyde (MDA)	

**Following researchers have investigated the antioxidant potency of Trigonella seed in different experimental models and suggested that its seeds rich in flavonoids and polyphenols can reduce oxidative stress.*

6.2.3.4 ANTICANCER EFFECT OF TRIGONELLA

Cancer is becoming a common cause of death in worldwide. Mostly therapeutic approach used but it only extends the patient's life period for few years or sometimes only for a few months. Conventional treatments have many side effects. In this view, alternative medicine are new way for treatment of cancer. Active ingredients of plant parts like fruits and vegetables are suggested for prevention of cancer. Increased uses of alternative medicine concepts are new way to prevent cancer. In this point of view, many researchers have focused on the role of fenugreek seeds in cancer prevention. They have confirmed the defensive effect of fenugreek seeds in cancer treatment through experimental models and by using cell lines (Table 6.4).

TABLE 6.4 Effect of Trigonella Extracts on Different Cell Lines

SL. No.	Name of Plant Part	Name of Cell Lines	Role/Result of Studies	References
1.	Whole plant alcoholic extracts	Neuroblastoma cell line, IMR-32 and HT29	Showed *in vitro* cytotoxicity	Verma et al. (2010)
2.	Ethanol extract of fenugreek seeds	MCF-7 cells	Showed decreased cell viability. Induced early apoptotic changes like movement of phosphatidylserine and degradation of DNA into small fragments. Decreased mitochondrial membrane potential. Arrest cell cycle at G2/M phase	Sebastian and Thampan (2007)
3.	Fenugreek seed extract	Breast, prostate, and pancreatic cancer cell lines.	Showed inhibitory effects on growth of breast, prostate, and pancreatic cancer cell lines. Induce cell death.	Shabbeer et al. (2009)

6.2.3.5 ANTI-INFLAMMATORY EFFECT OF TRIGONELLA

The plant showed noteworthy anti-inflammatory effect in various studies on experimental models. It increases the cell counts of macrophages which are the key role player in the inflammatory process. *Trigonella* extract also demonstrated increment in peritoneal cell exudate (Verma et al., 2010). In one study on cultured THP-1 cells, *Trigonella* methanolic seed extract showed inhibition in the production of tumor necrosis factor-α, which is one of the inflammatory cytokines (Kawabata et al., 2011). In another study on albino rats, arthritis was induced by adjuvants. The ethanolic extracts of *Trigonella* not only decreases the inflammatory interleukins level and reduces edema in the paw but also significantly influences the antioxidant machinery in tissue cartilages of the experimental model (Suresh et al., 2012).

6.3 MODES OF ACTION OF BIOACTIVE COMPOUNDS OF *TRIGONELLA FOENUM GRAECUM* (FENUGREEK)

The pharmacological and biological effects of the fenugreek plant are associated to the variety of components that exist in various tissues of the plant. It is mainly polyphenolic compounds, volatile substances, steroids, and different types of amino acids, etc. The chemical components are improved in the seeds and other parts of plants at the time of growing period and influenced by different cultural techniques like cell and tissue culture (suspension or static), post-harvest treatments, storage, and biological manipulation of yield (Nathiya et al., 2014).

Diosgenin, steroidal saponin is the most important pharmacological component derived from *Trigonella foenum graecum*. It is used in the treatment of several diseases like hypercholesterolemia, colon cancer, leukemia, and climacteric syndrome. Diosgenin is also useful as a substrate for the biosynthesis of hormones, oral contraceptives, and other steroidal compounds (Patel et al., 2012). Biosynthesis of diosgenin has not been clear yet. It is suggested that glycolytic, steroid, and mevalonate pathways are involved in this process. It is found that cholesterol is a precursor of diosgenin (Dewick, 2002). Diosgenin may be synthesized from squalene-2,3-oxide in two ways as suggested by Mehrafarin et al. (2010) and Vaidya et al. (2013) in their transcriptome sequencing studies on *T. foenum-graecum*:

- From lanosterol, it formed cholesterol; and
- From cycloartenol it formed sitosterol.

Triglyceride accumulation is inhibited by diosgenin as reported by Uemura et al. (2011). The expression of lipogenic genes in a hepatic cell line (HepG2 cells) are also influenced by diosgenin derived from Trigonella as it inhibits the transactivation of liver-X-receptor-α and thus decrease the hepatic lipid content. It is also reported that diosgenin is also involved in modulation of STAT3 signaling pathway in hepatocellular carcinoma. It effects the signaling pathway by suppressing the activation of JAK1, JAK2 and c-Src. Diosgenin is also involved in the regulation of various activities like cell proliferation inhibition and apoptotic effects of paclitaxel and doxorubicin.

6.4 POTENTIAL APPLICATIONS OF OMICS STUDIES AND SYSTEMS MEDICINE

Metabolomics is a comprehensive analysis of various biological components or metabolites to unravel the profile of metabolites modifications. Metabolites are low molecular weight biochemical compounds present in biological system. Metabolomics approaches provided novel strategies to explain the nature of nutritional and medicinal properties over the last decade. It is used in characterizing the metabolites of control and treatment groups (Kaddurah-Daouk et al., 2014, 2015). Several metabolic compounds may be present in the plant extract which may be either active or metabolized into bioactive form after processing. The bioavailability and compatibility with the natural system is also one of the criteria for any metabolite to be a suitable candidate for its therapeutic use. These different metabolites are evaluated as biomarkers during the metabolomics study and may be used to recognize targets. It is also used to study different pathways related to toxicity, side effects and efficacy. Metabolomics studies can also be used to investigate the mechanism and pharmacological effects of various herbs (Shi et al., 2016).

The strategies employed for metabolomics analysis involve *in-vivo* and *in-vitro* experimentation so as to investigate the change in metabolic profile and identification of antidiabetic bioactive compounds (Figure 6.2).

The analysis is done by either targeted or untargeted profiling strategies. The untargeted metabolomics leads to the study of a great number of chemically diverse metabolites and even metabolic profiling of some unknown metabolome. Techniques with high sensitivity and specificity such as LC-MS or NMR are required to be used for collecting the metabolic data. The complex data thus obtained is analyzed statistically by using multivariate statistical analyzes such as Principal Component Analysis (PCA), Partial Least Square Discriminant Analysis (PLS-DA), and Orthogonal Partial Least Square Discriminant Analysis (OPLS). The last step in this program is to discover the biomarker and identify the metabolic pathways which are malfunctioned in the diseased state. The targeted study involves much smaller number of metabolites usually less than 20 but it is a very specific, sensitive, and accurate analysis approach which focuses on a specific biological pathway such as enzyme inhibition and validate a particular biomarker precisely.

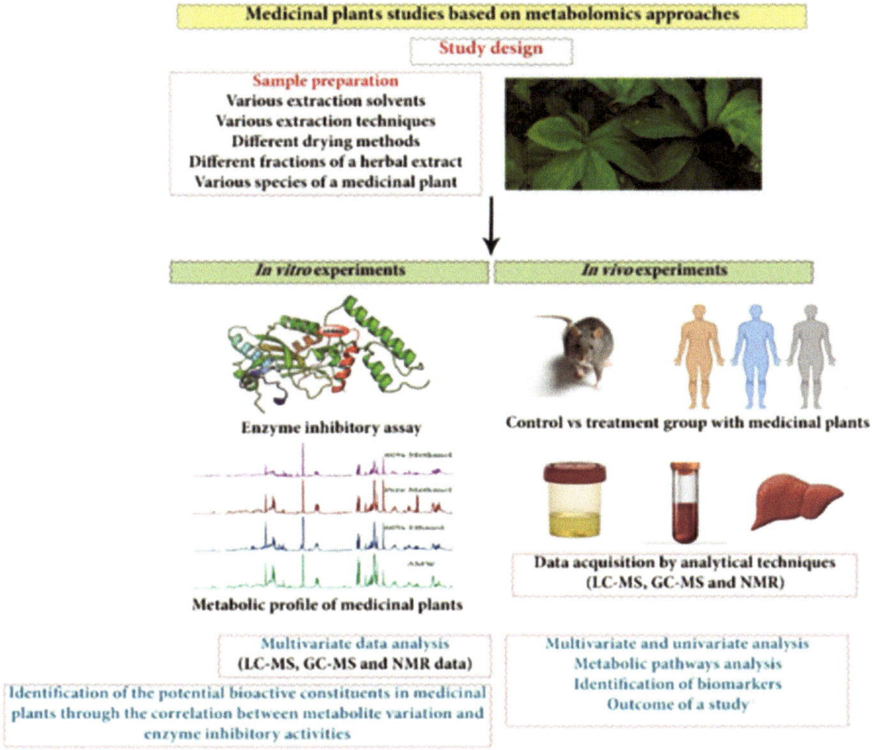

FIGURE 6.2 An overview of metabolomics approaches based on antidiabetic activity of medicinal plants.

Source: *Reprinted with permission from* Hasanpour et al. (2020). © Elsevier.

The antidiabetic potential of medicinal plants can be screened out using different experimental models such as animals, isolated tissue, and cell lines. Biochemical studies may be designed using computer-aided molecular docking to study the inhibitory effects of plant metabolites on carbohydrates, α-amylase, and α-glucosidase. The high throughput screening assays are becoming more popular these days as they offer fast and detailed analysis of plant active metabolites which have the potential to be developed into novel drugs for diabetes management (Lankatillake et al., 2019).

Apart from metabolomics, transcriptomic data of various medicinal plants are needed to be analyzed and profiled as suggested by Sangwan et al. in 2013:

- It helps to understand metabolic pathways under system biology with special reference to the metabolome;
- It helps us to compare the genes associated with biogeneration of the secondary metabolites under the similar group;
- It involves study of different sets of genes whose catalytic properties are same like under gene families and superfamilies;
- Explain regulation of metabolic pathways;
- Give details of functional genomics and significance of individual enzymes in enzymatic step.

Transcription profile can be generated by assembling the short reads generated through direct sequencing of targeted cDNAs sequences. RNA sequencing is a complete and well-organized way to explain transcriptomics on the basis of qualitatively and quantitatively characters. It is also helpful in the discovery of new exons and splicing variants. Target-based methods are used to study cDNA profiling and pattern changes in plants which is helpful in discovering new biosynthetic pathways. RNA-seq is an ultra-high-throughput DNA sequencing technology which provides a revolutionary advance in sequencing based on transcriptome scale and cost-effective approach. It is reported as a competitive technique to a microarray technology (Shi et al., 2011; Chikara et al., 2014).

In addition to above-mentioned studies and methods, drug design and revelation of drug-gene-disease associations are other approaches to analyze the medicinal potential of plant metabolites. Since a single compound sometimes is not efficient enough to treat the diseased condition system-level and network pharmacology is the new requirement which indicates network targets interacting with multi-components (Hutchinson and Kirk, 2011; Hopkins, 2007). The combinatorial strategies of various compounds working in synergy and their regulatory network pathways are required to be known through universal platforms for data integration and analysis (Li et al., 2011; Li and Zhang, 2013; Vitali et al., 2013; Kibble et al., 2015). Similar type of network pharmacological studies has been done on *Withania somnifera* and *Boswellia serrata* in order to understand their immunomodulatory mechanisms (Chandran and Patwardhan, 2017; Goswami et al., 2018).

The mechanistic landscape of Trigonella seeds have been laid down by using LC-MS analyzes. The study confirmed that Trigonella contains several phenolic compounds which are responsible for its hyperglycemic and hyperlipidemic properties. The bioactive substances like apigenin, calycosin, gallic acid, kaempferol, luteolin, orientin, pratensin, quercetin, tricin showed their active participation in insulin signaling pathway. They act as antioxidants as they can interact with carbonic anhydrase family of proteins. The plant extract is involved in EGFR/AKT/mTOR signaling cascade and show high enrichment for MAPK pathway, JAK-STAT pathway, Adipocytokine pathways thus showing cardioprotective potential (Banerjee et al.,).

The multi-approach study synergizing traditional knowledge, modern tools and techniques with advanced omics strategies may prove beneficial in understanding the mechanisms of action of a number of similar compounds in natural systems along with their combination synergy networks. Thus, this review explores the multi-dimensional analysis methods such as traditional data, experimental studies with advanced technology like metabolomics using LC-MS and network pharmacology which can have an important application in future drug development to manage metabolic diseases such as diabetes.

6.5 CONCLUSION

The properties of *Trigonella* studied in the present chapter suggests that it may prove to be a good substitute for antidiabetic medicines as it effectively regulates blood glucose levels and stops the aftermath of diabetes. The active principles of this plant get involved in different signaling mechanisms of body metabolism and thus control glycemic level. The advanced omics technologies help to discover and reveal the hidden biosynthetic pathways in which the plant's secondary metabolites get participates to show their hypoglycemic effects. Through metabolite profiling and metabolomic studies, the effect of biochemical or physiological disturbances on metabolite levels and their role in metabolic pathways in diseased conditions can be evaluated and then compared with the normal healthy state. This wealth of information linking the biochemical and physiological status and metabolite levels in health and diseased state facilitate the grouping of diseases, risk stratification, discovery of novel prognostic biomarkers, and make available new information on pathogenesis mechanisms and targets for therapy. The molecular mechanistic model of synergy of plant extracts may be further explored through suitable systems pharmacology experiments, and a detailed well-elaborated study is required to profile and analyze the transcriptional and translational data of antidiabetic effects of *Trigonella* plant metabolites.

KEYWORDS

- **antidiabetic**
- **bioactive compound**
- **diabetes mellitus**
- **glutathione peroxidase**
- **glutathione reductase**
- **glutathione-s-transferase**
- **malondialdehyde**
- **omics technologies**
- **orthogonal partial least square**
- **partial least square discriminant analysis**
- **principal component analysis**
- **superoxide dismutase**

REFERENCES

Aasim, M., Baloch, F. S., Nadeem, M. A., Bakhsh, A., Sameeullah, M., Day, S., Ozturk, M., et al., (2018). Fenugreek (*Trigonella Foenumgraecum* L.): An underutilized edible plant of modern world. In: *Global Perspectives on Underutilized Crops* (pp. 381–408). Springer: Cham.

Abdel-Barry, J. A., Abdel-Hassan, I. A., & Al-Hakiem, M. H., (1997). Hypoglycemic and antihyperglycemic effects of *Trigonella foenum-graecum* leaf in normal and alloxan-induced diabetic rats. *J. Ethnopharmacol., 58*, 149–155.

Afrisham, R., Aberomand, M., Ghaffari, M., Siahpoosh, A., & Jamalan, M., (2015). Inhibitory effect of *Heracleum persicum* and *Ziziphus jujuba* on activity of alpha-amylase. *J. Bot., 2015*, 824683.

Al-Jasass, F. M., & Al-Jasser, M. S., (2012). Chemical Composition and fatty acid content of some spices and herbs under Saudi Arabia conditions. *Sci. World J., 2012*, 1–5.

Annida, B., Stanely, M., & Prince, P., (2004). Supplementation of fenugreek leaves lower lipid profile in streptozotocin-induced diabetic rats. *J. Med. Food, 7*, 153–156.

Annida, B., Stanely, M., & Prince, P., (2005). Supplementation of fenugreek leaves reduces oxidative stress in streptozotocin-induced diabetic rats. *J. Med. Food, 8*, 382–385.

Arumugam, G., Manjula, P., & Paari, N., (2013). A review: Antidiabetic medicinal plants used for diabetes mellitus. *J. Acute Dis., 2*, 196–200. doi: 10.1016/S2221-6189(13)60126-2.

Baquer, N. Z., Kumar, P., & Taha, A., (2011). Metabolic and molecular action of *Trigonella foenum-graecum* (fenugreek) and trace metals in experimental diabetic tissues. *J. Biosci., 36*, 383–396.

Basch, E., Ulbricht, C., Kuo, G., Szapary, P., & Smith, M., (2003). Therapeutic applications of fenugreek. *Altern. Med Rev., 8*, 20–27.

Bordia, A., Verma, S. K., & Srivastava, K. C., (1997). Effect of ginger (Zingiber officinale Rosc.) and fenugreek (*Trigonella foenum-graecum* L.) on blood lipids, blood sugar and platelet aggregation in patients with coronary artery disease. *Prostaglandins Leukot. Essent. Fatty Acids, 56*, 379–384.

Campbell-Tofte, J. I., Mølgaard, P., & Winther, K., (2012). Harnessing the potential clinical use of medicinal plants as antidiabetic agents. *Botanics: Targets and Therapy, 2*, 7–19.

Chandran, U., & Patwardhan, B., (2017). Network ethnopharmacological evaluation of the immunomodulatory activity of *Withania somnifera. J. Ethnopharmacol., 197*, 250–256.

Chatterjee, S. P., Variyar, A., & Sharma, (2010). Bioactive lipid constituents of fenugreek. *Food Chem, 119*, 349–353.

Chikara, S. K., Pandey, M., Pandey, S., Vaidya, K., & Chaudhary, S., (2014). Next-generation sequencing: A revolutionary tool for plant variety improvement. *Am. J. Social Issues Humanities (AJSIH),* (Special Issue), 137–154.

Chintha, L., Tien, H., & Daniel, A. D., (2019). Understanding glycaemic control and current approaches for screening anti-diabetic natural products from evidencebased medicinal plants. *Plant Methods, 15*, 105.

Dangi, R. S., Lagu, M. D., Choudhary, L. B., Ranjekar, P. K., & Gupta, V. S., (2004). Assessment of genetic diversity in *Trigonella foenum-graecum* and *Trigonella caerulea* using ISSR and RAPD markers. *BMC Plant Biology, 4*(1), 13.

Dewick, P. M., (2002). *Medicinal Natural Products: A Biosynthetic Approach* (2nd edn.). Wiley, Chichester.

Dixit, P., Ghaskadbi, S., Mohan, H., & Devasagayam, T. P., (2005). Antioxidant properties of germinated fenugreek seeds. *Phytother. Res., 19*, 977–983.

Durazzo, A., D'Addezio, L., Camilli, E., Piccinelli, R., Turrini, A., Marletta, L., Marconi, S., et al., (2018). From plant compounds to botanicals and back: A current snapshot. *Molecules, 23*(8).

Feyzi, S., Varidi, M., Zare, F., & Varidi, M. J., (2015). Fenugreek (*Trigonella Foenum Graecum*) seed protein isolate: Extraction optimization, amino acid composition, thermo, and functional properties. *J. Sci. Food Agric., 95*, 3165–3176.

Gauttam, V. K., & Kalia, A. N., (2013). Development of polyherbal antidiabetic formulation encapsulated in the phospholipids vesicle system. *J. Adv. Pharm. Technol. Res., 4*, 108–117.

Goswami, D., Mahapatra, A. D., Banerjee, S., Kar, A., Ojha, D., Mukherjee, P. K., & Chattopadhyay, D., (2018). Boswellia serrata oleo-gum-resin and β-boswellic acid inhibits HSV-1 infection *in vitro* through modulation of NF-κB and p38 MAP kinase signaling. *Phytomedicine, 51*, 94–103.

Gupta, A., Gupta, R., & Lal, B., (2001). Effect of *Trigonella foenum-graecum* (fenugreek) seeds on glycaemic control and insulin resistance in type 2 diabetes mellitus: A double-blind placebo-controlled study. *The Journal of the Association of Physicians of India, 49*, 1057–1061.

Hannan, J. M. A., Ali, L., Rodeya, B., Khaleque, J., Alchter, M., Flatt, P. R., & Abdel, W. Y. H. A., (2006). Soluble dietary fibre fractions of *Trigonella foenumgraecum* (fenugreek) seed improves glucose homeostasis in animal models of type 1 and type 2 diabetes by delaying carbohydrate digestion and absorption and enhancing insulin action. *British Journal of Nutrition, 97*, 514–521.

Hannan, J. M., Rokeya, B., & Faruque, O., (2003). Effect of soluble dietary fibre fraction of *Trigonella foenum-graecum* on glycemic, insulinemic, lipidemic, and platelet aggregation status of type 2 diabetic model rat. *J. Ethnopharmacol., 88*, 73–77.

Hasanpoura, M., Iranshahyb, M., & Iranshahi, M., (2020). The application of metabolomics in investigating antidiabetic activity of medicinal plants. *Biomedicine & Pharmacotherapy, 128*, 110263.

He, Y., Ding, C., Wang, X., Wang, H., & Suo, Y., (2015). Using response surface methodology to optimize countercurrent chromatographic separation of polyphenol compounds from fenugreek (*Trigonella foenum-graecum* L.) seeds. *J. Liq. Chromatogr. Relat. Technol., 38*, 29–35.

Hooda, S., & Jood, S., (2004). Nutritional evaluation of wheat-fenugreek blends for product making. *Plant Foods Hum. Nutr., 59*, 149–154.

Hopkins, A. L., (2007). Network pharmacology. *Nat. Biotechnol., 25*, 1110–1111. https://doi. org/10.1038/nbt1007-1110.

Hutchinson, L., & Kirk, R., (2011). High drug attrition rates – where are we going wrong? *Nat. Rev. Clin. Oncol., 8*, 189–190. https://doi.org/10.1038/nrclinonc.2011.34.

Kaddurah, D. R., Weinshilboum, R., & Network, P. R., (2015). Metabolomic signatures for drug response phenotypes: Pharmacometabolomics enables precision medicine. *Clin. Pharmacol. Ther., 98*, 71–75.

Kaddurah-Daouk, R., Weinshilboum, R., & Network, P. R., (2014). Pharmacometabolomics: Implications for clinical pharmacology and systems pharmacology. *Clin. Pharmacol. Ther., 95*, 154–167.

Kaviarasan, S., Vijayalakshmi, K., & Anuradha, C. V., (2004). Polyphenol-rich extract of fenugreek seeds protects erythrocytes from oxidative damage. *Plant Foods Hum. Nutr., 59*, 143–147.

Kawabata, T., Cui, M. Y., & Hasegawa, T., (2011). Anti-inflammatory and anti-melanogenic steroidal saponin glycosides from fenugreek (*Trigonella foenum-graecum* L.) seeds. *Planta Med., 77*, 705–710.

Khan, V., Najmi, A. K., Akhtar, M., Aqil, M., Mujeeb, M., & Pillai, K. K., (2012). A pharmacological appraisal of medicinal plants with antidiabetic potential. *J. Pharm. Bioallied Sci., 4*, 27–42.

Kibble, M., Saarinen, N., Tang, J., Wennerberg, K., Mäkelä, S., & Aittokallio, T., (2015). Network pharmacology applications to map the unexplored target space and therapeutic potential of natural products. *Nat. Prod. Rep., 32*, 1249–1266 https://doi.org/10. 1039/c5np00005j.

Kooti, W., Moradi, M., Akbari, S., Sharafi, A. N., AsadiSamani, M., & Ashtary, L. D., (2015). Therapeutic and pharmacological potential of *Foeniculum vulgare* mill: A review. *J. HerbMed Pharm., 4*, 1–9.

Li, F., Fernandez, P. P., & Rajendran, P., (2010). Diosgenin, a steroidal saponin, inhibits STAT3 signaling pathway leading to suppression of proliferation and chemosensitization of human hepatocellular carcinoma cells. *Cancer Lett., 292*, 197–207.

Li, S., & Zhang, B., (2013). Traditional Chinese medicine network pharmacology: Theory, methodology, and application. *Chin. J. Nat. Med., 11*, 110–120 https://doi.org/10. 1016/ S1875-5364(13)60037-0.

Li, S., Zhang, B., & Zhang, N., (2010). Network target for screening synergistic drug combinations with application to traditional Chinese medicine. *BMC Syst. Biol., 5*(Suppl. 1), S10. https://doi.org/10.1186/1752-0509-5-S1-S10.

Maede, H., Milad, I., & Mehrdad, I., (2020). The application of metabolomics in investigating antidiabetic activity of medicinal plants. *Biomedicine & Pharmacotherapy, 128*.110263.

Meghwal, M., & Goswami, T. K., (2012). A review on the functional properties, nutritional content, medicinal utilization and potential application of fenugreek *J. Food Process Technol., 3*, 9.

Mehrafarin, A., Qaderi, A., Rezazadeh, S., Naghdi, B. H., Noormohammadi, G., & Zand, E., (2010). Bioengineering of important secondary metabolites and metabolic pathways in fenugreek (*Trigonella foenum-graecum* L.). *J. Med. Plants, 9*(35), 1–18.

Nagulapalli, V. K. C., Swaroop, A., Bagchi, D., & Bishayee, A., (2017). A small plant with big benefits: Fenugreek (*Trigonella foenum-graecum* Linn.) for disease prevention and health promotion. *Mol. Nutr. Food Res., 61*(1600950).

Nathiya, S., Durga, M., & Devasena, T., (2014). Therapeutic role of *Trigonella foenum-graecum* [Fenugreek]: A review. *Int. J. Pharm. Sci. Rev. Res., 27*(2), 74–80.

Passano, P., (1995). *The Many Uses of Methi Manushi, 2*, 31–34.

Patel, D. K., Prasad, S. K., Kumar, R., & Hemalatha, S., (2012). An overview on antidiabetic medicinal plants having insulin mimetic property. *Asian Pac. J. Trop. Biomed., 2*, 320–330.

Patel, K., Gadewar, M., Tahilyani, V., & Patel, D. K., (2012). A review on pharmacological and analytical aspects of diosgenin: A concise report. *Nat. Prod. Bioprospect., 2*, 46–52.

Prasanna, M., (2020). "Hypolipidemic effect of fenugreek: Diabetic patients: A clinical study." *Indian Journal of Pharmacology, 32*, 34–36.

Qamar, A. S., Zainab, R., Muhammad, H. A., Rizwan, S., Anum, I., Niaz, M., & Hafiz, U. U. R., (2020). Nutritional and therapeutic properties of fenugreek (*Trigonellafoenum-graecum*): A review. *International Journal of Food Properties., 23*(1), 1777–1791. doi: 10.1080/10942912.2020.1825482.

Raju, J., & Bird, R. P., (2006). Alleviation of hepatic steatosis accompanied by modulation of plasma and liver TNF-alpha levels by *Trigonella foenum-graecum* (fenugreek) seeds in Zucker obese (fa/fa) rats. *Int. J. Obese (Lond), 30*, 1298–1307.

Raju, J., Gupta, D., Araga, R. R., Yadav, P. K., & Boquer, N. Z., (2001). *Trigonella foenum-graecum* (fenugreek) seed powder improves glucose homeostasis in alloxan diabetic rat tissues by reversing the altered glycolytic, gluconeogenic, and lipogenic enzymes. *Molecular and Cellular Biochemistry, 224*, 45–51.

Rao, A. V., (2003). *Herbal Cure for Common Diseases Fusion Books*. New Delhi.

Reddy, R. L. R., Gowda, A. N. S., & Srinivasan, K., (2019). Antilithogenic and hypocholesterolemic effect of dietary fenugreek seeds (*Trigonella Foenum-graecum*) in experimental mice. *Medicinal Plants: Int. J. Phytomed. Related Ind., 11*, 145–154.

Rizvi, S. I., & Mishra, N., (2013). Traditional Indian medicines used for the management of diabetes mellitus. *J. Diabetes Res., 2013*, 712092.

Roberts, K. T., (2011). The potential of fenugreek (*Trigonella foenum-graecum*) as a functional food and nutraceutical and its effects on glycemia and lipidemia. *J. Med. Food, 14*, 1485–1489.

Sangwan, R. S., Tripathi, S., Singh, J., Narnoliya, L. K., & Sangwan, N. S., (2013). De novo sequencing and assembly of *Centella asiatica* leaf transcriptome for mapping of structural, functional, and regulatory genes with special reference to secondary metabolism. *Gene, 525*, 58–76.

Saravanan, G., Ponmurugan, P., Deepa, M. A., & Senthilkumar, B., (2014). Modulatory effects of diosgenin on attenuating the key enzymes activities of carbohydrate metabolism and glycogen content in streptozotocin-induced diabetic rats. *Can J. Diabetes, 38*, 409–414.

Sauvaire, Y., Baissac, Y., Leconte, O., Petit, P., & Ribes, G., (1996). Steroid saponins from fenugreek and some of their biological properties. *Adv. Exp. Med. Biol., 405*, 37–46.

Sebastian, K. S., & Thampan, R. V., (2007). Differential effects of soybean and fenugreek extracts on the growth of MCF-7 cells. *Chem. Biol. Interact., 170*, 135–143.

Shabbeer, S., Sobolewski, M., & Anchoori, R. K., (2009). Fenugreek: A naturally occurring edible spice as an anticancer agent. *Cancer Biol. Ther., 8*, 272–278.

Sharma, R. D., Raghuram, T. C., & Rao, N. S., (1990). Effect of fenugreek seeds on blood glucose and serum lipids in type I diabetes. *Eur. J. Clin Nutr., 44*, 301–306.

Shi, C. Y., Yang, H., Wei, C. L., Yu, O., Zhang, Z. Z., Jiang, C. J., Sun, J., et al., (2011). Deep sequencing of the *Camellia sinensis* transcriptome revealed candidate genes for major metabolic pathways of team-specific compounds. *BMC Genome, 12*, 131–150.

Shi, J., Cao, B., Wang, X. W., Aa, J. Y., Duan, J. A., Zhu, X. X., Wang, G. J., et al., (2016). *Metabolomics and its Application to the Evaluation of the Efficacy and Toxicity of Traditional Chinese Herb Medicines, 1026*, 204–216.

Snehlata, H. S., & Payal, D. R., (2012). Fenugreek (*Trigonella foenum-graecum* L.): An overview. *Int. J. Current Pharm. Rev. Res., 2*(4), 169–187.

Srichamroen, A., Thomson, A. B., Field, C. J., & Basu, T. K., (2009). *In vitro* intestinal glucose uptake is inhibited by galactomannan from Canadian fenugreek seed (*Trigonella foenum-graecum* L.) in genetically lean and obese rats. *Nutr. Res., 29*, 49–54.

Srinivasan, K., (2006). Fenugreek (*Trigonella foenum-graecum*): A review of health beneficial physiological effects. *Food Rev. Int., 22*(2), 203–224.

Srivastava, A., Singh, Z., Verma, V., & Choedon, T., (2020). Potential health benefits of fenugreek with multiple pharmacological properties. In: Mishra, N., (ed.), *Ethnopharmacological Investigation of Indian Spices* (pp. 137–153).

Suresh, P., Kavitha, C. N., & Babu, S. M., (2012). Effect of ethanol extract of *Trigonella foenum-graecum* (fenugreek) seeds on Freund's adjuvant-induced arthritis in albino rats. *Inflammation, 35*, 1314–1321.

Thirunavukkarasu, V., Anuradha, C. V., & Viswanathan, P., (2003). Protective effect of fenugreek (*Trigonella foenum-graecum*) seeds in experimental ethanol toxicity. *Phytother. Res., 17*, 737–743.

Tripathi, U. N., & Chandra, D., (2009). The plant extracts of *Momordica charantia* and *Trigonella foenum-graecum* have antioxidant and anti-hyperglycemic properties for cardiac tissue during diabetes mellitus. *Oxid. Med. Cell Longev., 2*, 290–296.

Uemura, T., Goto, T., & Kang, M. S., (2011). Diosgenin, the main aglycon of fenugreek, inhibits LXRα activity in HepG2 cells and decreases plasma and hepatic triglycerides in obese diabetic mice. *J. Nutr., 141*, 17–23.

Vaidya, K., Ghosh, A., Kumar, V., Chaudhary, S., Srivastava, N., Katudia, K., Tiwari, T., & Chikara, S. K., (2013). De novo transcriptome sequencing in *Trigonella foenum-graecum* L. to identify genes involved in the biosynthesis of diosgenin. *Plant Genome, 6*, 1–11.

Verma, S. K., Singh, S. K., & Mathur, A., (2010). *In vitro* cytotoxicity of *Calotropis procera* and *Trigonella foenum-graecum* against human cancer cell lines. *J. Chem. Pharm. Res., 2*, 861–865.

Vitali, F., Mulas, F., Marini, P., & Bellazzi, R., (2013). Network-based target ranking for polypharmacological therapies. *J. Biomed. Inform., 46*, 876–881 https://doi.org/10.1016/j.jbi.2013.06.015.

Wani, S. A., & Kumar, P., (2018). Fenugreek: A review on its nutraceutical properties and utilization in various food products. *J. Saudi Society Agri. Sci., 17*, 97–106.

Xue, W., Lei, J., Li, X., & Zhang, R., (2011). *Trigonella foenum-graecum* seed extract protects kidney function and morphology in diabetic rats via its antioxidant activity. *Nutr. Res., 31*, 555–562.

Yadav, U. C., Moorthy, K., & Baquer, N. Z., (2004). Effects of sodium-orthovanadate and *Trigonella foenum-graecum* seeds on hepatic and renal lipogenic enzymes and lipid profile during alloxan diabetes. *J. Biosci., 29*, 81–91.

Zia, T., Hasnain, S. N., & Hasan, S. K., (2001). Evaluation of the oral hypoglycemic effect of *Trigonella foenum-graecu,* L. (methi) in normal mice. *J. Ethnopharmacol., 75*, 191–195.

CHAPTER 7

Mulberry: From Root to Fruit with Antidiabetic Properties

MASARAT BASHIR[1] and ASHRAF DAR[2]

[1]*College of Temperate Sericulture, Sheri-Kashmir University of Agricultural Sciences and Technology, Srinagar, Jammu and Kashmir, India, E-mail: masarat.bashir4@gmail.com*

[2]*Department of Biochemistry, University of Kashmir, Srinagar, Jammu and Kashmir – 190006, India,*
E-mail: ashrafdar@kashmiruniversity.ac.in

ABSTRACT

Diabetes mellitus, more commonly known as diabetes, is a lifelong metabolic disease that is now prevalent in almost all populations across the world. This chronic disorder characterized by hyperglycemia (blood glucose $\geq$ 126 mg/dl after 8 hours of fasting) over a prolonged period of time eventually leads to damage of vital body organ systems. There is no cure for diabetes at present but it is managed by various drug therapies. During the last decade, there have been increased attempts for the use of herbal or plant-based treatments to manage diabetes. Several plants with significant medicinal values including mulberry have been studied for their antidiabetic properties. Different parts of the mulberry like leaf, root, fruit, and branches have shown strong hypoglycemic and antidiabetic properties in experimental animals and human subjects. The individual pharmaceutical components from mulberry have been isolated and their cellular targets involved in glucose metabolism pathways identified. The

Antidiabetic Plants for Drug Discovery: Pharmacology, Secondary Metabolite Profiling, and Ingredients with Insulin Mimetic Activity. Deepu Pandita, MPhil, Anu Pandita, MSc, & Chander Bhanu, MBBS, PGDDCE (Eds.)
© 2023 Apple Academic Press, Inc. Co-published with CRC Press (Taylor & Francis)

recent sequencing of the mulberry genome is expected to greatly facilitate our understanding of gene structures and metabolic pathways involved in the biosynthesis of active antidiabetic components in mulberry. Additionally, the mulberry genomic information could be exploited in future for sustained production of antidiabetic compounds by developing new transgenic mulberry varieties or by making use of recombinant DNA technology.

FIGURE 7.1 Mulberry plant.

Source: Luis Fernández García L. Fdez. / 2005-06-05 / Parque Ana Tutor, Madrid. https://creativecommons.org/licenses/by-sa/2.1/es/deed.en

7.1 INTRODUCTION

Mulberry (or Morus) belongs to the Moraceae family of flowering deciduous plants (Figure 7.1). The plant grows wild and is cultivated in many temperate zones of Asia, Southern Europe, and Northern America. Mulberry is cultivated on a large scale for commercial purposes and forms the backbone of the silk industry. The silk-producing worm or silkworm

(*Bombyx mori*) eats mulberry leaves as its sole source of food. Since ancient times, humans have also been consuming the fruit and leaf of mulberry for their great nutritive values.

There are about a dozen species belonging to the genus Morus (Venkatesh and Chauhan, 2008). Most of these mulberry species exhibit medicinal properties against various human disorders (Table 7.1). The three most common mulberry species including *M. alba, M. rubra, and M. nigra* have been comparatively well studied with respect to their anti-diabetic properties.

TABLE 7.1 Medicinal Properties of Different Species of Mulberry

Mulberry Species	Effective Against Diseases	References
M. alba	Diabetes, atherosclerosis, hyperglycemia, inflammation, high blood pressure, cough, neurodegeneration, cough, diarrhea, atherosclerosis, cancer	Singh et al. (2013)
M. nigra	Diabetes, cancer, microbial infection, obesity	Lim Ho and Choi Chang-Ik (2019)
M. rubra	Diabetes, constipation, intestinal worms, urinary tract problems, dysentery	Koca et al. (2008)
M. cathayana	Hypertension, inflammation	Duke et al. (1992)
M. laevigata	Blood problems, sores	Venkatesh and Chauhan (2008)
M. australis	Helminthic problems, skin problems, platelet disease	Chen et al. (1989)
M. serrata	Helminthic problems	Venkatesh and Chauhan (2008)
M. macroura	Oxidative stress, inflammation	Venkatesh and Chauhan (2008)
M. bomycis	Fever, inflammation, hypertension, anti-microbial, diabetes	Oh Na-Hyun et al. (2015)
M. indica	Hyperlipidemia, diabetes, oxidative stress	Andallu and Varadacharyulu (2003)

Over the years, mulberry has emerged as a plant with significant medicinal properties. Different parts of the plant, e.g., root bark, twigs, leaf, and fruit have been extensively studied for their benefits in human health and cure for diseases. These include antioxidant, antidiabetic, anticancer, anti-inflammatory, and antimicrobial properties. Furthermore, mulberry has proven to protect against various cardiovascular, neurological, gastrointestinal, and liver diseases. The leaf extract exhibits potential antiwrinkling, antihyperpigmentation, and anti-skin cancer effects in mice. In the traditional Chinese system of medicine, dried mulberry leaves have been used to treat frequent urination and increased thirst (symptoms of type 2 diabetes) (Wei et al., 2018). In Japan and Korea, diabetic patients are known to consume tea prepared with mulberry leaves to maintain their blood sugar (Chan et al., 2016). The recent research is suggesting that mulberry leaves, roots, branches, and fruits contain a range of bioactive molecules like alkaloids, flavonoids, polysaccharides, and amino acids with potential antidiabetic activities (Wei et al., 2018). This chapter elaborates upon antidiabetic values of different mulberry parts and isolated bioactive molecules, which have a great potential to be promoted in diabetic care.

7.2 ANTIDIABETIC PROPERTIES IN MULBERRY PARTS

Diabetes mellitus is a chronic disorder and is characterized by sustained hyperglycemia. The disease is associated with secondary metabolic complications like increased production of free radicals (e.g., reactive oxygen species or ROS, reactive nitrogen species or RNS, reactive chlorine species or RCS, etc.), and non-enzymatic glycosylated end products (e.g., glycated proteins/enzymes) in host. This creates a very high oxidative stress and damage to the blood vessels. In normal physiology, free radicals play an essential role in the functioning of MAPK (mitogen-activated protein kinase) and ERK (extracellular signal-related kinase) signaling pathways. These pathways are important for gene expression and cell proliferation. The endogenous antioxidant system comprising of antioxidant enzymes like glutathione (GSH), glutathione peroxidase (GSH-Px), catalase (CAT), superoxide dismutase (SOD), proteins, lipoic acid, Coenzyme Q, and uric acid, maintains the homeostasis of free radicals in the body. However, in type 2 diabetes, sustained hyperglycemia likely alters the structure and functioning of the endogenous antioxidant enzyme system by non-enzymatic glycation, thus promoting the accumulation of free radicals and consequently oxidative stress (Kumawat et al., 2013). When left untreated, diabetes leads to long-term complications including cardiovascular diseases, stroke, cognitive impairment and damage to nerves, kidneys, and eyes (Kharroubi and Darwish, 2015).

There are several commercial drugs available for the management of diabetes (Zierath, 2019). Due to the potential side effects and high cost of these antidiabetic drugs, increased need is felt across the world for safer and less expensive strategies to manage this chronic epidemic. Over the past decade, there has been increased attention on the identification and investigation of herbal medicines for the management of various chronic diseases, including diabetes mellitus (Han et al., 2019). Due to its use in traditional systems of medicine, mulberry has recently emerged as a strong antidiabetic medicinal plant (Tian et al., 2016). Several studies have now investigated antidiabetic properties in different parts of mulberry. These parts include leaf, fruit, root, and branches.

7.2.1 LEAF

From ancient times mulberry leaves (Figure 7.2) have been used in Asian countries for treatment of various diseases including type 2 diabetes. Various studies of mulberry leaf extracts in both animal and human diabetic models have shown significant benefit for type 2 diabetes. In experimental Wistar rats in which diabetes is induced by streptozotocin (STZ), treatment with mulberry leaves (*M. indica*) exhibit potential antihyperglycemic and antioxidant effects. There is a significant decrease in lipid peroxidation and CAT activity seen in erythrocytes of diabetic mice treated with mulberry leaves (Andallu et al., 2012). The dried leaf extracts of *M. alba* and *M. nigra* have a remarkable hypoglycemic effect in alloxan-induced diabetic mice. A single dose of leaf extract (500 mg/kg body weight) from both the mulberry species remarkably decreases blood glucose levels in these experimental mice (Hossienzadeh et al., 1999). The oral administration of *M. nigra* leaf extract (500 mg/kg body weight/day) in STZ-induced diabetic rats reduces the levels of blood glucose from 370 mg/dl to 154 mg/dl with a significant increase in the insulin level from 11.3 μU/ml to 14.6 μU/ml. These strong hypoglycemic effects of mulberry leaf extracts are apparently due to enhanced glucose uptake by muscle cells (El-Mawla et al., 2011; Chen et al., 1995). The hot-water extract of mulberry leaves also increases secretion of saliva, an indication for improvement in type 2 diabetes condition (Chen et al., 1995). It is important to mention that there is a reduced salivary flow in type 2 diabetic patients (Mata et al., 2004). The aqueous leaf extract of *M. rubra* has antioxidant activity and reduces diabetic complications in mice. This aqueous leaf extract may provide protection against hyperglycemia and dyslipidemia (Bala et al., 2010).

FIGURE 7.2 Mulberry leaf.

Source: Photo by 阿爾特斯. https://creativecommons.org/licenses/by-sa/3.0/deed.en

The leaf extract from *M. alba* has been shown to decrease weekly food intake in STZ-induced diabetic rats throughout a 5-week experimental window. In these animals, the observed hypoglycemic effect seen is probably a consequence of interference with intake of food or inhibition in absorption of glucose by the gastrointestinal system (Ojewole et al., 2006). Mice raised on *M. alba* leaf powder (6% extract) plus high-sucrose diets for 8 weeks show reduced insulin resistance and delayed onset of type 2 diabetes clinical symptoms (Tanabe et al., 2011).

Mulberry leaf powder significantly decreases the levels of total cholesterol, LDL (low density lipoprotein) and VLDL (very low-density lipoprotein) whereas it increases HDL (high density lipoprotein). The powder also decreases serum triglycerides, free fatty acids, and peroxides in plasma (Andallu et al., 2012). The dietary intake of mulberry leaf power may also protect against atherogenesis by significantly delaying lipoprotein oxidation (Harauma et al., 2007). Experiments in mice have shown that mulberry leaf powder decreases atherosclerotic lesion size (by 40%) in the aortae (Park et al., 2009).

Both *in vitro* and *in vivo* experiments suggest that mulberry leaf extracts (*M. alba*) have a potential postprandial hypoglycemic effect. Treatment of Caco-2 cells in culture with tea prepared from mulberry leaves reduces released glucose content in both apical and basal sides of cell monolayer. Although Caco2 cells are derived from colon, they functionally and morphologically resemble the enterocytes lining the small intestine. Pertinently, the hot water extract of mulberry leaves has an inhibitory effect on α-glucosidase, sucrase, and maltase enzymes (Hansawasdi et al., 2006). The oral administration of a single dose of mulberry leaf powder (~ 1.2 g, enriched with DNJ) significantly suppresses the elevation of postprandial blood glucose and insulin secretion in human subjects. Further, it is postulated that postprandial hypoglycemic effect observed with mulberry leaf water extract is caused by inhibition of glucose transport in the small intestine (Kimura et al., 2007; Park et al., 2009). These studies therefore suggest that mulberry leaf extracts may be promoted as a food supplement for management of postprandial blood glucose in patients with type 2 diabetes.

Hemoglobin, the oxygen carrying protein in RBC's, gets highly glycated (HbA1c) in type 2 diabetes. HbA1c is considered a biomarker for determination of chronic presence of elevated levels of glucose in serum. In STZ-induced diabetic rats, daily administration of ethanolic extract of mulberry leaf (1 g/kg body weight) for six weeks results in a decrease of blood glucose levels by 22% and HbA1c to 6.78 (from 9.02 in control) (Naowaboot et al., 2009).

Mulberry leaf extract has also been implicated in β-cell regeneration in the pancreas. The histological and biochemical examination of the pancreas from diabetic rats, after the animals were treated with mulberry leaf extract indicated regeneration of β-cells in the pancreas (Mohammadi et al., 2012). The results from a recent well-controlled study in which the use of mulberry leaf extract in human subjects suggests that mulberry leaf extract may be a useful complementary mealtime supplement for non-insulin dependent type 2 diabetic patients. The study proposes that mulberry leaf extract appears to be safe at a dose of 3 g/day in human subjects (Riche et al., 2017).

7.2.2 FRUIT

Mulberry fruits are fleshy, succulent, and delicious edible parts of the plant (Figure 7.3). They are low in calories and contain various phytonutrient compounds like vitamins, minerals, and polyphenols. The richness of mulberry fruits with these phytonutrients makes them ideal for promoting human health. In traditional Indian and Chinese systems of medicines, mulberry fruits have been used to enhance human health. Studies have shown that mulberry fruit extract has potent antidiabetic, antioxidant, antitumor, and neuroprotective activities. Moreover, hypolipidemic, and macrophage activating effects have been observed in mulberry fruit extracts (Kadam et al., 2019).

FIGURE 7.3 Mulberry fruit.

Source: Photo by Beko. https://creativecommons.org/licenses/by-sa/4.0/deed.en

The administration of mulberry fruits in the form of freeze-dried powder to experimental rats on a high-fat diet results in a significant decline in levels of serum and liver triglycerides, total cholesterol, and LDL, with a remarkable increase in levels of HDL. In red blood cells (RBCs) and liver, the lipid peroxidation decreases whereas the SOD and blood GSH-Px activities are markedly increased. There is also a significant decrease in atherogenesis in these experimental animals (Yang et al., 2010).

The ethyl acetate solubilized extract of mulberry fruit possesses potent α-glucosidase inhibitory activity and radical-scavenging activities against DPPH (1,1-diphenyl-2-picrylhydrazyl) and O_2 (superoxide) anion radicals *in vitro*. In STZ-induced diabetic experimental mice, ethyl acetate-solubilized extract of mulberry fruit decreases the fasting blood glucose levels and glycosylated serum proteins and increases antioxidant enzymatic activities like CAT, SOD, and GSH-Px, (Wang et al., 2013).

There are other studies that show ethanol extract of the mulberry fruit contains antioxidant activity (Bae and Suh, 2007) whereas the methanol extract possesses DPPH radical scavenging activity (Imran et al., 2010).

7.2.3 ROOT

Mulberry root (Figure 7.4) has been used in the traditional Chinese system of medicine against various human diseases. The roots are rich in many types of antioxidants, including flavonoids, guangsangons A-E, guangsangons K-N, astragalus, and polysaccharides. STZ-diabetic rats administered with ethanol extracts of root bark (*M. alba* for 10 days) show a significant decline in serum glucose level to a value of 155 mg/dl (similar to those of normal control rats, 143 mg/dl) from 379 mg/dl. The extracts further helped to improve insulin levels to 15.6 U/ml from 10.8 U/ml in these experimental rats. The lipid peroxide production significantly decreases in STZ-diabetic rats from 6.3 to 5.1 nmol TBARS/ml serum upon administration of ethanol extracts of root bark. In STZ minus control rats, the level of lipid peroxides observed is 5.3 nmol TBARS/ml serum (Singab et al., 2005). Thiobarbituric acid reactive species (TBARS) are a byproduct of lipid peroxidation and TBARS assay gives a measure of lipid peroxidation.

FIGURE 7.4 Mulberry root.
Source: Photo from https://permies.com/t/138285/Mulberries-Nitrogen

Besides antidiabetic properties of ethanol-soluble root extracts, the ethanol-insoluble fractions also could lower blood glucose levels in experimental animals, suggesting that different parts of root bark have antidiabetic potential (Chen et al., 1995). The administration of *M. nigra* bark extract to alloxan-induced experimental diabetic mice has been shown to effectively reduce the blood glucose levels (Hosseinzadeh et al., 1999). The mulberry root extracts also protect against the development of peripheral nervous lesions during the early stages of diabetes in alloxan-induced experimental rats. The extramedullary fiber and the areas of myelin sheath are significantly enlarged in alloxan-induced diabetic rats upon treatment with mulberry root extracts (Cui et al., 2006). These results suggest the neuroprotective properties embedded in mulberry root.

Protein tyrosine phosphatase 1B (PTP1B) has been implicated in obesity and type 2 diabetes. Inhibition of this enzyme could help in treatment of both obesity and diabetes. It has been found that ethyl acetate-soluble extract of the root bark of *Morus* sp. possesses inhibitory activities against PTP1B enzyme, suggesting a possible mechanism by which bioactive components in root bark may play a role in preventing both obesity and diabetes (Cui et al., 2006).

7.2.4 BRANCHES

The mulberry branches (Figure 7.5) have been traditionally wasted as agricultural waste or used as fire word. But recent studies have implicated mulberry branch bark extracts in a variety of pharmacological activities. Recently, it has been shown that mulberry branch extracts inhibit melanin synthesis and promote whitening. The alcoholic extracts of mulberry branches can also give some sort of protection to the kidneys in hyper-uricemic mice and protects against alcohol-induced liver injury and nerve damage. The branch extracts are now known to possess hypolipidemic, antioxidation, antibacterial, anticancer, anti-inflammatory, and antidia-betic properties (Hua-Yu et al., 2014).

FIGURE 7.5 Mulberry branch.

Source: Photo by Gaurav Dhwaj Khadka. https://creativecommons.org/licenses/by-sa/4.0/deed.en

The induction of diabetes in mice with STZ promotes polyuria, polyphagia, and weight loss. These mice also have enlarged liver and kidneys. The administration of branch bark extract slows weight loss in these STZ-induced diabetic mice. Examination of liver and kidneys shows that branch bark extract-treated STZ-induced diabetic mice have lesser enlargement of liver and kidneys compared to the untreated mice. In STZ-induced diabetic mice, the fasting glucose levels are high (21 mm/L at the end of 3rd week as compared to 5 mm/L in normal mice), which is decreased (~13 mm/L) upon treatment with branch bark extract (200 mg/kg body weight). Further, branch bark extract reduces the abnormally elevated levels of serum insulin effectively and improves insulin resistance in STZ-diabetic mice. These extracts (100 mg/kg body weight) also decrease mRNA expression of insulin 2 gene in diabetic mice, thus may help in overcoming insulin resistance. Moreover, the experiments indicate that branch bark extract can decrease the expression of glucose metabolism genes in diabetic mice. The extracts decrease the mRNA expression of Glucose 6 phosphatase, glucokinase, and phosphoenolpyruvate kinase genes. The decrease in expression of these genes is likely to slow down the gluconeogenesis reactions. The branch bark extract has also significant inhibitory effect on liver α-glycosidase enzyme, thus may reduce the post-prandial blood glucose. The experiments in STZ-induced diabetic mice show that branch bark extract administration can significantly reduce the levels of serum triglycerides and LDL. However, there is no change in HDL observed in these animals (Hua-Yu et al., 2014).

7.3 ANTI DIABETIC BIOACTIVE COMPONENTS IN MULBERRY

Mulberry is emerging as a source of high value-added bioactive components with pharmaceutical and nutraceutical applications. These bioactive components include anthocyanins, polyphenols, flavonoids, alkaloids, and others (Priya, 2012). The current trend in the food industry is supplementing foods with various bioactive ingredients for promoting health and prevention of different diseases in consumers. The bioactive compounds from various parts of mulberry that have been evaluated against diabetes are given in Table 7.2. These compounds may exert their antidiabetic effect by targeting different cellular enzymes and pathways involved in glucose metabolism.

7.3.1 SUGAR-MIMIC ALKALOIDS

Various parts of mulberry especially leaves contain a high concentration of sugar-mimicking alkaloids. For example 1,4-dideoxy-1,4-imino-D-arabinitol, 1-deoxynojirimycin and 1,4-dideoxy-1,4-imino-D-ribitol. These alkaloids show significant hypoglycemic properties and are known to inhibit the activities of all or some intestinal glucose metabolism enzymes. They regulate uptake of monosaccharides and are therefore promoted for therapeutic use in type 2 diabetes mellitus. 1,4-dideoxy-1,4-imino-D-arabinitol and 1-deoxynojirimycin have been thoroughly implicated in diabetic control in both animal and model systems.

TABLE 7.2 Antidiabetic Components in Various Parts of Mulberry

Antidiabetic Component	Source of Mulberry Part	Cellular Target	References
DAB	Root/branches/leaf/fruit	Glycogen phosphorylase	Fosgerau et al. (2000)
DNJ-1	Root/branches/leaf/fruit	α-glucosidase, GLUT4	Liu et al. (2014)
Fagomine-3	Root/branches/leaf/fruit	α-glucosidase	Taniguchi et al. (1998)
Flavonoids	Root/branches/leaf/fruit	α-glucosidase, PTP1B	Shibano et al. (2008)
Polysaccharides	Root/branches/leaf/fruit	α-glucosidase	Zhang et al. (2014)
Chlorogenic acid	Leaf	–	Hunyadi et al. (2012)

7.3.1.1 DAB (1, 4-DIDEOXY-1,4-IMINO-D-ARABINITOL)

The subcutaneous injection of DAB has been shown to improve glycemia in diabetic mice models. Besides being a α-glycosidase inhibitor, DAB has been shown to inhibit liver glycogen phosphorylase from various species including human. Liver glycogen phosphorylase is the rate-limiting enzyme in the glycogenolytic pathway. Glycogen phosphorylase catalyzes formation of glucose-1-phosphate from glycogen and is activated by glucagon, thus playing an essential role in the induction of hyperglycemia in type 2 diabetes. The *in vitro* experiments in primary hepatocytes have shown that DAB blocks glycogenolysis (glycogen breakdown) (Mackay et al., 2003).

7.3.1.2 DNJ (1-DEOXYNOJIRIMYCIN)

DNJ is a well-known anti diabetic compound from the mulberry. This analog of glucose was first isolated in 1976 from the mulberry root bark. In diabetic mice, administration of DNJ significantly reduces blood glucose levels by inhibiting α-glucosidase and the absorption of glucose in the intestinal brush border. Insulin plays a critical role in the maintenance of blood glucose homeostasis. During insulin-resistance state in Type 2 diabetes, insulin does not promote the uptake of glucose by skeletal muscles. As a consequence, insulin secretion increases excessively in blood as a compensation to maintain blood glucose homeostasis. The treatment of diabetic mice with DNJ significantly reduces insulin secretion (Liu et al., 2015).

DNJ also increases the expression of glucose transporter in skeletal muscles. GLUT4 transports glucose into the skeletal muscles and adipocytes and decreases hyperglycemia. Diabetic mice have significantly decreased levels of GLUT4 in muscle. Treating these mice with DNJ increases the expression of GLUT4 in skeletal muscles. Moreover, GLUT4 translocation to the plasma membrane of skeletal muscle cells is promoted by IRSβ-IRS-1-AKT-PI3K. In diabetic mice, the phosphorylation of all components of IRSβ-IRS-1-AKT-PI3K pathways is reduced. Treatment of diabetic rates greatly increases the phosphorylation of all components IRSβ-IRS-1-AKT-PI3K pathway and promotes GLUT4 translocation to plasma membrane (Liu et al., 2015). A mixture of DNJ and HDP polysaccharide from mulberry helps in regulating the expression of the hepatic gluconeogenic enzymes, glucokinase, PEP carboxykinase and glucose 6-phosphate. The polysaccharide in this mixture is able to protect pancreatic islets from alloxan induced damage, repair the destroyed pancreatic islets, upregulate the PDX-1, insulin-1 and insulin-2 expression in pancreas and normal secretion of insulin in serum (Li You-Gui et al., 2011).

In Otsuka Long-Evans Tokushima Fatty (OLETF) rats, DNJ treatment shows strong antidiabetic effects with improvement in fasting blood glucose levels, glucose tolerance, and insulin sensitivity. Furthermore, there is a substantial loss of body weight by DNJ seen in these experimental animals (Kong et al., 2008).

Oral administration of a single dose of 0.8 and 1.2 g of DNJ-enriched (1.5%) mulberry leaf powder significantly suppressed the elevation of postprandial blood glucose and secretion of insulin, revealing the physiological impact of mulberry DNJ in human volunteers. This study suggests that the newly developed DNJ-enriched powder can be promoted as a dietary supplement for preventing diabetes mellitus (Kimura et al., 2007).

7.3.1.3 FAGOMINE-3

Fagomine-3 is an iminosugar and is known to impart its antidiabetic activity by inhibiting intestinal α-glucosidase activity. In STZ-induced diabetic mice fagomine-3's antihyperglycemic activity is probably due to its potentiation of insulin release. Additionally, fagomine-3 has been shown to promote glyceraldehyde-induced insulin secretion (Taniguchi et al., 1998).

Several derivatives of fagomine-3 have been chemically synthesized which are much more potent in antidiabetic activity. For example, isofagomine works at an IC50 value of 0.7 μM compared to 200 μM for fagomine-3. Further, experiments in cultured hepatocytes show that isofagomine prevents basal and glucagon-stimulated glycogen degradation at an IC50 of 2–3 μM (Jakobsen et al., 2001).

7.3.2 FLAVONOIDS

The administration of mulberry leaf flavonoids is known to improve the diabetic symptoms in experimental mice. The animals show a significant reduction in blood glucose levels after administration of mulberry flavonoids. The levels of serum glycosylated proteins and glycated hemoglobin, which are the markers of immediate elevation in blood glucose, are markedly decreased. Examination of lipid profile in these animals show the significant decrease in levels of triglycerides, total cholesterol, LDL, and increase in the level of serum HDL. Mulberry flavonoids help to attenuate insulin resistance in these animals. The pancreatic islets which are almost lost in diabetic mice are partially restored with mulberry leaf flavonoids in these experimental mice. The levels of p-AMPK1/2, PGC-1α and

GLUT4 are significantly increased in diabetic mice with, indicating that AMPK-PGC-1α are possible targets of mulberry leaf flavonoids (Meng et al., 2020). Quercetin-19 and rutin-23 are effective inhibitors against **α**-glucosidase. Isoquercitrin-21 inhibits the activity of **α**-glucosidase inside rat intestines besides free radical scavenging activity and superoxide radical-scavenging activity (Shibano et al., 2008). Q3MG-24, a flavonol glycoside improves hyperglycemia in obese mice, and, in humans, may reduce oxidative stress in the liver after daily dietary intake (Katsube et al., 2010). Sanggenon C, sanggenon G, and kuwanon isolated from mulberry have been implicated in antidiabetic activity by their significant inhibitory action against PTP1B (Cui et al., 2006).

7.3.3 POLYSACCHARIDES

The various polysaccharides found in mulberry exert their potential anti-diabetic effect by inhibiting **α**-glucosidase enzyme. In alloxan-induced diabetic mice, the total polysaccharides from mulberry leaves (TPM) increases glucose tolerance and glycogen content. In normal rats, TPM has been found to increase blood insulin levels (Chen et al., 1996). In diabetic rats, treatment with a mulberry leaf polysaccharide (MLPII) inhibits apoptosis in pancreatic islet cells and ameliorates the insulin secretory capacity of pancreatic **β**-cells. (Zhang et al., 2014). The highly-purified polysaccharides α-arabinose, α-xylose, α-glucose, α-rhamnose, and α-mannose extracted from mulberry leaf are able to induce regeneration of pancreatic β-cells. These regenerated β-cells are able to effectively secrete insulin and reduce liver fat accumulation in diabetic rats. Glycosides isolated from a water extract of mulberry leaf are found to have potential type 2 antidiabetic activity (Li et al., 2011).

7.3.4 CHLOROGENIC ACID

Treating STZ-induced diabetic rats with 70% ethanolic fraction contacting chlorogenic acid shows strong antidiabetic activity in a dose dependent manner. The treatment has brought down the random plasma glucose level to approximately normal values in these experimental animals. Further studies are needed to find the cellular targets of these antidiabetic compounds (Hunyadi et al., 2012).

7.4 GENOMIC APPROACHES IN MULBERRY FOR ANTIDIABETIC DRUG DEVELOPMENT

Mulberry is rich in bioactive components of pharmacological importance. There is an increasing attention being paid towards characterization of pharmacological properties of these mulberry compounds. Some of these important bioactive components include DNJ, flavonoids, polysaccharides, and anthocyanins. Recently the whole genome sequencing of the mulberry plant (*Morus notabilis*) has been carried out. The total length of genome deciphered is 330 Mb. The repetitive sequence span over 128 Mb. There are 29,338 genes, out of which ~69% are have been supported by transcriptome sequencing (Hei et al., 2013). The genomic information can be exploited to understand various metabolic and biochemical pathways in mulberry. To make genomics information easier and more useful, an open-access mulberry genomic data base (MorusDB; https://morus.swu.edu.cn) has been established. This database not only provides the information about the mulberry gene sequences, gene structures, transposable elements but also provides information about the transcriptome and EST (expressed sequence tag) data (Li et al., 2014).

Although mulberry derived products are the rich source for antidiabetic drug development (Priya, 2012), the clinical potential and sustained supply of these drugs in the market will be compromised in the future because of low levels of production of these compounds in plant parts. With the sequencing of mulberry whole genome and the information gained from transcriptomic analysis, it would be possible to produce the various antidiabetic compounds in bacteria using recombinant DNA technology. Additionally, RNA interference-based gene silencing and virus-induced gene silencing technology can be used to exploit various biosynthetic metabolic pathways in mulberry (Hei et al., 2013). With the advent of genomics, transgenic mulberry plants can be developed which can accumulate greater quantities of antidiabetic compounds.

7.5 CONCLUDING REMARKS

Mulberry is an economically important plant and is cultivated across the temperate zones in the world. Forming a backbone of sericulture industry, mulberry has also been used for treatment of various ailments in Chinese traditional system of medicine and Indian Ayurveda (Priya, 2012). The recent studies and clinical trials of different mulberry parts (root, fruit, branches, leaf) and isolated compounds in the treatment of diabetes have greatly raised the prospectus of this plant in the area of herbal medicine. The experiments in mice, rats, and human subjects have shown that the hypoglycemic and antidiabetic effect in all parts of the plant is due to the presence of various bioactive compounds or phytochemicals like flavonoids, alkaloids, and polysaccharides (Wei et al., 2018).

Various studies have been performed to elucidate the molecular mechanism of action for mulberry-derived antidiabetic compounds and to identify their cellular targets. For example, DNJ, a polyhydroxylated mulberry alkaloid has been used in human trials and has shown promising antidiabetic results. The DNJ enriched mulberry leaf powder is on the market for the last few years for human use. Mulberry leaf tea, which was previously used in Korea and Japan to control hyperglycemia and gain other health benefits, is also available in the market. Further, studies might help to promote mulberry products or foods supplemented with mulberry compounds for people with type 2 diabetes in their daily glucose control prescription (Kimura et al., 2007).

Though the antidiabetic compounds are present in all parts of the plant, only leaves and fruits could provide the sustainable resource for extraction of those molecules. The recent whole-genome sequencing of mulberry genome has opened new avenues to understand the gene structures and metabolic pathways involved in synthesis of compounds with medicinal values. With the use of DNA recombinant technology, it would be possible to produce some of those compounds in bacterial systems at large scale. Furthermore, the information gained from mulberry genomic resource would be helpful to produce more transgenic varieties of mulberry, which will be able to produce medicinal compounds in larger quantities.

KEYWORDS

- **deoxynojirimycin**
- **flavonoids**
- **GLUT4**
- **hyperglycemia**
- **mulberry**
- **type 2 diabetes**
- **α-glucosidase**

REFERENCES

Andallu, B., & Varadacharyulu, N. C., (2003). Antioxidant role of mulberry (*Morus indica* L. cv. Anantha) leaves in streptozotocin-diabetic rats." *Clinica Chimica Acta: Int. J. Clin. Chem., 338*, 3–10.

Andallu, B., Kumar, A. V., & Varadacharyulu, N. C. (2012). Oxidative stress in streptozocin-diabetic rats: Amelioration by mulberry (*Morus indica* L.) leaves. *Chin. J. Integr. Med.*, 1–6.

Bae, S. H., & Suh, H. J., (2007). Antioxidant activities of five different mulberry cultivars in Korea. *LWT-Food Sci. Technol., 40*, 955–962.

Bala, S., Gupta, S., & Rani, U., (2010). Antidiabetogenic action of *Morus rubra* L. leaf extract in streptozotocin-induced diabetic rats. *J. Pharm. Pharmacol., 62*, 247–255.

Chan, E. W., Lye, P. Y., & Wong, S. K., (2016). Phytochemistry, pharmacology, and clinical trials of *Morus alba. Chin, J. Nat. Med., 14*, 17–30.

Chen, F., Lu, J., & Zhang, Y., (1996). Pharmacological studies on Morus (I): Effect of total polysaccharide of *Morus* (TPM) on carbohydrate metabolism in diabetic mice. *J. Shenyang Pharm. Univ., 13*, 24–26.

Chen, F., Nakashima, N., Kimura, I., Kimura, M., Asano, N., & Koya, S., (1995). Potentiating effects on pilocarpine-induced saliva secretion, by extracts and N-containing sugars derived from mulberry leaves, in streptozocin-diabetic mice. *Biol. Pharm. Bull., 18*, 1676–1680.

Cui, L., Na, M., Oh, H., Bae, E. Y., Jeong, D. G., Ryu, S. E., Kim, S., et al., (2006). Protein tyrosine phosphatase 1B inhibitors from Morus root bark. *Bioorg. Med. Chem. Lett., 16*, 1426–1429.

Duke, J. A., (1992). *Handbook of Edible Weeds* (p. 128). CRC Press Inc, USA.

El-Mawla, A. M. A. A., Mohamed, K. M., & Mostafa, A. M., (2011). Induction of biologically active flavonoids in cell cultures of Morus nigra and testing their hypoglycemic efficacy. *Sci. Pharm., 79*, 951–961.

Fosgerau, K., Westergaard, N., Quistorff, B., Grunner, N., Kristiansen, M., & Lundgren, K., (2000). *Arch Biochem Biophys., 380*, 274.

Han, D. G., Cho, S. S., Kwak, J. H., & Yoon, I. S., (2019). Medicinal plants and phytochemicals for diabetes mellitus: Pharmacokinetic characteristics and herb-drug interactions. *J. Pharma Inves., 49*, 603–612.

Hansawasdi, C., & Kawabata, J., (2006). α-Glucosidase inhibitory effect of mulberry (*Morus alba*) leaves on caco-2. *Fitoterapia., 77*, 568–573.

Harauma, A., Murayama, T., Ikeyama, K., Sano, H., Arai, H., Takano, R., Kita, T., et al., (2007). Mulberry leaf powder prevents atherosclerosis in apolipoprotein E-deficient mice. *Biochem. Biophys. Res. Commun., 358*, 751–756.

He, N., Zhang, C., Qi, X., Zhao, S., Tao, Y., Yang, G., Lee, T. H., et al., (2013). Draft genome sequence of the mulberry tree *Morus notabilis. Nat Commun., 4*, 2445.

Hosseinzadeh, A. S. H., (1999). Antihyperglycemic effects of Morus nigra and Morus alba in mice. *Pharm. Pharmacol. Lett., 9*, 63–65.

Hunyadi, A., Martins, A., Hsieh, T. J., Seres, A., & Zupko, I., (2012). Chlorogenic acid and rutin play a major role in the *in vivo* antidiabetic activity of *Morus alba* leaf extract on type II diabetic rats. *PLoS One, 7*, 50–61.

Imran, M., Khan, H., Shah, M., Khan, R., & Khan, F., (2010). Chemical composition and antioxidant activity of certain Morus species. *J. Zhejiang Univ. Sci. B., 11*, 973–980.

Jakobsen, P., Lundbeck, J. M., Kristiansen, M., Breinholt, J., Demuth, H., Pawlas, J., Candela, M. P., et al., (2001). Iminosugars: Potential inhibitors of liver glycogen phosphorylase. *Bioorg. Med. Chem., 9*, 733–744.

Kadam, R. A., Dhumal, N. D., & Khyade, V. B., (2019). The mulberry, *Morus alba* (L.): The medicinal herbal source for human health. *Int. J. Curr. Microbiol. App. Sci., 8,* 2941–2964.

Katsube, T., Yamasaki, M., Shiwaku, K., Ishijima, T., Matsumoto, I., Abe, K., & Yamasaki, Y., (2010). Effect of flavonol glycoside in mulberry (*Morus alba* L.) leaf on glucose metabolism and oxidative stress in liver in diet-induced obese mice. *J. Sci. Food Agric., 90*, 2386–2392.

Kharroubi, A. T., & Darwish, H. M., (2015). Diabetes mellitus: The epidemic of the century. *World J. Diabetes, 6*, 850–867.

Kimura, T., Nakagawa, K., Kubota, H., Kojima, Y., Goto, Y., Yamagishi, K., Oita, S., Oikawa, S., & Miyazawa, T., (2007). Food-grade mulberry powder enriched with 1-deoxynojirimycin suppresses the elevation of postprandial blood glucose in humans. *J. Agric. Food Chem., 55*, 5869–5874.

Koca, I., Ustun, N. S., Koca, A. F., & Karadeniz, B., (2008). Chemical composition, antioxidant activity and anthocyanin profiles of purple mulberry (Morus rubra) fruits. *J. Food Agri. and Environment, 6*, 39–42.

Kong, W. H., Oh, S. H., Ahn, Y. R., Kim, K. W., Kim, J. H., & Seo, S. W., (2008). Antiobesity effects and improvement of insulin sensitivity by 1-deoxynojirimycin in animal models. *J. Agric. Food Chem., 56*, 2613–2619.

Kumawat, M., Sharma, T. K., Singh, I., Singh, N., Ghalaut, V. S., Vardey, S. K., & Shanker, V., (2013). Antioxidant enzymes and lipid peroxidation in type 2 diabetes millatus patients with and without nephropathy. *North Am. J. Med. Sci., 5*, 213–219.

Li, T., Qi, X., Zeng, Q., Xiang, Z., & He, N., (2014). MorusDB: A resource for mulberry genomics and genome biology. *Database, 2014*. bau054.

Li, Y. G., Dong-Feng, J., Zhong, S., Lv, Z. Q., Lin, T. B., Chen, S., & Hu, G. Y., (2011). Hybrid of 1-deoxynojirimycin and polysaccharide from mulberry leaves treat diabetes mellitus by activating PDX-1/insulin-1 signaling pathway and regulating the expression of glucokinase, phosphoenol-pyruvate carboxykinase and glucose-6-phosphatase in alloxan-induced diabetic mice. *J. Ethnopharmacol., 134,* 961–970.

Lim, S. H., & C-Ik. C., (2019). Properties of *Morus nigra* L. (black mulberry) as a promising nutraceutical resource. *Nutrients, 11*, 437.

Liu, H. Y., Fang, M., & Zhang, Y. Q. (2014). *In vivo* hypoglycaemic effect and inhibitory mechanism of the branch bark extract of the mulberry on STZ-induced diabetic mice. *The Scientific World Journal, 2014*, pp. 1–11.

Liu, Q., Li, X., Li, C., Zheng, Y., & Peng, G., (2015). 1-deoxynojirimycin alleviates insulin resistance via activation of insulin signaling PI3K/AKT pathway in skeletal muscle of dB/dB mice. *Molecules, 20*, 21700–21714.

Mackay, P., Ynddal, L., Andersen, J. V., & McCormack, J. G., (2003). Pharmacokinetics and anti-hyperglycemic efficacy of a novel inhibitor of glycogen phosphorylase,1,4-dideoxy-1,4-imino-Darabinitol, in glucagon-challenged rats and dogs and in diabetic OB/OB mice. *Diabetes, Obesity, and Metabolism, 5*, 397–407.

Mata, A. D., Marques, D., Rocha, S., Francisco, H., Santos, C., Mesquita, M. F., & Singh, J., (2004). Effects of diabetes milletus on salivary secretion and its composition in the human. *Mol. Cell Biochem, 261*, 137–142.

Meng, Q., Qi, X., Fu, Y., Chen, Q., Cheng, P., Yu, X., Sun, X., et al., (2020). Flavonoids extracted from mulberry (*Morus alba* L.) leaf improve skeletal muscle mitochondrial function by activating AMPK in type 2 diabetes. *J. Ethnopharmacol., 10*, 112–326.

Mohammadi, J., & Naik, P. R., (2012). The histopathologic effects of Morus alba leaf extract on the pancreas of diabetic rats. *Turk. J. Biol., 36*, 211–216.

Naowaboot, J., Pannangpetch, P., Kukongviriyapan, V., Kukongviriyapan, U., Nakmareong, S., & Itharat, A., (2009). Mulberry leaf extract restores arterial pressure in streptozotocin-induced chronic diabetic rats. *Nutr. Res., 29*, 602–608.

Oh, N. H., Han, J. W., Shim, D. W., et al., (2015). Anti-inflammatory properties of Morus bombycis Koidzumi via inhibiting IFN-β signaling and NLRP3 inflammasome activation. *J. Ethnopharmacol., 24*, 424–428.

Ojewole, J. A., Musabayane, C. T., & Bwititi, P. T., (2006). Effects of *Persea americana* mill (Lauraceae) ethanolic leaf extract on blood glucose and kidney function in streptozotocin-induced diabetic rats and on kidney cell lines of the proximal (LLCPK1) and distal tubules (MDBK). *Methods Find. Exp. Clin. Pharmacol., 28*, 2–23.

Park, J. M., Bong, H. Y., Jeong, H. I., Kim, Y. K., Kim, J. Y., & Kwon, O., (2009). Postprandial hypoglycemic effect of mulberry leaf in goto-kakizaki rats and counterpart control Wistar rats. *Nutr. Res. Pract., 3*, 272–278.

Priya, S., (2012). Medicinal values of mulberry: An overview. *Journal of Pharmacy Research, 5*, 3588–3596.

Riche, D. M., Riche, K. D., East, H. E., Barrett, E. K., & May, W. L., (2017). Impact of mulberry leaf extract on type 2 diabetes (Mul-DM): A randomized, placebo-controlled pilot study. *Complementary Therapies in Med., 32*, 105–108.

Shibano, M., Kakutani, K., Taniguchi, M., Yasuda, M., & Baba, K., (2008). Antioxidant constituents in the dayflower (*Commelina communis* L.) and their α-glucosidase-inhibitory activity. *J. Nat. Med., 62*, 349–353.

Singab, A. N. B., El-Beshbishy, H. A., Yonekawa, M., Nomura, T., & Fukai, T., (2005). Hypoglycemic effect of Egyptian *Morus alba* root bark extract: Effect on diabetes and lipid peroxidation of streptozotocin-induced diabetic rats. *J. Ethnopharmacol., 100*, 333–338.

Singh, R., Bagachi, A., Semwal, A., & Bharadwaj, A., (2013). Traditional uses, phytochemistry, and pharmacology of *Morus alba* Linn.: A review. *J. Med. Plants Res., 7*, 461–469.

Tanabe, K., Nakamura, S., Omagari, K., & Oku, T., (2011). Repeated ingestion of the leaf extract from Morus alba reduces insulin resistance in KK-Ay mice. *Nutr. Res., 31*, 848–854.

Taniguchi, S., Asano, N., Tomino, F., & Miwa, I., (1998). Potentiation of glucose-induced insulin secretion by fagomine, a pseudo-sugar isolated from mulberry leaves. *Horm. Metab. Res., 30*, 679–683.

Tian, S., Tang, M., & Zhao, B., (2016). Current antidiabetes mechanisms and clinical trials using *Morus alba* L. *J. Trad. Chin. Med. Sci., 3*, 3e8.

Venkatesh, K. R., & Chauhan, S., (2008). Mulberry: Life enhancer. *J. Med. Plants Res., 2*, 271–278.

Wang, Y., Xiang, L., Wang, C., Tang, C., & He, X., (2013). Antidiabetic and antioxidant effects and phytochemicals of mulberry fruit (Morus alba L.) polyphenol enhanced extract. *PLoS One, 8*, 1–10.

Wei, H., Liu, S., Liao, Y., Ma, C., Wang, D., Tong, J., Feng, J., et al., (2018). A systematic review of the medicinal potential of mulberry in treating diabetes mellitus. *The Am. J. Chin. Med., 46*, 1743–1770.

Yang, X., Yang, L., & Zheng, H., (2010). Hypolipidemic and antioxidant effects of mulberry (*Morus alba* L.) fruit in hyperlipidaemia rats. *Food Chem. Toxicol., 48*, 2374–2379.

Zhang, Y., Ren, C., Lu, G., Mu, Z., Cui, W., Gao, H., & Wang, Y., (2014). Antidiabetic effect of mulberry leaf polysaccharide by inhibiting pancreatic islet cell apoptosis and ameliorating insulin secretory capacity in diabetic rats. *Int. Immunopharmacol., 22*, 248–257.

Zierath, J. R., (2019). Major advances and discoveries in diabetes-2019 in review. *Current Diabetes Rep.,* 19.

CHAPTER 8

The Antidiabetic Effect of *Nigella sativa* L. with Respect to Its Phytochemicals and the Mechanism of Action

IFFAT ZAREEN AHMAD

Natural Products Laboratory, Department of Bioengineering, Integral University, Dasauli, Kursi Road, Lucknow – 226026, Uttar Pradesh, India, E-mail: iffat@iul.ac.in

ABSTRACT

Nigella sativa is a very important seed with immense pharmacological activity. It possesses enormous antidiabetic activity because of the bioactive compounds it harbors. *N. sativa* is a very popular spice which is used in many Indian cuisines therefore it is a very common nutraceutical and the exploration of its antidiabetic activity could be an excellent contribution to the field of medical science. Because of the toxicity and adverse repercussions of chemical drugs, formulations based on herbal drugs will be a useful alternative for the management of diabetes after deducing its action mechanism and thorough clinical trials. The main antidiabetic compounds of *N. sativa* which have shown antidiabetic properties include thymoquinone, thymol, thymohydroquinone, dithymoquinone, nigellone, alpha-hederin, carotenoids, flavonoids, and fatty acids. *N. sativa* oil, the seed extracts and purified bioactive compounds have shown remarkable antidiabetic activity in both the *in vitro* and *in vivo* research.

The antidiabetic effect can be attributed to the strong antioxidant activities of its phytochemicals, decreased gluconeogenesis, and mimic

Antidiabetic Plants for Drug Discovery: Pharmacology, Secondary Metabolite Profiling, and Ingredients with Insulin Mimetic Activity. Deepu Pandita, MPhil, Anu Pandita, MSc, & Chander Bhanu, MBBS, PGDDCE (Eds.)
© 2023 Apple Academic Press, Inc. Co-published with CRC Press (Taylor & Francis)

activity similar to insulin at both the cell and molecular levels in different organs. The study has been undertaken using omics approach in which the exploration of enzyme dipeptidyl peptidase-4 inhibitors was done by intrinsic inhibitors. The mechanism of the glucose-lowering effects of thymoquinone using *in silico* tools has been studied. *N. sativa* is a promising medicinal seed and its bioactive compounds have the prospective of being important leads to synthesize drugs having antidiabetic activities.

8.1 INTRODUCTION

Diabetes mellitus is the most prevalent ailments in which metabolic processes are disturbed. It affects individuals of every sex, age, and race is affected. It has resulted into morbidity and mortality due to various complications (Patel et al., 2011a). Human bodies have antioxidative defense mechanisms comprising strategies which may or may not involve enzymes. It reduces the production of reactive oxygen species, accountable for numerous deteriorating ailments together with diabetes (Patel et al., 2011b). Diabetes is quickly increasing throughout the world. Person affected with diabetes have high glucose level in blood as a consequence of deficit of the insulin (Ponnusamy et al., 2011). The prevalent form of the disease is Type 2 diabetes or noninsulin-dependent diabetes mellitus and it is responsible for 90%–95% of cases (Li et al., 2004). WHO states that by the year 2025, the patients with diabetes are expected to rise up to 300 million (Sy et al., 2005).

Herbal medicines that are derived from medicinal plants have gained significance since they are effective with minimum side effects and less explored scientifically. Nigella sativa belongs to the Ranunculaceae family and has a wide array of therapeutic properties which it possesses. The common name of *N. sativa* is a black seed, and it is native to Southwest Asia, North Africa, and Southern Europe and is cultured in Saudi Arabia, the Middle Eastern Mediterranean region, India, Pakistan, South Europe, Turkey, and Syria (Khare, 2004). It is used globally as a flavoring agent in cuisines and as a spice (Al-Ali et al., 2008). It has found applications in conventional medicine systems like Unani, Ayurveda, Homeopathy, and Siddha (Sharma et al., 2005; Goreja et al., 2003). Seeds and oil are generally used in medicines and food. Its reference has been given in In Islamic literature as one of the supreme forms of medicinal herb to the extent that it is said to be the cure of every disease except death as mentioned in Tibb-e-Nabwi, Prophetic Medicine (Al-Bukhari, 1976). It has an array of pharmacological activities including antihypertensive, diuretics, digestive, anti-diarrheal, antimicrobial, bronchodilator, immunomodulator, anti-inflammatory, analgesic, antidiabetic, anticancer, hepatoprotective, gastro-protective, Kidney protective, antioxidant, and in treating skin disorders (Abel-Salam, 2012; Khaled, 2009; Assayed, 2010; Abdel-Zaher et al., 2011; Boskabady et al., 2010; Goreja, 2003).

8.2 STUDIES ON ANTIDIABETIC ACTIVITIES OF *NIGELLA SATIVA*

The study by Al-Awadi and Gumma (Al-Awadi and Gumma, 1987) have studied a mixture comprising of *N. sativa*, gum, myrrh, aloe, and asafetida on diabetic rats in Kuwait on bringing down the glucose level and showed its effectiveness. The mixture comprising *N. sativa* showed that bringing down of blood glucose level was because of the retardation of liver gluconeogenesis and the mixture of plant extracts may be beneficial in curing non-insulin-dependent diabetes mellitus (Al-Awadi et al., 1991; Mohamed et al., 2009). The *N. sativa* oil showed a considerable hypoglycemic effect on both normal as well as diabetic rabbits induced by alloxan with no modification in insulin concentrations (Al-Hader et al., 1993). The study of Matira et al. (2008) established the ameliorative effects of raw concentrate and n-hexane distillate of *N. sativa* seed against diabetes. The clinical study showed considerable betterment with respect to cumulative cholesterol, low density lipoprotein (LDL) cholesterol and fasting blood glucose (FBG) demonstrating effectiveness against insulin resistance disorder (Najmi et al., 2008). Nadia and Taha (2009) studied *N. sativa* oil (NSO) and thymoquinone (TQ) for their effect on oxidative stress and neuropathy in streptozotocin (STZ)-induced diabetic rats. The data of the study showed considerable increment in concentrations of norepinephrine and dopamine and a considerable reduction in levels of serotonin in comparison to the control.

The crude distillate of *N. sativa* inhibited the electrogenic in vitro uptake of sugar in intestinal (Meddah et al., 2009). The therapy with *N. sativa* plus TQ lessened the blood sugar and enhanced serum insulin in STZ-induced diabetic rats. The treatment also affected the insulin immunoreactive beta-cells markedly (Kanter, 2008). Ethanol extract of *N. sativa* seeds considerably decreased the blood sugar levels, fats, plasma insulin and enhanced changed amount of products of lipid peroxidation and antioxidant enzymes including superoxide dismutase (SOD), catalase (CAT), reduced glutathione (GSH) and glutathione peroxidase (GPx) in liver and kidney in STZ-induced diabetic rats when given orally (Kaleem et al., 2006; Kanter et al., 2003).

Noticeable fall in the level of blood sugar and enhancement in serum insulin concentration were seen after administering NSO for four weeks. The pancreas of *N. sativa* oil-treated group was studied after the immune histochemical staining. The large regions showed positive immunoreactivity for the occurrence of insulin. The current research was carried out to explore the probable insulinotropic activities of NSO in STZ and nicotinamide (NA)-induced diabetes in hamsters. NA was given by injection little earlier than STZ in the hampers were administered with *N. sativa* oil orally for four weeks after inducing them with diabetes. The level of serum insulin was evaluated by enzyme immunoassay. anti-insulin monoclonal antibody was used to stain Islets. Following the administration of NSO for four weeks considerable decrement in the amount of plasma glucose and marked rise in the amount of serum insulin were seen. Big regions showing positive immuno-reactivity for the occurrence of insulin were seen in the pancreases after immunohistochemical staining. Thus, the results revealed that the glucose-lowering activity of NSO in STZ plus NA diabetic hamsters was because of the induction of β cell function with subsequent enhancement in serum insulin level (Fararh et al., 2002).

To reveal the mechanism of the glucose-lowering effect of *N. sativa* oil on STZ-induced diabetic hamsters, by evaluating the glucose level in the liver, and to explore the probable immune-stimulating influence of NSO on peritoneal macrophages, the present study was undertaken. After administering with NSO for six weeks following onset of disease, liver cells were taken up using collagenase enzyme to evaluate glucose production in hepatic cells. The activity of phagocytes was evaluated by injecting fluorescent latex followed by collection of peritoneal macrophages after 24 h. It was seen that the activity of phagocytes and its index of peritoneal macrophages and lymphocyte number were increased considerably. *N. sativa* oil reduces the glucose level in blood. Also, the gluconeogenic pathway was suppressed and the production of glucose from precursors was less in liver cells in treated hamsters. The results showed that the hypoglycemic activity of NSO is because of reduction in liver gluconeogenesis, and the immunostimulatory response of NSO is arbitrated by providing stimulus to the activity of phagocytic macrophage (Fararh et al., 2004).

The aim of this investigation was to study the promising defensive activity of *N. sativa* towards β-cell damage from STZ-induced diabetic rats. *N. sativa* was given for three days before the administration of STZ, and were continuously given during the four-week period. Oxidative stress is supposed to have a part in the disease progression of diabetes mellitus. To estimate the alterations in the antioxidative defense mechanism at the cell level, the activities of antioxidative enzymes were estimated in pancreas. Also, nitric oxide in the serum and MDA levels in erythrocyte and pancreas to check the disparity between the levels of oxidants and antioxidants. The immunohistochemical methods were used to observe the pancreatic β-cells. It was seen that STZ has induced a considerable amount of lipid peroxidation products and NO levels in serum, and it has reduced the activities of antioxidant enzymes. NS administration was seen to offer a protecting activity by decreasing lipid peroxidation and NO level in serum, and enhancing the activity of antioxidant enzymes. In immunohistochemical staining, the disintegration of Islet cells and less insulin staining was seen in rats with STZ-induced diabetes. Amplified insulin staining intensity and conservation of beta-cell were obvious in diabetic rats administered with *N. sativa* (Kanter et al., 2004).

Amongst many bioactive compounds present in *N. sativa*, Thymoquinone (TQ), which is a volatile mono oxygenated terpene has been reported to possess promising hypoglycemic activity. Besides TQ, thymol, thymohydroquinone, dithymoquinone, nigellone, alpha-hederin, carotenoids, flavonoids, and fatty acids are also responsible for the activity of *N. sativa* against diabetes (Figure 8.1). The curative activity of α-lipoic acid, L-carnitine, and *N. sativa* or their mixture in glucose and fat metabolism was studied in STZ-induced diabetic rat model. For assessment of carbohydrate metabolism, the activity pyruvate dehydrogenase along with FBG, insulin, and its sensitivity, HOMA, C-peptide were estimated. Amalgamation of α-LA, L-carnitine, and *N. sativa* contributed markedly in monitoring of the glucose metabolism in diabetic rats (Salama, 2011). The effects of aqueous extract of *N. sativa*, its oil and TQ was studied on serum insulin and sugar levels in STZ-induced rats with diabetes. Administration of rats with *N. sativa* extract, oil, and TQ markedly reduced the MDA in tissue and serum glucose resulted due to diabetes and markedly increased insulin level in serum and tissue SOD. In histopathology study, it was shown that TQ amended the toxic effects of STZ to a great extent followed by water extract of *N. sativa*. The NSO reinstated standard insulin levels, but not successful in decreasing serum glucose concentration. The data showed that *N. sativa* extract and TQ displayed protection against STZ-induced diabetes by reducing oxidative stress and restoring β-cell function in pancreas (Abdelmeguid et al., 2010). The defensive property of NSO on insulin immunoreactivity and pancreatic β-cells as seen by ultrastructural changes in STZ-induced diabetic rats was described (Kanter et al., 2009). Insulin was stained intensely and β-cell titer was preserved in the diabetic rats provided with *N. sativa*. The protecting activity of *N. sativa* was obvious by a reasonable rise in the granular secretory vesicles and also minor damage with cristae damage in the β-cell mitochondria of diabetic rats (Kanter et al., 2009). The antidiabetic property of TQ was studied on the main enzymes of sugar metabolism and their activity of in STZ-nicotinamide (NA)-induced diabetic rats. Dosage of TQ improved the glucose levels in STZ-NA induced diabetic rats when given orally. The concentration of insulin, hemoglobin (Hb) rose with considerable reduction in glucose and HbA levels. The modified activities of enzymes of sugar metabolism were reinstated to almost standard. These data showed that TQ was linked to positive alterations in liver enzymatic activities and thus employs possible anti-hyperglycemic properties (Pari and Sankaranarayanan, 2009). The *N. sativa* exhibited the interactive influence with parathyroid hormone of human in the betterment of bone mass, connectivity, biomechanical behavior, and power in insulin-dependent diabetic rats and was shown to be highly efficient in comparison to the usage with *N. sativa* or human parathyroid hormone separately (Altan et al., 2007). In a study with clinical samples, the adjuvant activity of NSO on insulin resistance syndrome were investigated by taking different clinical and biochemical factors. NSO was seen to be active as a supplementary medication in patients with insulin resistance syndrome. NSO has a considerable effect in patients having diabetes and dyslipidemic syndrome (Najmi et al., 2010). *N. sativa* is of enormous medicinal advantage in persons having diabetes and those dealing with intolerance to glucose as it stresses on production of insulin-induced by glucose in addition to suppressing absorption of glucose from the mucosa of the intestine (Kapoor, 2009). The response of TQ on STZ-induced diabetic rats was explored. The response of seeds of *N. sativa* on the management of sugar in patients with type 2 diabetes mellitus was explored. *N. sativa* seeds were used as a supplementary treatment in persons with type 2 diabetes mellitus in addition to their antidiabetic medicines. When administered with the dosage of 2 gm/day, *N. sativa* showed marked decrease in FBG, 2 h after the meal, and glycosylated HbA1c with no substantial variation in body weight (BW). The data revealed that this dose of *N. sativa* might be an advantageous supplementary treatment to medicines given orally to diabetic patients (Bamosa et al., 2010). The antidiabetic property of *N. sativa* ethanol extract of seed (NSE) was studied *in vivo* in diabetic *Meriones shawi*. Plasma lipid profile, insulin, leptin, and adiponectin parameters were measured. ACC phosphorylation and Glut4 protein concentration were evaluated in hepatic and skeletal muscle. The rats displayed an improvement in glycemia when administered with *N. sativa*. It was also shown that treatment with *N. sativa* applies an insulin-stimulating action by inducing ACC phosphorylation, a main constituent of the insulin-independent AMPK signaling pathway, and by rising concentration of muscle Glut4 (Benhaddou-Andaloussi et al., 2011).

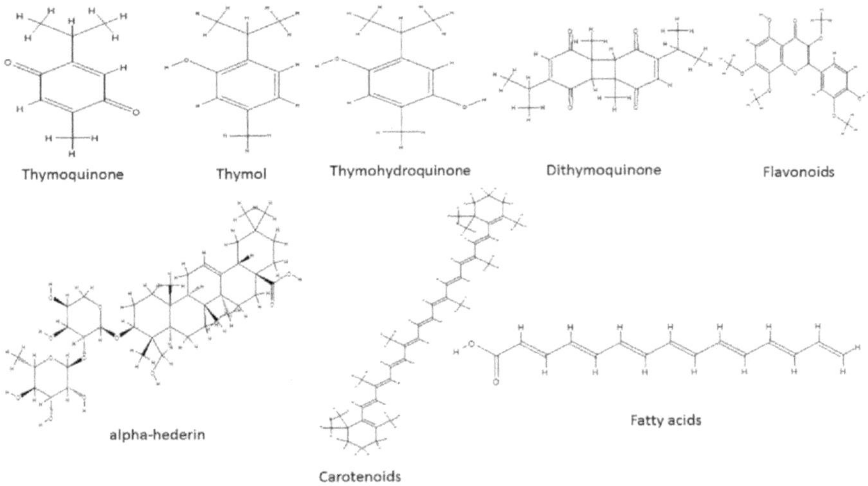

FIGURE 8.1 Chemical structures of the antidiabetic phytochemicals of *N. sativa*.

TQ is a constituent of the NSO. It is one of the potent constituents showing antidiabetic property. But the antidiabetic property has also been reported in its water and also extract of *N. sativa* which is free of fats. *N. sativa* may be advantageous in people diabetes and those having intolerance to glucose as it decreases hunger, absorption of intestinal glucose, liver gluconeogenesis, blood sugar level, cholesterol, triglycerides, body weight and induces insulin secretion induced by glucose in pancreatic β-cells; mends tolerance to glucose as proficiently as metformin. Also, it has not presented considerable side reactions and is very less toxic. In a study comprising of STZ-induced diabetes in rats it partially regenerated pancreatic β-cells, enhanced the serum insulin level, and decreased the raised serum glucose. The antioxidative effect and defensive function of TQ in the progression of type I diabetes may be via inhibitory pathway NO. It also applies an insulin-sensitizing act in liver cells (Murli et al., 2011).

The hydroalcoholic distillate of *N. sativa* was evaluated for its effect on hypoglycemic and hypolipidemic ailment in STZ-induced diabetes in rats. It was observed that *N. sativa* considerably reduced the level of sugar in blood in comparison to diabetic group. The serum glutamic-pyruvic transaminase (SGPT), serum glutamic-oxaloacetic transaminase (SGOT) and C-reactive protein (CRP) were also reduced markedly. It can be said that *N. sativa* might be active in treating liver disfunction. A suggestive, antilipidemic activity was also seen as total cholesterol (TC), total glyceride (TG), LDL, very-low-density lipoprotein (VLDL) displayed substantial reduction whereas HDL showed considerable increase by *N. sativa* treatment (Asaduzzaman et al., 2015).

The current study investigated the permanent glucose-lowering activity of *N. sativa* in persons with type 2 diabetes mellitus taking antidiabetic medicine orally and to explore its role on oxidative status of such persons. There was a considerable decrease in FBG, HbA1c and TBARS but a considerable rise in total antioxidant capacity (TAC), SOD, and glutathione in *N. sativa* treated persons in comparison to control. In this group, insulin resistance was markedly reduced, while activity of β-cell was markedly greater than the standard digits throughout the complete tenure of therapy (Katabi et al., 2015).

The characteristics of *N. sativa* on glucose level balance, fat profile and metabolic parameters in diabetes mellitus were reviewed. The particular studies showed that *N. sativa* can reduce blood glucose and lipid profile malfunction with different potent strategies which involves its antioxidative properties and role in secretion of insulin, absorption of glucose, gluconeogenesis, and expression of gene. Few researchers made a comparison of the properties of different types of *N. sativa* samples, and they described dissimilar features with different kinds of *N. sativa* seed. *N. sativa* can recover blood glucose level and lipid profile in diabetic models (Heshmati and Namazi, 2015).

The research was conducted to study the response of hydroalcoholic distillate of *N. sativa* on the concentrations of glucose in STZ-induced diabetes in rats. The rats were evaluated after 32 days of administration with *N. sativa* extract to estimate its response on FBG. At the completion of the work, the animals were on fasting overnight, then given anesthesia and sacrificed to obtain samples of liver and pancreas tissues. The islets number and cells were totaled and the diameters of the islet were calculated in micrometer. The level of glycogen was estimated in the hepatic by Periodic Acid-Schiff (PAS) staining. The data showed a considerable increase in the BW after treating with *N. sativa* extract and the reduction in FBG level. Histopathological study showed the partial recovery of glycogen level in the liver and the protection of the pancreatic islet cells (Alimohammadi et al., 2013).

The other study showed the medicinal properties of *N. sativa* and *Propolis* methanol extract was compared on STZ-induced diabetes in male rats and considering diabetic nephropathy. Administering *N. sativa* methanol extract to the diabetic rats with transformed biochemical and pathological investigations reaching the negative control (Haddad et al., 2015).

The study was done to explore the hypoglycemic property of *N. sativa* water extract. The extract exhibited no changes in urea and albumin level at all dosages. But, in diabetic rats, the extract considerably reduced blood sugar, TG, T-cholesterol, LDL-c, and TBARS and exhibited a reinstated insulinemia and a substantial rise in HDL-c. The data on the liver showed a reduction in fats and a probable glycogenogenesis (Bensiameur-Touati et al., 2017).

The research work was performed to explore the response of *N. sativa* extract having TQ on female diabetic rats. The data has shown a substantial decrease in blood sugar in case of high dose of treatment group with *N. sativa* extract as compared to low dose. Also, very high reduction in body weight and insulin levels was seen. The given data suggested a glucose-lowering effect of *N. sativa* extract may be ascribed through a reduction in liver gluconeogenesis (Abduallah, 2017).

The subacute toxicity of *N. sativa* aqueous extract (AqE.NS) was explored and the lowest dose was selected to study its antidiabetic activity. After giving different dosages of the water extract to mice, biochemical parameters were evaluated and histological study was performed on liver and kidney. After that, diabetes was induced using alloxan and the rats were administered water extract for 8 weeks. Sera were collected for the evaluation of metabolic parameters. AqE.NS exhibited no changes in urea and albumin at all the dosage, but liver toxicity was established by histopathological micrographs of hepatic cells. In rats having diabetes, AqE.NS considerably reduced glucose level, TG, T-cholesterol, LDL-c, and TBARS and displayed a reinstated insulinemia and a marked enhancement in HDL-c. The data on the liver showed a decreased lipid profile and a probable glycogenogenesis (Bensiameur-Touati et al., 2017).

The antidiabetic activity of NSO was studied on male Wistar rats on STZ-induced diabetes model. The rats were administered with NSO for 30 days orally. Fasting glucose level, insulin, and lipid contents were estimated. Antioxidant activity in terms of catalase and GSH were estimated in pancreas and liver tissues. The tissues were observed under microscope for histopathological studies, glycogen contents in liver, immunohistochemistry of insulin and morphometry of pancreatic islet were carried out. NSO has markedly decreased, fasting sugar content in blood and increased insulin concentration in rats with diabetes when compared with diabetic rats without medication. Lipid profile was also significantly improved in diabetic rats receiving oil. The antioxidant enzyme activities were enhanced in pancreas and liver both. The histopathology also showed improved picture in *N. sativa* oil-treated groups. It decreased haptic glucose content, immunoreactive% of insulin, and mean diameter of pancreatic islet (Abdelrazek et al., 2018).

In a study, *N. sativa* and *Cinnamomum cassia's* extract separately and in synergism with metformin was studied for its prospective in alleviating signs of diabetes mellitus, and as a supplementary medicine with metformin was carried out. Combinatorial effect of *N. sativa* and *Cinnamomum cassia* extracts considerably controlled glucose levels in blood, fat profile and kidney function parameters in comparison to the control group. Also, considerable reverse in the pancreatic cell impairment was seen on management with the mixture of extracts (Kaur et al., 2018).

The study was planned to estimate the glucose-lowering efficacy of hydro-alcoholic seed extract of *N. sativa* in rat model. The body weight, Oral Glucose Tolerance Test (OGTT), blood sugar content, SGOT, and SGPT contents were estimated. The data showed significant positive antidiabetic effect in dose-dependent manner (Khan and Saxena, 2018).

In a research work, *N. sativa* was explored as a single medication for freshly detected type 2 diabetes mellitus (T2DM) persons. The patients were arbitrarily allocated to take metformin tablets or NSO capsules separately in a hospital. The two groups of T2DM patients were given treatment for three months. *N. sativa* was not better than metformin in lowering fasting blood glucose (FBG) in newly diagnosed T2DM patients, 2 h pp, and A1C or rising secretory function. But it was equivalent to metformin in reducing weight, waist circumference (WC), and body mass index (BMI) considerably. NS was equally efficient as metformin with respect to its activity on fasting insulin, insulin sensitivity, insulin resistance (IR), alanine transaminase (ALT), total cholesterol (TC), LDL, HDL, total glyceride (TG) and total antioxidant capacity (TAC). Metformin displayed considerable increment in aspartate aminotransferase (AST) and creatinine which was kept in *N. sativa* group (Moustafa et al., 2019).

N. sativa was reported to have potential control on diabetes under laboratory conditions. After dosing with *N. sativa*, there was a substantial decrement in FBG, blood glucose level 2 h after the meals, glycated Hb, and insulin resistance, and an increment in serum insulin. The data proposed that *N. sativa* could be applied as a supplementary for hypoglycemic drugs taken orally in controlling diabetes (Hamdan et al., 2019).

In research, the influence of NSO on alloxan-induced diabetic rabbits was explored. The dosage considerably decreased sugar levels in blood and fat profile, but augmented the mean body weight, HDL-C, and vitamin C levels of diabetic rabbits. It also markedly reduced CAT activity, TC, TGs, LDL-C, and VLDL-C levels, but regulated bilirubin levels in diabetic rabbits (Table 8.1) (Akhtar et al., 2020).

TABLE 8.1 The Antidiabetic Activities of *N. sativa* and its Phytochemicals on Different Models of Study

SL. No.	Extract/ Compound	Model	Effects	References
1.	*N. sativa*, gum, myrrh, aloe, and asafetida	Rat	Increased glucose tolerance	Al-Awadi and Gumma (1987)
2.	*N. sativa* seeds	STZ-induced diabetic rat	Blood glucose and fructosamine ↓ Hb and albumin levels ↓ Hepatic glycolytic enzymes ↑ Gluconeogenic enzyme ↓ Antioxidant activity ↑	Mohamed et al. (2009)

TABLE 8.1 *(Continued)*

SL. No.	Extract/ Compound	Model	Effects	References
3.	NSO	Alloxan-induced diabetes in rabbits	FBG ↓	Al-Hader et al. (1993)
4.	Raw and n-hexane extract of *N. sativa*	STZ-induced diabetic male rat	Serum glucose and cholesterol concentrations ↓ Serum triglyceride concentrations ↓ Pancreatic GSH ↑ Reappearance of Î² cells in pancreas.	Matira (2008)
5.	NSO	60 patients	Total cholesterol ↓ LDL cholesterol ↓ Fasting glucose level in blood ↓	Najmi et al. (2008)
6.	NSO and TQ	STZ-induced diabetic male rat	Muscle and brain types cardiac creatine kinase ↓ Improvement in GST, GSH, and catalase ↑ NO and malondialdehyde concentrations ↓	Nadia and Taha (2009)
7.	Crude aqueous extract of *N. sativa* seeds	Rat	Electrogenic glucose absorption in intestine *in vitro* ↓ Sodium-dependent glucose transport in separated rat jejunum ↓ Better glucose tolerance ↑	Meddah et al. (2009)
8.	NS and TQ	STZ-induced diabetic rats	Serum glucose ↓ Serum insulin concentrations ↑ Area of insulin immunoreactive β-cells ↑	Kanter (2008)

TABLE 8.1 *(Continued)*

SL. No.	Extract/ Compound	Model	Effects	References
9.	*N. sativa* seeds	STZ-induced diabetic rats	↓ Blood glucose, lipids, plasma insulin, lipid peroxidation products ↑ SOD, CAT, GSH, and GPx in liver and kidney ↑	Kaleem et al. (2006)
10.	*N. sativa*	STZ-induced diabetic rats	Blood glucose level ↓ Serum insulin level ↑ Rejuvenation of β-cells in pancreas partially	Kanter et al. (2003)
11.	*N. sativa*	STZ and nicotinamide induced diabetic hamsters	Serum glucose ↓ Blood insulin level ↑ Large areas showing positive immuno-reactivity for the existence of insulin	Fararh et al. (2002)
12.	NSO	STZ-induced diabetic hamsters	Glucose synthesis in liver from gluconeogenic precursors ↓ Phagocytic activity and index of peritoneal macrophages ↑ Lymphocyte titer in peripheral blood ↑	Fararh et al. (2004)
13.	*N. sativa*	STZ-induced diabetes in rats	Lipid oxidation and serum NO ↓ Activity of antioxidant enzymes ↑ Staining intensity for insulin ↑ β-cell numbers conservation ↑	Kanter et al. (2004)

TABLE 8.1 *(Continued)*

SL. No.	Extract/ Compound	Model	Effects	References
14.	*N. sativa* water extract, NSO, TQ	STZ-induced diabetic rats	MDA in tissue ↓ Serum glucose ↓ SOD in tissue and serum insulin ↑ Improvement of the toxic effects of STZ by TQ	Abdelmeguid et al. (2010)
15.	α-lipoic acid (α-LA), L-carnitine, *N. sativa*	STZ-induced diabetes in rats	Blood glucose level ↓ Insulin and C-peptide contents ↑ Activity of antioxidant enzymes ↑ Oxidative damage of DNA ↓ Increase of HOMA-β in G3 and G6 ↑	Salama (2011)
16.	NSO	STZ-induced diabetic rats	Staining intensity of insulin and conservation of β-cell count ↑ Less granular secretory vesicles ↑ Cristae damage in the mitochondria of β-cell ↓	Kanter et al. (2009)
17.	TQ	(STZ)-nicotinamide (NA)-induced diabetic rats	Insulin and Hb levels ↑ Glucose and HbA(1C) levels ↓ The modified carbohydrate metabolism enzyme activities were reinstated to near normal.	Pari and Sankaranarayanan (2009)

TABLE 8.1 *(Continued)*

SL. No.	Extract/ Compound	Model	Effects	References
18.	NS and human parathyroid hormone (hPTH)	(STZ)-induced diabetic rats	Insulin immunoreactive β-cells area ↑	Altan et al. (2009)
19.	*N. sativa* capsule	94 patients	FBG, 2hPG, and HbA1 ↓	Bamosa et al. (2010)
			Fasting blood glucose was reduced ↓	
			HbAlC was reduced ↓	
			Insulin resistance calculated by HOMA2 ↓	
			β-cell function ↑	
20.	*N. sativa* seed ethanol extract	*Meriones shawi*	Regularization of glycemia in animals	Benhaddou-Andaloussi et al. (2011)
			Insulinemia and HDL-cholesterol ↑	
			Leptin and adiponectin were unchanged.	
			OGTT, hepatic, and muscle triglyceride levels ↓	
			Muscle and liver ACC phosphorylation and muscle Glut4 ↑	
21.	*N. sativa* seed hydroalcoholic extract	STZ-induced diabetes in rats	Blood glucose level ↓	Asaduzzaman et al. (2015)
			SGPT, SGOT, and CRP ↓	
			TC, TG, LDL, VLDL ↓	
			HDL ↑	

TABLE 8.1 *(Continued)*

SL. No.	Extract/ Compound	Model	Effects	References
22.	*N. sativa*	114 type 2 diabetic patients	FBG, HbA1c, TBARS ↓ TAC, SOD, and glutathione ↑ Insulin resistance ↓ Activity of β-cell ↑	Kaatabi et al. (2015)
23.	*N. sativa* hydroalcoholic extract	STZ-induced diabetes in rats	BW ↑ FBG level ↓ Partial recovery of hepatic glycogen content Protection of the pancreatic islet cells.	Alimohammadi et al. (2013)
24.	AqE.NS	Alloxan-induced diabetes in rats	TG, T-cholesterol, LDL-c, and TBARS ↓ HDL-c ↑ Lipids ↓	Bensiameur-Touati et al. (2017)
25.	*N. sativa* extract	STZ-induced diabetic rodentsis	Blood glucose ↓ BW and insulin content ↓	Abduallah et al. (2017)
26.	AqE.NS	*Mus musculus* mice	TG, T-cholesterol, LDL-c, and TBARS ↓ HDL-c ↑	Bensiameur-Touati et al. (2017)
27.	NSO	STZ-induced diabetic rats	FBG ↓ Insulin levels ↑ CAT and GSH in pancreas and liver ↑ Mean pancreatic islet diameter and insulin immunoreactive parts% ↑	Abdelrazek et al. (2018)

TABLE 8.1 *(Continued)*

SL. No.	Extract/ Compound	Model	Effects	References
28.	*Nigella sativa and Cinnamomum cassia* extracts	STZ-induced diabetic rats	Considerable regularization of blood glucose levels, fat profile and renal function factors	Kaur et al. (2018)
29.	Hydro-alcoholic extract of seeds of *N. sativa*	STZ-induced diabetic rats	Normalized the increasing blood glucose level ↓ SGOT and SGPT concentrations ↓	Khan and Saxena (2018)
30.	*Nigella sativa*	T2DM patients	FBG, 2 h pp, and A1C ↑ %B weight, WC, and BMI ↓ %S, IR, ALT, TC, LDL, HDL, TG, and TAC ↓	Moustafa et al. (2019)
31.	NSO	Alloxan-induced diabetes in rabbits	Blood glucose and lipid contents ↓ Mean BW, HDL-C, and vitamin C contents ↑ CAT activity, TC, TGs, LDL-C, and VLDL-C levels ↓ Regularized bilirubin content	Akjtar et al. (2020)
32.	TQ	STZ-induced diabetes in hamsters	Blood glucose and glycated Hb contents ↓ Glucose production ↓	Fararh et al. (2005)
33.	NSO, nigellone, and TQ	STZ-induced diabetes in rats	Extra pancreatic actions	El-Dakhakhny et al. (2002)

TABLE 8.1 *(Continued)*

SL. No.	Extract/ Compound	Model	Effects	References
34.	*N. sativa* seeds methanol extract and *Propolis* methanol extract	STZ-induced diabetic rats	FBG ↓ MDA content in serum and tissue homogenate of and kidney ↓ Serum CAT, SOD, and GST ↑ IL-6, IgG, IgA, and IgM ↓ Percentage of CML ↓ Urea, creatinine, and uric acid contents ↓ Urinary albumin ↓ Creatinine in urine ↑ Serum electrolytes levels ↑ BWG, BWG%, FER, FER% ↓	El Rabey et al. (2017)
35.	Aqueous extract of *N. sativa*	Isolated rat jejunum	Glucose transport across isolated rat jejunum ↓	Meddah et al. (2009)
36.	TQ	Mice	SOD, CAT, and GSH-Px activities in liver, SOD activity in heart ↓ Hepatic and cardiac lipid peroxidation ↓ Heart and kidney DT-diaphorase activity ↑	Mansour et al. (2002)

Note: ↑: increase; ↓: decrease.

Abbreviations: Glutathione S-transferase (GST); body weight gain percentage (BWG%); food efficiency ratio (FER); body mass index (BMI); secretory function (%S).

8.3 MECHANISMS UNDERLYING THE ANTIDIABETIC EFFECTS OF *NIGELLA SATIVA*

N. sativa is a very promising antidiabetic plant and has the potential to develop into the future drug. Therefore, it becomes important to know its underlying function mechanism. The strategy of blood lowering activity of *N. sativa* is multifaceted, which includes enhancing insulin concentration, reducing insulin resistance, stimulation of β cells activity, direct insulin-like effect and reducing absorption of glucose in the intestine.

The response of *N. sativa* on the key signaling pathways which participate in regulating glucose balance and metabolism, and also on the function of mitochondria in insulin-sensitive cells was determined. $C2C1_2$ myotubes, hepatocytes H4IIE and 3 T3-L1 adipocytes were given a dosage of ethanol extract of *N. sativa*. The main insulin proteins (AKT, ERK1/2) and AMPK (AMPK, ACC) pathways were studied by Western blot analysis. To reveal the response of *N. sativa* on respiration and oxidative phosphorylation isolated liver mitochondria were used. It was seen that *N. sativa* activated AKT and ERK1/2 in $C2C1_2$ to almost 50% more than the DMSO vehicle. Similarly, *N. sativa* enhanced phosphorylation of AMPK and ACC in a way comparable to the positive control AICAR. In H4IIE, *N. sativa* raised the phosphorylation of AKT, yet decreased the stimulation of ERK1/2. The AMPK metabolic pathway of the liver was also activated. In comparison, *N. sativa* did not show any effect on the insulin or the AMPK pathway in 3 T3-L1 adipocytes. To conclude, *N. sativa* decreased the consumption of oxygen of separated liver mitochondria on the basis of dose. The antidiabetic activity of *N. sativa* can be accredited to an enhanced sensitivity to insulin and an increase of the insulin-dependent AMPK pathway that can be associated with the negative impacts on the respiratory chain of mitochondria. Remarkable differences particular to the cell in action also persist. These results support the earlier study on the stimulation of glucose transport by *N. sativa* (Benhaddou et al., 2008).

The hypoglycemic activity of *N. sativa* seed extracts and TQ has been studied in many models of animal and also on humans, including clinical trials. All studies showed that the mechanism of hypoglycemic activity largely occurs by decrementing blood glucose levels and incrementing insulin concentration. In addition to this, few studies also reported loss of appetite, weight loss, and reduction in glyceride levels and partial regeneration of pancreatic beta cells. Randomized clinical trials of *N. sativa* in diabetic patients have led to a decrease in levels of FBG, postprandial glucose, LDL cholesterol, triglycerides, and BW (Rashid and Dwivedi, 2019).

The mechanism which is the basis for the lowering of glucose by TQ in STZ-induced diabetes in hamsters on the basis of liver glucose synthesis was revealed. After the treatment with TQ, blood sugar and glycated Hb concentrations were considerably decreased in animals. After 30 days of TQ treatment, glucose synthesis in isolated hepatocytes was determined. Glucose synthesis was noticeably lesser in hamsters after treatment with TQ after keeping the separated liver cells with precursors of gluconeogenesis as compared to controls (Fararh et al., 2005).

The response of NSO on blood sugar levels was evaluated in STZ-induced diabetes in rats. Further, the response of NSO, nigellone, and TQ were evaluated on the secretion of insulin in separated pancreatic islets of rat in the presence of glucose. NSO considerably reduced plasma sugar concentration in diabetic rats. The hypoglycemic activity of NSO was, however, not in coordination with the induction of insulin production under the influence of NSO, nigellone or TQ. The results showed that the hypoglycemic activity of NSO may be ascribed to extra-pancreatic activity and not to induced release of insulin (El-Dakhakhny et al., 2002).

The research was performed out to see the effects of *N. sativa* aqueous distillate, its oil, and the bioactive compound TQ on serum insulin content and sugar levels in STZ-induced diabetic rats. The tissue malondialdehyde (MDA) and serum glucose concentrations were increased and insulin and SOD content were reduced due to diabetes. Management with *N. sativa* extract, its oil and TQ to rats considerably reduced the tissue MDA and serum sugar which were enhanced as a consequence of diabetes and markedly enhanced serum insulin level and tissue SOD content. TQ reduced the harmful impacts of STZ, which included isolated nucleoli, heterochromatin clusters and vacuolization and fragmentation in mitochondria as seen in ultrastructural study. Somewhat, the water distillate of *N. sativa* reversed these toxic outcomes of STZ. The NSO reinstated regular insulin levels, but could not reduce serum glucose levels to standard. The glucose-lowering activity witnessed might be because of the improvement of β-cell, thus resulting into raised insulin concentrations. Thus, *N. sativa* and thymoquinone provided the protection to the β-cells against oxidative stress (Abdelmeguid et al., 2010).

The hypoglycemic activity of seed of *N. sativa* was attributed to insulin-like activation of glucose uptake by tissues of muscle and adipose tissue (Benhaddou-Andaloussi et al., 2010). Another reason might be the retardation of absorption of glucose in the intestine (Meddah et al., 2009). Another study ascribed the hypoglycemic activity of the seed to the normalization of insulin resistance (Alsaif et al., 2009) better extrapancreatic activities of insulin 38 and the presence of antioxidant constituents in the seed (Balbaa et al., 2017; El Rabey et al., 2017; Mansour et al., 2002; Pari et al., 2009) (Figure 8.2).

In the given research work, STZ was injected to induce hyperglycemia in rat model. Since, liver is the chief origin of circulating IGF-1, it was used for evaluating the expression of this gene. The data showed an insignificant induction of STZ of IGF-1 gene expression which established the previous research (El-Bahr et al., 2014). On the contrary, earlier findings (Li et al., 2004) pointed out a considerable suppression of IGF-1 in diseased rats followed by alloxan treatment. This might be because of usage of another diabetogenic factor, difference in time of experiment and dissimilar RT-PCR methodology.

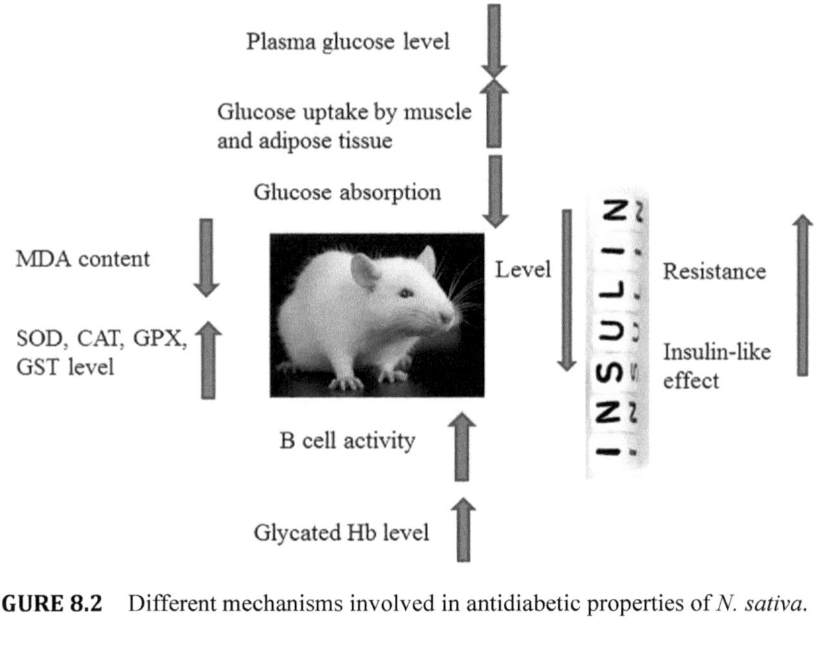

FIGURE 8.2 Different mechanisms involved in antidiabetic properties of *N. sativa*.

The Bcl-2 has been shown to reduce the synthesis of ROS by attaching to cytochrome c or stopped its entry to the cytosol (Shimizu et al., 1995). An enhanced expression of Bcl-2 induced raised expression of antioxidative enzymes and raised concentration of GSH in cells (Amstad et al., 2001). The marked-up regulation in the expression of Bcl-2 in rats with diabetes might be to neutralize the oxidative stress induced by diabetes.

This data is in agreement with those reported earlier in STZ-induced diabetic rats (El-Bahr et al., 2013). An expression of Bcl2 gene in STZ-induced diabetic rats administered with *N. sativa* seed was seen to be increased. The up regulation of Bcl2 gene in diseased rats administered with seed might be attributed to the anti-apoptotic and antioxidant effect of Bcl2 in rats with diabetes (El-Bahr et al., 2013).

The studies also showed the overexpression of SOD and GPX gene in STZ-induced diabetes in rats. On the other hand, no changes have been observed in the expression of mRNA concentrations of SOD and GPX 22. Diabetes reduced the activities and expression of protein in case of SOD and GPX enzymes. The study suggested that STZ induced overexpression of SOD, GPX, CAT, and GST.

8.4 ANTIDIABETIC ACTIVITY OF *NIGELLA SATIVA* USING *IN SILICO* APPROACHES

The bioinformatics tools used in *in silico* approaches including ADMET analysis, 2D-, 3D-, and 4D-QSAR, QM/MM evaluation, quantum chemical calculations, Molecular docking, simulations, and dynamics studies have been used in the study of drug discovery, delivery, and targeting. These *in silico* approaches facilitate to envisage the type of interactions of a ligand and its receptor. Synthetic as well as natural inhibitors have been used.

The research was done to deduce the mechanism of the hypoglycemic activity of TQ in which *in silico* tools using glycogen phosphorylase enzyme assay were applied. The obtained data showed that thymoquinone is associated with possible antihyperglycemic properties. Additionally, antidiabetic effects of were not associated with inhibition of glycogen phosphorylase (El-Ameen et al., 2015).

In the present study, exploratory research on dipeptidyl peptidase-4 enzyme inhibitors by natural inhibitors has been carried out involving the salvation parameter. For this, another molecular modeling technique was used. The molecular dynamics study was performed to find out the best natural inhibitor by analyzing the molecular docking simulation data. The compound having the minimum binding energy scores for ligands was the best inhibitor. The interactions of the ligand and the receptor were further assessed using the molecular dynamics approach. The data attained after docking and dynamics of both techniques led to the conclusion that dithymoquinone, a component of NSO exhibited better interaction of DDP-4 enzyme and therefore, it can be said to be the greatest inhibitor candidate for future *in vitro* and *in vivo* studies (Mesli et al., 2019).

KEYWORDS

- **antidiabetic**
- **bioactive compounds**
- **catalase**
- **fasting blood glucose**
- **glutathione**
- ***in silico***
- **insulin**
- **low-density lipoprotein**
- **nicotinamide**
- ***Nigella sativa***
- **streptozotocin**
- **superoxide dismutase**
- **thymoquinone**

REFERENCES

Abdelmeguid, N. E., Fakhoury, R., Kamal, S. M., & Al Wafai, R. J., (2010). Effects of *Nigella sativa* and thymoquinone on biochemical and subcellular changes in pancreatic β-cells of streptozotocin-induced diabetic rats. *J. Diabetes, 2*(4), 256–266.

Abdelrazek, H. M. A., Kilany, O. E., Muhammad, M. A. A., Tag, H. M., & Abdelazim, A. M., (2018). Black seed thymoquinone improved insulin secretion, hepatic glycogen storage, and oxidative stress in streptozotocin-induced diabetic male Wistar rats. *Oxid. Med. Cell Longev., 2018*, 8104165. doi: 10.1155/2018/8104165. PMID: 29686746.

Abdel-Zaher, A. O., Abdel-Rahman, M. S., & Elwasei, F. M., (2011). Protective effect of *Nigella sativa* oil against tramadol-induced tolerance and dependence in mice: Role of nitric oxide and oxidative stress. *Neurotoxicology, 32*(6), 725–733.

Abduallah, A. M., (2017). The effect of *Nigella sativa* extract (Thymoquinone) on glucose-insulin levels and body weight of induced diabetic female rats. *American Journal of Life Sciences, 5*(2), 52–56.

Abel-Salam, B. K., (2012). Immunomodulatory effects of black seeds and garlic on alloxan-induced diabetes in albino rat. *Allergol. Immunopathol. (Madr)*, *40*(6), 336–340.

Akhtar, M. T., Qadir, R., Bukhari, I., Ashraf, R. A., Malik, Z., Zahoor, S., Murtaza, M. A., et al., (2020). Antidiabetic potential of *Nigella sativa* L seed oil in alloxan-induced diabetic rabbits. *Tropical Journal of Pharmaceutical Research, 19*, 283–289.

Al-Ali, A., Alkhawajah, A. A., Randhawa, M. A., & Shaikh, N. A., (2008). Oral and intraperitoneal LD50 of thymoquinone, an active principle of *Nigella sativa*, in mice and rats. *J. Ayub. Med. Coll. Abbottabad, 20*(2), 25–27.

Al-Awadi, F. M., & Gumma, K. A., (1987). Studies on the activity of individual plants of an antidiabetic plant mixture. *Acta. Diabetol. Lat., 24*(1), 37–41.

Al-Awadi, F. M., Fatania, H., & Shamte, U., (1991). The effect of a plant mixture extract on liver gluconeogenesis in streptozotocin-induced diabetic rats. *Diab. Res., 18*(4), 163–168.

Al-Bukhari, M. I., (1976). In: Al-Bukhari, S., (ed.), *The Collection of Authentic Sayings of Prophet Mohammad (Peace be Upon Him), Division 71 on Medicine* (2nd edn.). Ankara: Hilal Yayinlari.

Al-Hader, A., Aqel, M., & Hasan, Z., (1993). Hypoglycemic effect of volatile oil of *Nigella sativa* seeds. *Pharmaceutical Biology, 31*(2), 96–100.

Ali, B., Louis, M. C., Diane, V., Yara, H., & Pierre, H. S., (2008). Antidiabetic effects of *Nigella sativa* are mediated by activation of insulin and AMPK pathways, and by mitochondrial uncoupling. *Canadian Journal of Diabetes, 32*(4), 333.

Alimohammadi, S., Hobbenaghi, R., Javanbakht, J., Kheradmand, D., Mortezaee, R., Tavakoli, M., Khadivar, F., & Akbari, H., (2013). Protective and antidiabetic effects of extract from *Nigella sativa* on blood glucose concentrations against streptozotocin (STZ)-induced diabetic in rats: An experimental study with histopathological evaluation. *Diagn Pathol., 8, 11*, 125, 137. doi: 10.1186/1746-1596-8-137.

Alsaif, M. A., (2008). Effect of *N. sativa* oil on impaired glucose tolerance and insulin insensitivity induced by high-fat-diet and turpentine-induced trauma. *Pak. J. Biol. Sci., 11*, 1093–1099.

Altan, M. F., Kanter, M., Donmez, S., Kartal, M. E., & Buyukbas, S., (2007). Combination therapy *Nigella sativa* and human parathyroid hormone on bone mass, biomechanical behavior and structure in streptozotocin-induced diabetic rats. *Acta Histochem., 109*(4), 304–314.

Amstad, P. A., Liu, H., Ichimiya, M., Berezesky, I. K., Trump, B. F., Buhimschi, I. A., & Gutierrez, P. L., (2001). BCL-2 is involved in preventing oxidant-induced cell death and in decreasing oxygen radical production. *Redox Rep., 6*, 351–362.

Asaduzzaman, M., Nahar, L., Sha, M. F. R., & Roy, M. N., (2015). Hypoglycemic and hypolipidemic potential of *N. sativa* L. seed extract in streptozotocin (STZ)-induced diabetic rats. *J. Plant Biochem. Physiol., 3*(4).

Assayed, M. E., (2010). Radioprotective effects of black seed (*Nigella sativa*) oil against hemopoietic damage and immunosuppression in gamma-irradiated rats. *Immunopharmacol. Immunotoxicol., 32*(2), 284–296.

Balbaa, M., Abdulmalek, S. A., & Khalil, S., (2017). Oxidative stress and expression of insulin signaling proteins in the brain of diabetic rats: Role of *Nigella sativa* oil and antidiabetic drugs. *PloS One, 12*(5), e0172429.

Bamosa, A. O., Kaatabi, H., Lebdaa, F. M., Elq, A. M., & Al-Sultanb, A., (2010). Effect of *Nigella sativa* seeds on the glycemic control of patients with type 2 diabetes mellitus. *Indian J. Physiol. Pharmacol., 54*(4), 344–354.

Benhaddou-Andaloussi, A., Martineau, L. C., Vallerand, D., Haddad, Y., Afshar, A., Settaf, A., & Haddad, P. S., (2010). Multiple molecular targets underlie the antidiabetic effect of Nigella sativa seed extract in skeletal muscle, adipocyte, and liver cells. *Diabetes Obes. Metab., 12*, 148–157.

Benhaddou-Andaloussi, A., Martineau, L., Vuong, T., Meddah, B., Madiraju, P., Settaf, A., & Haddad, P. S., (2011). The *in vivo* antidiabetic activity of *Nigella sativa* is mediated through activation of the AMPK pathway and increased muscle glut4 content. *Evid. Based Complement. Alternat. Med., 2011*, 538671.

Bensiameur-Touati, B., Kacimi, G., El-Mehdi, H., Berdja, S., &Aouichat-Bouguerra, S., (2017). *In vivo* subacute toxicity and antidiabetic effect of aqueous extract of *Nigella sativa*. *Evidence-Based Complementary and Alternative Medicine, 2017*, 13. 8427034.

Boskabady, M. H., Mohsenpoor, N., & Takaloo, L., (2010). Antiasthmatic effect of *Nigella* sativa in airways of asthmatic patients. *Phytomedicine, 17*(10), 707–713.

El Rabey, H. A., Al-Seeni, M. N., & Bakhashwain, A. S., (2017). The antidiabetic activity of *Nigella sativa* and propolis on streptozotocin-induced diabetes and diabetic nephropathy in male rats. *Evidence-Based Complementary Alter. Med., 2017*. 10.1155/2017/5439645.

El-Ameen, N. M. H., Taha, M. M. E., Abdelwahab, S. I., Khalid, A., Elfatih, F., Kamel, M. A., & Sheikh, B. Y., (2015). Antidiabetic properties of thymoquinone is unassociated with glycogen phosphorylase inhibition. *Pharmacogn. J., 7*, 406–410.

El-Bahr, S. M., & El-Sabagh, I. M., (2014). Hepatic gene expression of insulin like growth factor and selected antioxidants in diabetic rats treated with turmeric or black cumin. *Br. Biotechnol. J., 4*, 778–793.

El-Bahr, S. M., (2013). Curcumin regulates gene expression of insulin-like growth factor, B-cell CLL/lymphoma 2 and antioxidant enzymes in streptozotocin-induced diabetic rats. BMC Complement. *Altern. Med., 13*. 10.1186/1472-6882-13-368.

El-Dakhakhny, M., Mady, N., Lembert, N., & Ammon, H. P., (2002). The hypoglycemic effect of *Nigella sativa* oil is mediated by extrapancreatic actions. *Planta Med., 68*(5), 465–466.

Farah, K. M., Atoji, Y., Shimizu, Y., & Takewaki, T., (2002). Insulinotropic properties of *Nigella sativa* oil in streptozotocin plus nicotinamide diabetic hamsters. *Res. Vet. Sci., 73*, 279–282.

Fararh, K. M., Atoji, Y., Shimizu, Y., & Takewaki, T., (2002). Isulinotropic properties of *Nigella sativa* oil in streptozotocin plus nicotinamide diabetic hamster. *Res. Vet. Sci., 73*(3), 279–282.

Fararh, K. M., Atoji, Y., Shimizu, Y., Shiina, T., Nikami, H., & Takewaki, T., (2004). Mechanisms of the hypoglycaemic and immunopotentiating effects of *Nigella sativa* L. oil in streptozotocin-induced diabetic hamsters. *Res. Vet. Sci., 77*(2), 123–129.

Fararh, K. M., Shimizu, Y., Shiina, T., Nikami, H., Ghanem, M. M., & Takewaki, T., (2005). Thymoquinone reduces hepatic glucose production in diabetic hamsters. *Res. Vet. Sci., 79*(3), 219–223.

Goreja, W. G., (2003). *Black Seed: Nature's Miracle Remedy*. New York, NY 7 Amazing Herbs Press.

Hamdan, A., Haji, I. R., & Mokhtar, M. H., (2019). Effects of *Nigella sativa* on type-2 diabetes mellitus: A systematic review. *Int. J. Environ. Res. Public Health, 16*(24), 4911.

Heshmati, J., & Namazi, N., (2015). Effects of black seed (*Nigella sativa*) on metabolic parameters in diabetes mellitus: A systematic review. *Complementary Therapies in Medicine, 23*(2), 275–282.

Kaatabi, H., Bamosa, A. O., Badar, A., Al-Elq, A., Abou-Hozaifa, B., Lebda, F., Al-Khadra, A., & Al-Almaie, S., (2015). *Nigella sativa* improves glycemic control and ameliorates oxidative stress in patients with type 2 diabetes mellitus: Placebo-controlled participant blinded clinical trial. *PLOS One, 10*(2), e0113486.

Kaleem, M., Kirmani, D., Asif, M., Ahmed, Q., & Bano, B., (2006). Biochemical effects of *Nigella sativa* L seeds in diabetic rats. *Indian J. Exp. Biol., 44*(9), 745–748.

Kanter, M., (2008). Effects of *Nigella sativa* and its major constituent, thymoquinone on sciatic nerves in experimental diabetic neuropathy. *Neurochem. Res., 33*(1), 87–96.

Kanter, M., Akpolat, M., & Aktas, C., (2009). Protective effects of the volatile oil of *Nigella sativa* seeds on beta-cell damage in streptozotocin-induced diabetic rats: A light and electron microscopic study. *J. Mol. Histol., 40*(5, 6), 379–385.

Kanter, M., Coskun, O., Korkmaz, A., & Oter, S., (2004). Effects of *Nigella sativa* on oxidative stress and beta-cell damage in streptozotocin-induced diabetic rats. *Anat. Rec. A Discov. Mol. Cell. Evol. Bioli., 279*(1), 685–691.

Kanter, M., Meral, I., Yener, Z., Ozbek, H., & Demir, H., (2003). Partial regeneration/proliferation of the beta-cells in the islets of Langerhans by *Nigella sativa* L. in streptozotocin-induced diabetic rats. *Tohoku. J. Exp. Med., 201*(4), 213–219.

Kapoor, S., (2009). Emerging clinical and therapeutic applications of *Nigella sativa* in gastroenterology. *World J. Gastroenterol., 7*, 2170–2171.

Kaur, G., Invally, M., Khan, M. K., & Jadhav, P., (2018). A nutraceutical combination of *Cinnamomum cassia* & *Nigella sativa* for type 1 diabetes mellitus. *Journal of Ayurveda and Integrative Medicine, 9*(1), 27–37.

Khaled, A. A. S., (2009). Gastroprotective effects of *Nigella Sativa* oil on the formation of stress gastritis in hypothyroidal rats. *Int. J. Physiol. Pathophysiol. Pharmacol., 1*, 143–149.

Khalid, O. M., Eltayeb, M., Ali, I. A., & Musa, O. A., (2019). Effect of oral *Nigella sativa* on fasting blood glucose in adults. *Int. J. Diabetes Metab. Disord., 4*(1), 1–3.

Khan, M., & Saxena, V., (2018). Screening of Antidiabetic potential of hydro-alcoholic seeds extract of *Nigella sativa* in streptozotocin-induced diabetic rats. *IOSR Journal of Pharmacy, 8*(4), 73–78.

Khare, C. P., (2004). *Encyclopedia of Indian Medicinal Plants*. New York: Springes-Verlag Berlin Heidelberg.

Li, J. B., Wang, C. Y., Chen, J. W., Feng, Z. Q., & Ma, H. T., (2004). Expression of liver insulin-like growth factor 1 gene and its serum level in rats with diabetes. *World J. Gastroenterol., 10*, 255–259.

Li, W. L., Zheng, H. C., Bukuru, J., & De Kimpe, N., (2004). Natural medicines used in the traditional Chinese medical system for therapy of diabetes mellitus. *J. Ethnopharmacol., 92*(1), 1–21.

Mansour, M. A., Nagi, M. N., El-Khatib, A. S., & Al-Bekairi, A. M., (2002). Effects of thymoquinone on antioxidant enzyme activities, lipid peroxidation and DT-diaphorase in different tissues of mice: A possible mechanism of action. *Cell Biochem. Funct., 20*, 143–151.

Matira, K., & Zesmin, F. D., (2008). Effects of the crude and the n-hexane extract of *Nigella sativa* Linn. (kalajira) upon diabetic rats. *Bangladesh J. Pharmacol., 4*, 17–20.

Meddah, B., Ducroc, R., Faouzi, M. E. A., Eto, B., Mahraoui, L., Benhaddou-Andaloussi, A., Martineau, L. C., et al., (2009). *Nigella sativa* inhibits intestinal glucose absorption and improves glucose tolerance in rats. *Journal of Ethnopharmacology, 121*(3), 419–424.

Mesli, F., Daoud, I., & Ghalem, S., (2019). Antidiabetic activity of *Nigella sativa* (Black seed) by molecular modeling elucidation, molecular dynamic and conceptual DFT investigation. *Pharmacophore, 10*(5), 57–66.

Mohamed, A. M., EL-Sharkawy, F. Z., Ahmed, S. A. A., Aziz, W. M., & Badary, O. A., (2009). Glycemic control and therapeutic effect of *Nigella sativa* and *Curcuma longa* on rats with streptozotocin-induced diabetic hepatopathy. *J. Pharmacol. Toxicol., 4*(2), 45–57.

Moustafa, H. A. M., El Wakeel, L. M., Halawa, M. R., Sabri, N. A., El-Bahy, A. Z., & Singab, A. N., (2019). Effect of *Nigella sativa* oil versus metformin on glycemic control and biochemical parameters of newly diagnosed type 2 diabetes mellitus patients. *Endocrine, 65*(2), 286–294.

Murli, M., Gaura, J., Sharma, R., & Haldiya, K. R., (2011). *Antidiabetic Properties of a Spice Plant Nigella Sativa, 1*(1), 1–8.

Nadia, M. H., & Taha, R. A., (2009). Effects of *Nigella sativa* oil and thymoquinone on oxidative stress and neuropathy in streptozotocin-induced diabetic rats. *Pharmacology, 84*, 127–134.

Najmi, A., Haque, S. F., Naseeruddin, M., & Khan, R. A., (2008). Effect of *Nigella sativa* oil on various Clinical and biochemical parameters of metabolic syndrome. *Int. J. Diab. Dev. Ctries., 16*, 85–87.

Najmi, A., Nasiruddin, M., Khan, R. A., & Haque, S. F., (2008). Effect of *Nigella sativa* oil on various clinical and biochemical parameters of insulin resistance syndrome. *Int. J. Diab. Dev. Ctries., 28*, 11–14.

Pari, L., & Sankaranarayanan, C., (2009). Beneficial effects of thymoquinone on hepatic key enzymes in streptozotocin-nicotinamide induced diabetic rats. *Life Sci., 85*(23–26), 830–834.

Patel, D. K., Kumar, R., Prasad, S. K., & Hemalatha, S., (2011). *Pedalium murex Linn* (Pedaliaceae) fruits: A comparative antioxidant activity of its different fractions. *Asian Pac. J. Trop. Biomed., 1*(5), 395–400.

Patel, D. K., Kumar, R., Prasad, S. K., Sairam, K., & Hemalatha, S., (2011). Antidiabetic and *in vitro* antioxidant potential of *Hybanthus enneaspermus* (Linn) F. muell in streptozotocin-induced diabetic rats. *Asian Pac. J. Trop. Biomed., 1*(4), 316–322.

Ponnusamy, S., Ravindran, R., Zinjarde, S., Bhargava, S., & Kumar, A. R., (2011). Evaluation of traditional Indian antidiabetic medicinal plants for human pancreatic amylase inhibitory effect *in vitro*. *Evid. Based Complement. Alternat. Med., 2011*, 515647.

Rashid, M., & Dwivedi, S. D., (2019). *Nigella sativa* as highly potent hypoglycemic agent. *International Journal of Applied Research, 5*(3), 247–255.

Salama, R. H., (2011). Hypoglycemic effect of lipoic acid, carnitine, and *Nigella sativa* in diabetic rat model. *Int. J. Health Sci. (Qassim), 5*(2), 126–134.

Sharma, P. C., Yelne, M. B., & Dennis, T. J., (2005). *Database on Medicinal Plants Used in Ayurveda* (pp. 420–440). New Delhi.
Shimizu, S., Eguchi, Y., Kosaka, H., Kamiike, W., Matsuda, H., & Tsujimoto, Y., (1995). Prevention of hypoxia-induced cell death by Bcl-2 and Bcl-xL. *Nature, 374*, 811–813.
Sjogren, K., Sheng, M., Moverare, S., Liu, L. S., Wallenius, K., Törnell, J., Isaksson, O., et al., (2002). Effects of liver derived insulin like growth factor I on bone metabolism in mice. *J. Bone Mineral Res., 17*, 1977–1987.
Sy, G. Y., Cissé, A., Nongonierma, R. B., Sarr, M., Mbodj, N. A., & Faye, B., (2005). Hypoglycaemic and antidiabetic activity of acetonic extract of *Vernonia colorata* leaves in normoglycaemic and alloxan-induced diabetic rats. *J. Ethnopharmacol., 98*(1, 2), 171–175.

CHAPTER 9

Aegle marmelos (L.) as a Functional Food for Managing Diabetes Mellitus

KHYATIE ANJARIA, HIMANSHU DAWDA, KARISHMA RAJBHAR, and USHA MUKUNDAN

Department of Botany, Ramniranjan Jhunjhunwala College, Ghatkopar (W), Mumbai – 400 086, Maharashtra, India, E-mail: usha.rjc@gmail.com

ABSTRACT

Although *Aegle marmelos* (L.) or Bael is indigenous to North India, it has been grown extensively across India. In recent times, the consumption of nutrient-rich foods with therapeutic properties is increasingly being emphasized in the long-term management of major lifestyle diseases such as diabetes mellitus. Diabetes mellitus results from disturbances in the homeostasis of carbohydrate metabolism due to inadequate insulin levels. Insulin deficiency causes various metabolic aberrations. Furthermore, high uric acid, urea, and creatine levels are observed in patients with diabetes. Moreover, increased oxidative stress causes additional complications observed in these patients. This chapter reviews the potential of *A. marmelos,* particularly the leaves and fruits, as functional foods for reducing blood sugar levels, which is an essential aspect of managing diabetes mellitus and associated complications. In the discussion, the phytochemical bioactive profiles of the *A. marmelos* leaves and fruits as well as their antidiabetic and antioxidant functions have been highlighted. The review confirms that the extracts of *A. marmelos* leaves and fruits can be consumed regularly as functional foods.

Antidiabetic Plants for Drug Discovery: Pharmacology, Secondary Metabolite Profiling, and Ingredients with Insulin Mimetic Activity. Deepu Pandita, MPhil, Anu Pandita, MSc, & Chander Bhanu, MBBS, PGDDCE (Eds.)
© 2023 Apple Academic Press, Inc. Co-published with CRC Press (Taylor & Francis)

9.1 INTRODUCTION

Aegle marmelos, which has numerous uses in daily life, belongs to the family Rutaceae. *A. marmelos* is indigenous to India and possesses several medicinal properties. *A. marmelos* is commonly called the Bael tree. Interestingly, almost every part of the Bael tree is a component of at least one herbal medicine. In addition to the uses of the Bael tree in Indian ethnomedicine, it is considered an auspicious tree in Hinduism and is used extensively in religious ceremonies (Sharma et al., 2013).

9.2 DISTRIBUTION

A. marmelos is a subtropical plant growing at an altitude of approximately 1,200 m above mean sea level. The plant easily adapts to various habitats; consequently, it can be planted worldwide. According to prehistoric data from 800 B.C., *A. marmelos* originated from the eastern ghats and central India. Owing to the importance of the Bael tree in Hinduism, *A. marmelos* is primarily grown in temple courtyards and in the vicinity of temples and shrines throughout India. The dry hilly forests and dry plains of central and southern India, sub-Himalayan tracts between the Jhelum and West Bengal, east coast, Deccan plateau, and Andaman Islands are highly suited to the growth of *A. marmelos*. Bael trees are found along the entire north-south extent of India; accordingly, these trees are found in Andhra Pradesh, Bihar, Himachal Pradesh, Jammu and Kashmir, Karnataka, Kerala, Madhya Pradesh, Maharashtra, Punjab, Rajasthan, Tamil Nadu, Uttar Pradesh, and West Bengal (Sharma et al., 2013). It grows at altitudes of 1,200 m in regions like Punjab, where temperatures range from 48.89°C during summer to −6.67°C during winter. In addition, prolonged drought is thought to facilitate the fruiting of *Aegle* because studies have reported the absence of fruiting in the absence of a long dry season. Reports indicate that *A. marmelos* grows easily on rich, well-drained soil as well as on swampy, alkaline, or stony soils with pH 5 to 8 (Patel et al., 2012; Sawale et al., 2018).

9.3 DESCRIPTION OF PLANT

A. marmelos is a medium-sized tree that grows slowly to the height of 12–15 m. The trunk is generally short, and sometimes the branches are drooping. Some varieties exhibit the presence of spines. The bark of the Bael tree is thick, flaking, soft, and spreading. A clear or transparent gummy sap is exuded from wounded branches of the Bael tree. The sap gradually solidifies and appears similar to gum Arabica. Bael tree sap or gum is initially sweet to taste later irritates the throat. The leaves of Bael plant are of the deciduous type. They mostly occur singly or in groups of 2 or 3 members. The leaves are oval, alternate, and pointed, and the leaflets are shallow toothed. The leaf is 4–10 cm long and 2–5 cm wide, and the terminal leaf has a long petiole. New leaves are pinkish maroon in color and can appear glossy. Mature leaves that are bruised or damaged emit a disagreeable odor. Young branchlets have 4–7 clusters of fragrant flowers. The varieties of *A. marmelos* are based on the structure of fruit. The fruit diameter varies between 5 and 20 cm and are round, pyriform, oval, or oblong. The fruit may have a hard, thin, woody, or soft rind (Patel et al., 2012; Sharma et al., 2013; Sawale et al., 2018).

9.4 EFFECTS OF LIFESTYLE DISEASES

In recent times, the incidence of overweight and obesity in India is increasing at an alarming rate. The body mass index (BMI) of overweight individuals is higher than 25. According to a survey, approximately 31% of urban Indians are either overweight or obese. The risk of heart disease in obese individuals is three times higher than that of normal-weight individuals. Furthermore, the additional weight increases the incidence of cardiovascular disease, hypertension, diabetes mellitus (type 2), stroke, dyslipidemia, osteoarthritis, and some cancers. Work-related hypertension and stress are risk factors for heart ailments, which increase the risk by 50%. Nearly 40% of urban Indians exhibit high lipid levels as well as high levels of cholesterol and triglycerides. Bad cholesterol and triglycerides are crucial risk factors for heart disease. An unhealthy diet, inadequate physical inactivity, and stress related to psychosocial pressures are considered the main causes of an increase in the incidence of heart disease. Data reports that Indians have smaller sized coronaries when compared with Caucasians. Rate of narrowing of blood vessels are faster and start at a younger age. Indians have a high level of genetic inclination for cholesterol boost and diabetes, and when this is coupled with bad alterations in lifestyle, it produces an epidemic of heart disease. India has maximum patient's occurrence with early disease of coronary artery taking place, nearly 10 years before the other populations in the entire world. Mean age of first heart attack in Indians is 53, but now more and more youths in 30s are seen experiencing acute heart attacks. The strong positive association between excess weight obesity and other numerous well-established risk issues for morbidity and mortality have made an urgent priority of reversing the obesity epidemic. Obesity and type 2 diabetes are preventable. Thus, adjustments in the everyday lifestyle are useful in preventing as well as obesity in the high-risk adults who also have impaired glucose tolerance (Rajbhar et al., 2016).

9.5 ADVERSE EFFECTS OF A NON-NUTRITIOUS DIET

9.5.1 CHEMICAL CHANGES

Studies have revealed that the consumption of foods containing high levels of sugar and fat alter chemical activity within the brain. The changes in chemical activity increase the dependency of the brain on such foods for regulation of mood. For example, as described by Marriott in 1994, that carbohydrate cravers who overconsume carbohydrate-rich food for getting relief from depression symptoms.

9.5.2 DEPRESSION

A non-nutritious diet can contribute to depression in multiple ways. The primary factors are the type and quantity of fats found in many non-nutritious foods. Recent evidence reveals a link between low levels of serotonin and suicidal tendencies. Furthermore, low levels of neurotransmitters are partly responsible for complete insensitivity to consequences as well as triggering dangerous, thoughtless, and aggressive actions that may result in suicide (Rao et al., 2008).

9.5.3 FATIGUE

The consumption of non-nutritious food has been reported to cause fatigue. Furthermore, the consumption of non-nutritious foods has been associated with sleeplessness, excessive exertion, jet lag, overeating, and aging. Azzolino et al. (2020) have described how altering food intake as well as body composition control fatigue perception, possibly through underlying mitochondrial dysfunction.

9.5.4 FLUCTUATIONS IN BLOOD SUGAR LEVELS

Physical and emotional stress, which trigger the release of hormones causing high blood sugar levels, have been associated with the consumption of non-nutritious foods. In women, hormonal changes that occur during menstruation and menopause affect blood sugar levels (Wilcox, 2005).

9.5.5 HEART DISEASES

A non-nutritious diet is the major cause of heart diseases. Several studies have shown that acute myocardial infarction is caused by plaque development in arteries, which causes an additional strain on the heart to pump blood downstream. During upstream flow, a shortfall occurs in the blood returning to the heart; consequently, the heart is damaged in two ways. First, the heart becomes fatigued by the constant additional strain during pumping and second, the heart receives a limited oxygen supply. The commencing and the remedy mutually take a prolonged time and a huge determination to prevail (Riccardi et al., 2011; Wan et al., 2017).

9.5.6 HIGH CHOLESTEROL

In addition to causing the formation of plaques and narrowing arteries, cholesterol adversely affects the liver, where it is metabolized. High levels of cholesterol caused by the consumption of non-nutritious food and strain the liver and eventually damage it. Furthermore, liver damage is one of the long-term effects of cardiovascular problems like arterial plaque (Wilcox, 2005).

9.5.7 SUSCEPTIBILITY TO TYPE 2 DIABETES

Consumption of non-nutritious food can result in overweight and obesity. Studies have shown that being overweight increases the risk of diabetes. Diabetes is a type of disease which stays for lifetime, thus proper therapy approaches with special importance on diet is a way to control the diabetes. Reducing the symptoms and preventing the appearance of the complications can also be controlled with a functional diet. Physical inactivity together with overweight or obesity ranks among the top risk factors for prediabetes and diabetes Type 2 (Qin et al., 2010; Sami et al., 2017).

9.5.8 INCREASED RISK OF CANCER

Consumption of non-nutritious food can cause overweight and obesity, which are associated with a significantly higher risk of cancer of the kidney, esophagus, pancreas, colon, and rectum, endometrium, thyroid gland, liver, and gallbladder. Around 20% of obesity are responsible for all malignancies. Obesity influences are gender and site-specific. Association between the obesity and cancer risk is mostly because of anthropometric parameters and due to lifestyle influences. Anthropometric parameters include BMI, weight increase, and amount of body fat mostly visceral fat. Obesity is considered a primary factor contributing to cancer-related mortality (Dhaka et al., 2011; De Pergola et al., 2013).

9.5.9 INCREASED RISK OF DEMENTIA

An excessive intake of non-nutritious food, fatty food and sweets substantially increase the concentration of insulin in the body and reduces memory and the ability to think, which increases the risk of dementia (Gómez-Pinilla et al., 2008).

9.5.10 IMPATIENCE AND UNCONTROLLABLE CRAVINGS

The intake of sugary products like cupcakes and doughnuts can temporarily cause a spike in blood sugar levels, which promotes feelings of happiness and satisfaction. However, when the sugar levels return to normal, irritability, and confusion set in (París et al., 2013).

9.5.11 LACK OF ENERGY

The lack of energy is an extensively described temporary but unfavorable effect of consuming non-nutritious foods. non-nutritious foods do not provide vital nutrients; however, one may feel extremely satiated after consuming these foods (Romieu et al., 2017; Azzolino et al., 2020).

9.5.12 REDUCED APPETITE CONTROL

The consumption of excess trans fats that are present in processed and non-nutritious foods deliver mixed signals to the brain, which create confusion about the intensity of hunger; consequently, people tend to overeat (Gómez-Pinilla et al., 2008; Dhaka et al., 2011).

9.5.13 LIVER DISEASE/LIVER FAILURE

Generally, the symptoms of fatty liver syndrome are visible; however, diabetes, poor nutrition and obesity cause progressive liver damage with or without visible symptoms. Morbid obesity may contribute to severe liver damage with all the symptoms of alcoholic hepatitis. Furthermore, morbid obesity may result in cirrhosis and liver failure (Bhatt et al., 2015).

9.5.14 MALNUTRITION

Studies have shown that insufficient consumption of calories and protein cause malnutrition. Furthermore, the intake of non-nutritious food has been associated with a high caloric intake but not high protein intake. The excessive consumption of calories may cause obesity, while the ingestion of excess sodium may contribute to interdialytic weight gain (Romieu et al., 2017).

9.5.15 MEMORY AND LEARNING PROBLEMS

Diet and exercise with environment are likely to alter our brain wellbeing and mental function. The frequent intake of non-nutritious food adversely affects memory. The capacity of diet to regulate cognitive abilities have long-term consequences considering current studies which indicate that nutritional influences may be transmitted over age group by affecting epigenetic events. Research have also indicated that an unnecessarily consuming calories may undo the positive results of some diets (Gómez-Pinilla et al., 2008).

9.5.16 OBESITY

The most common health problem associated with eating non-nutritious food or overeating is obesity. Obesity is a result of multiple diseases and deficiencies of essential nutrients. Non-nutritious foods, such as fried, processed, and fast food are highly popular. Currently, people prefer foods that are full of fats and non-essential constituents, which leads to obesity (Wilcox, 2005).

9.5.17 POOR CONCENTRATION

The inability to concentrate is another crucial consequence of consuming non-nutritious food. non-nutritious food has been reported to exhibit either immediate or moderately long-term effects. Consuming non-nutritious meals that are rich in oil induces a feeling of drowsiness, which results in failure to concentrate. The consumption of non-nutritious food over prolonged periods results in fat build-up and a reduction in blood circulation. An insufficient supply of oxygen and crucial nutrients, particularly proteins, adversely affects the functioning of brain cells.

9.5.18 SKIN PROBLEMS

The type of food consumed frequently affects the appearance of skin, particularly facial skin. Carbohydrate-rich foods lead to sudden increases or spikes in blood sugar levels, and a sudden spike in sugar levels in blood may trigger acne.

9.5.19 TIREDNESS

Excessive eating results in obesity, which significantly reduces energy levels and results in tiredness. Individuals with unhealthy eating preferences, particularly those with a preference for salty foods, are more likely to feel tired throughout the day (Yau et al., 2013).

9.5.20 TOOTH DECAY

Non-nutritious foods adversely affect oral health because of the added salt and sugar they contain. non-nutritious food debris is highly liable to sticking to teeth. This type of food debris permits acid accumulation on the superficial portion of teeth and ultimately leads to tooth decay and cavities (Gupta et al., 2013).

9.5.21 DIGESTIVE PROBLEMS

Disturbances in the equilibrium between good and bad intestinal bacteria can be caused by the use of antibiotics, excessive consumption of non-nutritious food and stress. Alterations in gut flora result in gas, bloating, constipation, and irritable bowel syndrome (IBS). Recent studies have reported an association between high counts of bad bacteria and obesity.

9.5.22 VITAMIN DEFICIENCIES

Some studies have revealed that non-nutritious diets are associated with blindness. Poor eating habits are considered to be a primary cause of an eye condition known as nutritional optic neuropathy. This condition is generally related to a deficiency of folic acid and vitamin B complex, according to reports published by the National Institutes of Health (Sawicka-Pierko et al., 2014).

9.5.23 WEAKNESS

Processed and non-nutritious food increase the risk of obesity, diabetes, and heart disease. Food directly affects mental and physical well-being and overall health. The consumption of non-nutritious food does not provide adequate quantities of vital nutrients. that the consumption of non-nutritious food has been shown to complicate an individual's health condition (Anand et al., 2015).

9.6 DIABETES

Diabetes has become a common lifestyle-related disease throughout the world. Most patients with diabetes fall into two categories, namely type 1 and type 2 diabetes. A majority of the patients have type 2 diabetes or diabetes mellitus. Type 1 diabetes is considered an autoimmune disorder and involves genetic factors. Progressive damage of beta cells, which results in complete insulin deficiency, is a characteristic of Type 1 diabetes. Type 2 diabetes results from interactions between genetic factors and lifestyle-related factors and typically manifest after years or decades. A gradual rise increase in insulin requirement is a characteristic of Type 2 diabetes. In addition to gradual beta-cell failure, type 2 diabetes results in high levels of glucose in the blood. Type 2 diabetes typically occurs with other metabolic disorders such as hypertension and dyslipidemia. The body is unable to produce sufficient insulin; hence, the concentration of blood glucose increases. Antidiabetic drugs are designed to reduce the concentration of blood glucose and inducing the production of a sufficient quantity of insulin (Wilcox, 2005; Lindström, 2011).

9.7 HEALTHY FOODS AND FUNCTIONAL FOOD

Healthy foods are nutrient-rich foods; they contain vitamins, minerals, and phytochemicals that are beneficial to the human body. Healthy foods are considered valuable to health. Regulatory authorities, such as the U.S. Food and Drug Administration, have not provided an accurate authoritative description of functional food. Foods that are considered healthy may be either natural, organic, or whole foods. Dietary supplements are also considered healthy foods. These types of products are marketed in health food stores and in the 'health' or 'organic' segments of supermarkets.

Non-nutritious or junk food is usually poor in nutrients because of its components and preparation techniques; non-nutritious food most often has an abundance of calories, which are unnecessary.

Earlier, most foods were considered functional food to varying degrees because they provided taste, aroma, as well as a high level of nutrition. However, now foods are being analyzed extensively for additional physiological benefits, which can diminish the risk of chronic disease and enhance overall health. Recently, the interest in research on a type of foods that are known as 'functional foods' is increasing globally. Functional food is still a novel and undefined concept in most of the world. However, Japan had originally introduced functional foods as a concept in the 1980s in response to escalating health care expenses. Ministry of Health and Welfare introduced a monitoring system to certify certain foods with verified health benefits to promote the wellbeing of the nation's elderly population. These foods are certified and are marked using a distinct seal that reads Food for Specified Health Use (FOSHU). In Japan, since July 2002, approximately 300 manufactured foods have been accorded FOSHU status. In the United States, functional foods have not been recognized by governing bodies concerned with public health and nutrition. Nevertheless, numerous groups have recommended descriptions intended for this type of new food classification. National Academy of Science Food and Nutrition Board in 1994 have defined functional food as any modified food or food ingredient that may provide a health benefit beyond the traditional nutrients it contains. International Life Sciences Institute defines these foods as a food that, by virtue of the presence of physiologically active components, provide a health benefit beyond basic nutrition. Another term that is frequently used to describe functional foods is nutraceuticals. The term nutraceuticals were coined in 1991 by Foundation for Innovation in Medicine for practically describing bioactive constituents that provide health benefits (Hasler, 2002).

9.8 *A. MARMELOS* AS FUNCTIONAL FOOD

Any food should exhibit efficacy in clinical studies and meet safety requirements in clinical and non-clinical investigations to qualify as a functional food. Furthermore, the active or effective ingredient should be identified and quantified. At Functional Food Center 10th International Conference held at Santa Barbara, CA, in 2012 functional foods were defined as natural or processed foods that contain known or unknown biologically active compounds, which is defined amounts provide a clinically proven and recognized health benefit for the prevention, management, or even treatment of chronic disease (Riccardi et al., 2011; Martirosyan et al., 2015).

A. marmelos fruits are rich in nutrients, which are listed below (Singh et al., 2014; Murthy et al., 2020):

Nutrient in Bael Fruit	Amount per 100 g of Pulp
Calcium	85 mg
Calories	137 kcal
Carbohydrate	31.8 g
β-Carotene	55 mg
Copper	0.21 mg
Fat	0.39 g
Iron	0.6 mg
Minerals	1.7 g
Niacin	1.1 mg
Phosphorus	31.8 mg
Potassium	600 mg
Protein	1.8 g
Riboflavin	0.03–1.19 mg
Thiamine	0.13 mg
Total minerals	1.70 g
Vitamin A	0.05 mg
Vitamin B2	1.2 mg
Vitamin C	8 mg
Water	61.5 g

Bael fruit consists of numerous functional bioactive compounds. It also contains an indigenous natural antioxidant, large amounts of dietary fiber and high levels of carotenoid, phenolic, flavonoid, ascorbic acid, and alkaloids with strong antioxidant activities. In addition, the Bael fruit has a visually attractive yellowish-orange pulp that is fragrant and flavorful. Most of the volatile compounds in the fruit are monoterpenes and sesquiterpenes. Marmelosin, luvangetin, aurapten, marmelide, and tannins are well known phytoconstituents of Bael fruit. The bioactive constituents of Bael, which include alkaloids, coumarins, polysaccharides, and essential oils that are present in relatively large quantities, exhibit different medicinal properties (Table 9.1) (Riccardi et al., 2011; Martirosyan et al., 2015).

TABLE 9.1 List of Reported Bioactive Constituents Detected in Bael Fruits and Leaves

Bioactive Compound Class	Common Name of Bioactive Compound	IUPAC Name of Bioactive Compound	Molecular Formula	Molecular Weight (g/mol)
Alkaloid	Aegeline	N-(2-Hydroxy-2-(4-methoxyphenyl) ethyl) cinnamamide	$C_{18}H_{19}NO_3$	297.3
Alkaloid	Aegelinine (Halfordinol)	4-(2-pyridin-3-yl-1,3-oxazol-5-yl) phenol	$C_{14}H_{10}N_2O_2$	238.24
Alkaloid	Fragin	(Z)-hydroxyimino-[3-methyl-1-(octanoylamino) butan-2-yl]-oxidoazanium	$C_{13}H_{27}N_3O_3$	273.37
Alkaloid	O-Methylhalforodinine	5-[4-(3-methylbut-2-enoxy) phenyl]-2-pyridin-3-yl-1,3-oxazole	$C_{19}H_{18}N_2O_2$	306.4
Alkaloid	O-Isopentanylhalfordinol	5-[4-(3-methylbut-2-enoxy) phenyl]-2-pyridin-3-yl-1,3-oxazole	$C_{19}H_{18}N_2O_2$	306.40
Alkaloid	N-2-[4-(3,'3'-Dimethylallyloxy) Phenyl]Ethyl Cinnamide	(Z)-N-[2-[4-(3-methylbut-2-enoxy)phenyl] ethyl]-3-phenylprop-2-enamide	$C_{22}H_{25}NO_2$	335.4
Alkaloid	O-(3,3-Dimethylallyl) Halfordinol	5-[4-(3-methylbut-2-enoxy) phenyl]-2-pyridin-3-yl-1,3-oxazole	$C_{19}H_{18}N_2O_2$	306.4
Alkaloid	Alkaloid B or Ethyl Cinnamide	(E)-N-[2-ethoxy-2-(4-methoxyphenyl) ethyl]-3-phenylprop-2-enamide	$C_{20}H_{23}NO_3$	325.4
Alkaloid	Alkaloid D or Ethyl Cinnamide	(E)-N-[2-methoxy-2-(4-methoxyphenyl) ethyl]-3-phenylprop-2-enamide	$C_{19}H_{21}NO_3$	311.4
Carotene	Carotenoids	(2S,5S,6R)-2-[(6E,8E,10E,12E,14E,16E,18E,20E,22E,24E)-2,6,10,14,19,23-hexamethyl-25-(2,6,6-trimethylcyclohexen-1-yl) pentacosa-6,8,10,12,14,16,18,20,22,24-decaen-2-yl]oxy-6-(hydroxymethyl) oxane-3,4,5-triol	$C_{46}H_{68}O_6$	717

TABLE 9.1 *(Continued)*

Bioactive Compound Class	Common Name of Bioactive Compound	IUPAC Name of Bioactive Compound	Molecular Formula	Molecular Weight (g/mol)
Coumarin	Marmelosin or Marmelide or Imperatorin	9-(3-methylbut-2-enoxy)furo[3,2-g]chromen-7-one	$C_{16}H_{14}O_4$	270.28
Coumarin	Marmesin	(2S)-2-(2-hydroxypropan-2-yl)-2,3-dihydrofuro[3,2-g]chromen-7-one	$C_{14}H_{14}O_4$	246.26
Coumarin	Marmin	7-[(E,6R)-6,7-dihydroxy-3,7-dimethyloct-2-enoxy]chromen-2-one	$C_{19}H_{24}O_5$	332.4
Coumarin	Alloimperatorin	9-hydroxy-4-(3-methylbut-2-enyl)furo[3,2-g]chromen-7-one	$C_{16}H_{14}O_4$ 270.28	270.28
Coumarin	Methyl Ether	methoxymethane	C_2H_6O or CH_3OCH_3	46.07
Coumarin	Xanthotoxol	9-hydroxyfuro[3,2-g]chromen-7-one	$C_{11}H_6O_4$	202.16
Coumarin	Scoparone	6,7-dimethoxychromen-2-one	$C_{11}H_{10}O_4$	206.19
Coumarin	Scopoletin	7-hydroxy-6-methoxychromen-2-one	$C_{10}H_8O_4$	192.17
Coumarin	Umbeliferone	7-hydroxychromen-2-one	$C_9H_6O_3$	162.14
Coumarin	Marmenol	7-(2,6-dihydroxy-7-methoxy-7-methyloct-3-enoxy)chromen-2-one	$C_{19}H_{24}O_6$	348.4
Coumarin	Psoralen	furo[3,2-g]chromen-7-one	$C_{11}H_6O_3$	186.16
Flavonoid	Rutin	2-(3,4-dihydroxyphenyl)-5,7-dihydroxy-3-[(2S,3R,4S,5S,6R)-3,4,5-trihydroxy-6-[[(2R,3R,4R,5R,6S)-3,4,5-trihydroxy-6-methyloxan-2-yl]oxymethyl]oxan-2-yl]oxychromen-4-one	$C_{27}H_{30}O_{16}$	610.5

TABLE 9.1 *(Continued)*

Bioactive Compound Class	Common Name of Bioactive Compound	IUPAC Name of Bioactive Compound	Molecular Formula	Molecular Weight (g/mol)
Flavonoid	Flavone	2-phenylchromen-4-one	$C_{15}H_{10}O_2$	222.24
Flavonoid	Flavan-3-ols	2-(4-hydroxyphenyl)-3,4-dihydro-2H-chromene-3,5,7-triol	$C_{15}H_{14}O_5$	274.27
Flavonoid	Flavonol 3-O-D-glycosides	2-phenyl-3-[(2*S*,3*R*,4*S*,6*R*)-3,4,5-trihydroxy-6-(hydroxymethyl)oxan-2-yl]oxychromen-4-one	$C_{21}H_{20}O_8$	400.4
Polysaccharide	Galactose	(3R,4S,5R,6R)-6-(hydroxymethyl)oxane-2,3,4,5-tetrol	$C_6H_{12}O_6$	180.16
Polysaccharide	Arabinose	(3R,4S,5S)-oxane-2,3,4,5-tetrol	$C_5H_{10}O_5$	150.13
Polysaccharide	Glucuronic Acid	(2S,3S,4S,5R)-3,4,5,6-tetrahydroxyoxane-2-carboxylic acid	$C_6H_{10}O_7$	194.14
Polysaccharide	L-Rhamanose	(2R,3R,4R,5R,6S)-6-Methyloxane-2,3,4,5-tetrol	$C_6H_{12}O_5$	164.157
Phenylpropanoids	Hydroxycoumarin	3-hydroxychromen-2-one	$C_9H_6O_3$	162.14
Phenylpropanoids	Lignan	ethyl 6,7-dimethoxy-3-methyl-4-oxo-1-(3,4,5-trimethoxyphenyl)-2,3-dihydro-1H-naphthalene-2-carboxylate	$C_{25}H_{30}O_8$	458.5
Phenylpropanoids	2-Phenylpropene	prop-1-en-2-ylbenzene	C_9H_{10}	118.18
Minor constituent	Ascorbic acid	(2*R*)-2-[(1*S*)-1,2-dihydroxyethyl]-3,4-dihydroxy-2*H*-furan-5-one	$C_6H_8O_6$ or $HC_6H_7O_6$	176.12

TABLE 9.1 *(Continued)*

Bioactive Compound Class	Common Name of Bioactive Compound	IUPAC Name of Bioactive Compound	Molecular Formula	Molecular Weight (g/mol)
Minor constituent	Sitosterol	(3*S*,8*S*,9*S*,10*R*,13*R*,14*S*,17*R*)-17-[(2*R*,5*R*)-5-ethyl-6-methylheptan-2-yl]-10,13-dimethyl-2,3,4,7,8,9,11,12,14,15,16,17-dodecahydro-1*H*-cyclopenta[a]phenanthren-3-ol	$C_{29}H_{50}O$	414.7
Minor constituent	A-Amyrin	(3*S*,4*aR*,6*aR*,6*bS*,8*aR*,11*R*,12*S*,12*aR*, 14*aR*,14*bR*)-4,4,6*a*,6*b*,8*a*,11,12,14*b*-octamethyl-2,3,4*a*,5,6,7,8,9,10,11,12,12*a*,14,14*a*-tetradecahydro-1*H*-picen-3-ol	$C_{30}H_{50}O$	426.7
Tannin	Skimmianine	4,7,8-trimethoxyfuro[2,3-b]quinoline	$C_{14}H_{13}NO_4$	259.26
Terpenoid	α-Phellandrene	2-methyl-5-propan-2-ylcyclohexa-1,3-diene	$C_{10}H_{16}$	136.23
Terpenoid	p-cymene	1-methyl-4-propan-2-ylbenzene	$C_{10}H_{14}$ or $CH_3C_6H_4CH(CH_3)_2$	134.22
Terpenoid	d-Limonene	(4*R*)-1-methyl-4-prop-1-en-2-ylcyclohexene	$C_{10}H_{16}$	136.23
Terpenoid	p-Menth-1-en-3,6-diol	2-methyl-5-propan-2-ylcyclohex-2-ene-1,4-diol	$C_{10}H_{18}O_2$	170.25
Terpenoid	Cineol	1,3,3-trimethyl-2-oxabicyclo[2.2.2]octane	$C_{10}H_{18}O$	154.25
Terpenoid	Citronellal	3,7-dimethyloct-6-enal	$C_{10}H_{18}O$	154.25
Terpenoid	Citral	(2*E*)-3,7-dimethylocta-2,6-dienal	$C_{10}H_{16}O$	152.23
Terpenoid	Cuminalhehyde	4-propan-2-ylbenzaldehyde	$C_{10}H_{12}O$	148.2

Source: Patker et al. (2012); Sawale et al. (2018); Murthy et al. (2020); and https://pubchem.ncbi.nlm.nih.gov.

9.9 MEDICINAL PROPERTIES

Since the 1500s, this fruit was utilized by the East Indians and Portuguese as well as British colonials to cure dysentery and diarrhea. According to modern medicine, half-ripe or unripe Bael fruit is used as a treatment for prolonged diarrhea and dysentery without fever. *A. marmelos* is used for controlling diabetes mellitus in Indian medicinal systems such as Ayurveda, Unani, and Siddha. Studies have reported the cytotoxic or anticancer activity of *A. marmelos*. The anticancer activity of Bael leaves was reported in a hydroalcoholic extract in an animal model; an analysis of the mechanism of action revealed that the initiation of apoptosis was caused by skimmianine present in the hydroalcoholic extract.

9.10 ANTIOXIDANT SHIELD

Antioxidants are the first line of defense against free radical activity and are crucial for preserving health. When the availability of antioxidants is limited, cumulative oxidative damage is observed. The requirement of antioxidants increases with exposure to free radicals. Several mechanisms exist in tissues to protect against damage caused by free radicals and reactive oxygen species (ROS). These mechanisms have different targets and function in distinct cellular compartments. An important line of defense is enzyme systems such as superoxide dismutase, catalase, and glutathione peroxidase, which regulate the levels of the most damaging ROS in tissues (Percival, 1998).

Another mechanism for defense against the damage induced by free radicals in the presence of antioxidant chemicals. An antioxidant molecule can be characterized as a molecule that is sufficiently stable to donate an electron to a free unhindered radical for neutralizing it. Neuralization reduces its capacity to cause damage. Some antioxidants consisting of glutathione, uric acid and ubiquinol are produced during normal metabolic processes. Others are acquired through the regular diet. Among 4,000 known antioxidants, the most effective antioxidants are carotenoids, vitamin C and vitamin E. Countless additional food substances, such as phenolics and polyphenolic complexes, exhibit antioxidant properties because they are significant for health. The third line of antioxidative defense is a complex group of enzymes designed to repair oxidative damage in DNA, lipids, and proteins and limit the continuous spread of peroxyl lipid radicals through chain reactions. These enzymes restore the damage done to biomolecules and reconstruct damaged cell membranes. These enzymes include lipase, protease, DNA repair enzymes, methionine sulfoxide reductase and transferases (Table 9.2) (Percival, 1998).

TABLE 9.2 Main Cellular Antioxidants

First Line of Defense (Antioxidant Enzymes)	Second Line of Defense (Radical Scavengers)	Third Line of Defense (Repair Enzymes)
Superoxide dismutase	Glutathione	Proteases
Catalase	Ascorbic acid	DNA repair enzymes
Glutathione peroxidase	α-tocopherol	Methionine sulfoxide reductase
Glutathione reductase	Uric acid	Transferases

Superoxide dismutase is foremost protective redox enzyme in cells and serves as a crucial antioxidant. Superoxide dismutase and related enzymes are the most plentiful antioxidant enzymes in animals. Cellular concentrations of superoxide dismutase relative to metabolic activity is a reliable predictor of life span.

Catalase is an enzyme present in all aerobic eukaryotes, and it catalyzes the disintegration of hydrogen peroxide to form gaseous oxygen and water. Catalase has maximum turnover number when compared to other enzymes. One unit of catalase can convert millions of units of hydrogen peroxide into water and oxygen per second. Glutathione peroxidase is a common name of the enzyme group with peroxidase activity, which is the primary biological means of protecting targets biomolecules from oxidative damage (Muller et al., 2007).

Vitamins are known for reducing molecular damage caused by ROSs and reactive nitrogen species (RNSs). Vitamin C, a vital water-soluble antioxidant, is an important micronutrient for metabolic activities in the body. It also helps in the regeneration of vitamin E with glutathione or compounds that can easily donate reducing equivalents. This vitamin is crucial in neutrophil function. It readily scavenges ROS as well as NO and hypochlorous acid. It can protect lipids and lipoproteins against oxidative damage (Frei, 1991, 1994; Noroozi et al., 1998).

Vitamin E is the most effective fat-soluble antioxidant. A phenolic hydroxyl unit of tocopherol reacts with organic peroxyl radicals to create corresponding hydroperoxides and tocopheryl radicals. Thus, vitamin E functions like an antioxidant that breaks the chain during the process of lipid peroxidation in cell membrane layers. The generated hydroperoxides are subsequently detoxified through a non-radical reaction. However, tocopherol may be restored by reduction through interaction with other antioxidants. The tocopherol radical is rendered ineffective for more lipid peroxidation termination. It is regenerated in its condensed form by other antioxidants (Burton et al., 1982).

Glutathione is an endogenous intracellular thiol-containing tripeptide consisting of cysteine, glycine, and glutamic acid. Glutathione is an essential cellular antioxidant involved in many cellular processes, such as transport of amino acid, protection against oxidative stress, detoxification of xenobiotics and drug metabolism (Wang et al., 1998).

The products of metabolism are free radicals, which are potentially harmful to cells. Damage caused by free radicals is minimized by antioxidants and repair mechanisms existing inside cells. Hence, the extent of oxidative damage in metabolically functional tissues of healthy individual with a sufficient dietary intake is usually insignificant, and if damage occurs, most of it is repaired (Halliwell et al., 1992).

9.11 ANTIOXIDATIVE AND ANTIDIABETIC PROPERTIES OF *A. MARMELOS*

Normal metabolic activities in the human body produce free radicals. Free radicals, such as ROSs, cause oxidative stress. ROSs are considered harmful to the human body because they damage macromolecules including DNA, proteins, and lipids. Thus, antioxidants are beneficial because they scavenge free radicals such as ROSs and reduce oxidative stress in the human body. Literature revealed that Bael fruits exhibit significant amounts of antioxidant activity. Charoensiddhi et al. in 2010 and Patel et al. in 2012 reported the antioxidant properties of Bael fruits and leaves. The antioxidant activity of aqueous extracts of *A. marmelos* leaves was tested in male albino rats (Patkar et al., 2012). They estimated the amounts of glucose, urea, glutathione-S-transferase (GST) in plasma, malondialdehyde (MDA) and glutathione (GSH) quantities in erythrocytes in the male albino rats. They also reported a decrease in concentrations of blood glucose and malondialdehyde and an increase in the concentrations of erythrocyte glutathione and plasma glutathione-S-transferase compared with the control at the end of the last week. Thus, they concluded that the aqueous extract of *A. marmelos* leaves is useful for long-term control of diabetes. Simultaneously, Patel et al. (2012) reported the antioxidant and hepatoprotective activities of *A. marmelos*. Nigam et al. (2015) reviewed literature and reported that extracts of *A. marmelos* efficiently reduced alloxan-induced oxidative stress and caused a decrease in blood sugar levels. This report was contrary to the results obtained in a study on the effect of methanolic leaf extracts of *A. marmelos* on antioxidants such as reduced glutathione, glutathione reductase, glutathione peroxidase, catalase, and superoxide dismutase (SOD). The study showed that the activity was dose related; furthermore, an increase in antioxidant activity and decrease in lipid peroxidation were observed after treatment with *A. marmelos* leaf extract. They also reported a substantial decline in alloxan-induced lipid peroxidation and hydroperoxide levels in the serum and liver. The aqueous extract of *A. marmelos* leaves showed hypoglycemic and antioxidant activities. The concentrations of glucose, urea, and GST (glutathione-S-transferase) in the plasma and glutathione (GSH) and malondialdehyde (MDA) levels in the erythrocytes of alloxan-induced diabetic rats were investigated. Glutathione (GSH) levels declined in erythrocytes, whereas glutathione-S-transferase (GST) and malondialdehyde (MDA) levels in plasma increased in diabetic male albino rats; however, the aforementioned level was restored to normal with the administration of *A. marmelos* leaf extract, which indicated the antioxidant potential of *A. marmelos* leaves.

Nigam et al. in 2015 confirmed that eugenol and marmesinin are responsible for the antioxidant activity of Bael because these compounds have independently shown their activity against oxidative stress. The aforementioned study also investigated the *in vitro* antioxidant activity of the methanolic extract of *A. marmelos* leaf by testing DPPH scavenging activity. The results data revealed that *A. marmelos* methanolic extracts showed reasonable antioxidant activity with an IC50 value of 23 ± 0.08; hence, the methanolic leaf extract may be used as an inhibitor of free radicals. Phytochemical screening of methanolic and water extracts of *A. marmelos* leaves revealed that they are rich in ascorbic acid, flavonoids, glutathione, polyphenols, saponins, reducing sugars and terpenoids. The water extract showed higher antioxidant activity

(92%) than the standard antioxidant BHT (81%). The substantial radical scavenging activity can be attributed to high concentrations of ascorbic acid, glutathione, and tocopherol. The study concluded that treatment with *A. marmelos* leaf extracts at doses of 100 and 200 mg/kg of body weight for approximately 35 days substantially reduced lipid peroxidase activity. When the levels of lipids, lipoproteins, and antioxidant enzymes in isoproterenol-treated rats were analyzed, the effect of *A. marmelos* leaf extract at 200 mg/kg of body weight was observed to be equivalent to effect of alpha-tocopherol at a dose of 60 mg/kg of body weight.

Abdallah et al. (2017) showed that alloxan treatment resulted in a considerable increase in the amount of MDA and a considerable decline in the activities of GSH and SOD compared with the control group. Treatment of the diabetic rats with an extract of *A. marmelos* fruit resulted in remarkable amelioration of the adverse effects of diabetes; moreover, a significant improvement in the levels of GSH and MDA and SOD activities were observed once compared with the untreated diabetic rat. The fruit extract was found to be extremely effective at a dose of 500 mg/kg of body weight; it significantly inhibited oxidative stress in the diabetic group. The treatment of alloxan-induced diabetes in mice by using *A. marmelos* fruit extract considerably reduced the MDA level and increased the GSH and SOD levels. Furthermore, treatment with *A. marmelos* fruit extract considerably reduced the amount of thiobarbituric acid, a reactive substance, and hydroperoxides in the diabetic rats; however, the treatment increased the levels of SOD and GSH in the tissues and plasma. Treatment with *A. marmelos* fruit extract before irradiation notably reduced lipid peroxidation and considerably increased the level of GSH in mouse tissues. The antioxidative effect of *A. marmelos* fruit extract has a potent *in vitro* antioxidant activity, which may be attributed to its content of bioactive compounds. The amelioration of lipid peroxidation by fruit extract of *A. marmelos* was attributable to antioxidative phytochemicals present in it, mainly flavonoids.

Sonar et al. (2020) reported that *A. marmelos* extract possessed potent antioxidant activity because of the presence of phytochemicals such as tannins, phenolics, and flavonoids. The overall antioxidant activity of the extract is high because of the presence of electron-donating molecules because the molecules reacted with free radicals and interrupted the chain reaction of the radicals. Some reports also revealed that consuming Bael extract at a high dose, approximately 25% of body weight, increases the activity of ROS scavengers such as glutathione peroxidase, catalase, glutathione reductase, and SOD; furthermore, the results are better than those obtained with glibenclamide. Antioxidant activity has been attributed to alkaloids, flavonoids, flavonoid glycosides, phlobatannins, sterols, and tannins (Patel et al., 2012; Murthy et al., 2020).

Marmelosin in *A. marmelos* was quantified using HPLC analysis. Marmelosin is an alkaloid and is a potential antioxidant and an antidiabetic substance extracted from the underutilized Bael fruit and leaves. The maximum amount of marmelosin was found in sample 04, which had 0.0055% marmelosin in the leaves and 1.336% in the fruit.

Plant	**Leaf Extract (%)**	**Fruit Extract (%)**
A. marmelos 01	0.0045	1.260
A. marmelos 02	0.0040	1.258
A. marmelos 03	0.0050	1.300
A. marmelos 04	0.0055	1.336

Several studies have reported that aqueous extracts of *A. marmelos* seeds reduce blood glucose concentration; consequently, it is considered useful for treating patients with acute diabetes. Bael extract when consumed at a high dose of 25% of body weight is more effective than the antidiabetic drug glibenclamide. The antidiabetic effect of the extract is because of bioactive compounds in the fruit; these compounds stimulate beta cells of islet of Langerhans for producing insulin (Anjaria et al., 2014).

A study by Nigam et al. in 2015 demonstrated the hypoglycemic effect of *A. marmelos* extract in rats with glucose-stimulated hyperglycemia for 7 successive days at an oral dose of 250 mg/kg of body weight. Treatment with the extract substantially improved the capability of the rats to utilize an external glucose burden. The standard blood glucose concentration was reduced by 67% after *A. marmelos*. The efficacy of the *A. marmelos* extract was 71% of that of glibenclamide. The effect of *A. marmelos* extracts prepared in water, alcohol, and petroleum ether of were tested for hypoglycemic activity. The aqueous and alcoholic extracts caused hypoglycemia at 500 mg/kg in the normal fasted rabbits; however, the petroleum ether extract did not affect blood sugar levels.

The results of a clinical trial of a trademarked formulation containing *A. marmelos* leaves showed that the formulation effectively treated patients with slight to moderate diabetes in addition to insulin-dependent patients. The treatment increases insulin secretion from the pancreas. Treatment of diabetic pancreas with *A. marmelos* leaf extract showed an improvement in the efficiency of the pancreatic beta cells. This study indicates the hypoglycemic effect of *A. marmelos* leaf extract as well as its role in the restoration of damaged pancreas.

The potential of *A. marmelos* leaf extract as an antidiabetic drug in rats with streptozotocin-induced diabetes was assessed. The extract exhibited an insulin-like effect by restoring of the blood glucose; moreover, it restored body weight to normal levels. An aqueous extract of *A. marmelos* leaves in a powdered form at doses equivalent to 1 g/kg/day produced significant ($p < 0.01$) anti-hyperglycemic effects within 3 days in rabbits with alloxan-stimulated diabetes. Similarly, in normal rabbits, the treatment reduced blood glucose levels by up to 35.3% after 4 hours of administration; furthermore, moderate hypoglycemics effects were recorded after 12 hours. The potential antidiabetic effects of the *A. marmelos* leaf extract in diabetic rats was explored. In one group of diabetic animals, insulin was administered through injection, and in the other group *A. marmelos* leaf extract was administered orally. This study confirmed that the active principle in *A. marmelos* leaf extract has hypoglycemic activity analogous to that of insulin.

9.12 CONCLUSION

Foods are assessed extensively for their additional physiological benefits, such as reducing the risk of some diseases and improving health. The literature clearly indicates that weight reduction in overweight patients reduces the risk of several associated diseases and improves their life expectancy at all ages. Functional and antidiabetic foods like *A. marmelos* fruits are remarkably effective for prevention and/or treatment of diabetes because they prevent overweight and obesity. If consumed as a functional food *A. marmelos* can affect the regulation of the energy balance equation by controlling energy intake and/or the amount of energy dissipated as heat (thermogenesis); this regulation is manifested as the antidiabetic effect of *A. marmelos*.

KEYWORDS

- ***Aegle marmelos***
- **antidiabetic**
- **body mass index**
- **food for specified health use**
- **functional food**
- **glutathione**
- **glutathione-s-transferase**
- **irritable bowel syndrome**
- **malondialdehyde**
- **obesity**
- **reactive nitrogen species**
- **reactive oxygen species**

REFERENCES

Abdallah, I. Z., Salem, I., El-Salam, A., & Nayrouz, A. S., (2017). Evaluation of antidiabetic and antioxidant activity of *Aegle marmelos* L. correa fruit extract in diabetic rats. *The Egyptian Journal of Hospital Medicine, 67*(2), 731–741.

Anand, S. S., Hawkes, C., De Souza, R. J., Mente, A., Dehghan, M., Nugent, R., & Kromhout, D., (2015). Food consumption and its impact on cardiovascular disease: Importance of solutions focused on the globalized food system: A report from the workshop convened by the World Heart Federation. *Journal of the American College of Cardiology, 66*(14), 1590–1614.

Anjaria, K., Mukundan, U., & Dawda, H., (2014). *In Vitro Culture & Efficacy Studies of two Antidiabetic Plants* (pp. 1–244). Lambert Academic Publishing House.

Azzolino, D., Arosio, B., Marzetti, E., Calvani, R., & Cesari, M., (2020). Nutritional status as a mediator of fatigue and its underlying mechanisms in older people. *Nutrients, 12*(2), 444.

Bhatt, H. B., & Smith, R. J., (2015). Fatty liver disease in diabetes mellitus. *Hepatobiliary Surgery and Nutrition, 4*(2), 101–108.

Burton, G. W., Joyce, A., & Ingold, K. U., (1982). First proof that vitamin E is major lipid-soluble, chain-breaking antioxidant in human blood plasma. *Lancet (London, England), 2*(8293), 327–327.

Charoensiddhi, S., & Anprung, P., (2010). Characterization of Bael fruit (*Aegle marmelos* [L.] correa) hydrolysate as affected by enzyme treatment. *Journal of Food Biochemistry, 34*(6), 1249–1267.

De Pergola, G., & Silvestris, F., (2013). Obesity as a major risk factor for cancer. *Journal of Obesity*, 1–11.

Dhaka, V., Gulia, N., Ahlawat, K. S., & Khatkar, B. S., (2011). Trans fats—sources, health risks and alternative approach: A review. *Journal of Food Science and Technology, 48*(5), 534–541.

Frei, B., (1991). Ascorbic acid protects lipids in human plasma and low-density lipoprotein against oxidative damage. *The American Journal of Clinical Nutrition, 54*(6), 1113S–1118S.

Frei, B., (1994). Reactive oxygen species and antioxidant vitamins: Mechanisms of action. *The American Journal of Medicine, 97*(3), S5–S13.

Gómez-Pinilla, F., (2008). Brain foods: The effects of nutrients on brain function. *Nature Reviews Neuroscience, 9*(7), 568–578.

Gupta, P., Gupta, N., Pawar, A. P., Birajdar, S. S., Natt, A. S., & Singh, H. P., (2013). Role of sugar and sugar substitutes in dental caries: A review. *International Scholarly Research Notices*, 1–5.

Halliwell, B. A. R. R. Y., Gutteridge, J. M., & Cross, C. E., (1992). Free radicals, antioxidants, and human disease: Where are we now? *The Journal of Laboratory and Clinical Medicine*, *119*(6), 598–620.

Hasler, C. M., (2002). Functional foods: Benefits, concerns, and challenges—a position paper from the American council on science and health. *The Journal of Nutrition, 132*(12), 3772–3781.

Lindström, J., & Virtanen, S. M., (2011). Functional foods and prevention of diabetes. In: *Functional Foods* (pp. 261–276). Woodhead Publishing.

Marriott, B. M., (1994). Effects of nutrients on neurotransmitter release. In: *Food Components to Enhance Performance: An Evaluation of Potential Performance-Enhancing Food Components for Operational Rations*. National Academies Press (US).

Martirosyan, D. M., & Singh, J., (2015). A new definition of functional food by FFC: What makes a new definition unique? *Functional Foods in Health and Disease, 5*(6), 209–223.

Muller, F. L., Lustgarten, M. S., Jang, Y., Richardson, A., & Van, R. H., (2007). Trends in oxidative aging theories. *Free Radical Biology and Medicine, 43*(4), 477–503.

Murthy, H. N., Bhat, M. A., & Dalawai, D., (2020). Bioactive compounds of Bael (*Aegle marmelos* (L.) Correa). *Bioactive Compounds in Underutilized Fruits and Nuts,* 459–486.

Nigam, V., & Nambiar, V. S., (2015). Therapeutic potential of *Aegle marmelos* (L.) Correa leaves as an antioxidant and antidiabetic agent: A review. *International Journal of Pharma Sciences and Research, 6*(3), 611–621.

Noroozi, M., Angerson, W. J., & Lean, M. E., (1998). Effects of flavonoids and vitamin C on oxidative DNA damage to human lymphocytes. *The American Journal of Clinical Nutrition, 67*(6), 1210–1218.

París, A. S., Carceller, D. B., Lacleta, I. M., & Gamboa, R. A., (2013). Sugar and diabetes; international recommendations. *Nutr. Hosp., 28*(4), 72–80.

Patel, P. K., Sahu, J., Sahu, L., Prajapati, N. K., & Dubey, B. K., (2012). Aegle marmelos: A review on its medicinal properties. *International Journal of Pharmaceutical and Phytopharmacological Research, 1*(5), 332–341.

Patkar, A. N., Desai, N. V., Ranage, A. A., & Kalekar, K. S., (2012). A review on *Aegle marmelos*: A potential medicinal tree. *International Research Journal of Pharmacy, 3*(8), 86–91.

Percival, M., (1998). Antioxidants clinical nutrition insights. *Advanced Nutrition, 31*, 201–205.
Qin, L., Knol, M. J., Corpeleijn, E., & Stolk, R. P., (2010). Does physical activity modify the risk of obesity for type 2 diabetes: A review of epidemiological data. *European Journal of Epidemiology, 25*(1), 5–12.
Rajbhar, K., Dawda, H., & Mukundan, U., (2016). Tea polyphenols: A preventive measure to attenuate obesity and type-2 diabetes. *World Journal of Pharmaceutical Sciences, 4*(3), 346–350.
Rao, T. S., Asha, M. R., Ramesh, B. N., & Rao, K. J., (2008). Understanding nutrition, depression, and mental illnesses. *Indian Journal of Psychiatry, 50*(2), 77–82.
Riccardi, G., Capaldo, B., & Vaccaro, O., (2011). Functional foods for diabetes and obesity. *Nutritional and Metabolic Bases of Cardiovascular Disease,* 138–146.
Romieu, I., Dossus, L., Barquera, S., Blottière, H. M., Franks, P. W., Gunter, M., & Willett, W. C., (2017). Energy balance and obesity: What are the main drivers. *Cancer Causes & Control, 28*(3), 247–258.
Sami, W., Ansari, T., Butt, N. S., & Ab Hamid, M. R., (2017). Effect of diet on type 2 diabetes mellitus: A review. *International Journal of Health Sciences, 11*(2), 65.
Sawale, K. R., Deshpande, H. W., & Kulkarni, D. B., (2018). Bael (*Aegle marmelos*) a super fruit of an hour: A review. *Int. J. Chem. Studies, 6*(3), 1720–1723.
Sawicka-Pierko, A., Obuchowska, I., & Mariak, Z., (2014). Nutritional optic neuropathy. *Klinika Oczna, 116*(2), 104–110.
Sharma, N., & Dubey, W., (2013). History and taxonomy of *Aegle marmelos*: A review. *International Journal of Pure and Applied Bioscience, 1*(6), 7–13.
Singh, A., Sharma, H. K., Kaushal, P., & Upadhyay, A., (2014). Bael (*Aegle marmelos* Correa) products processing: A review. *African Journal of Food Science, 8*(5), 204–215.
Sonar, M. P., & Rathod, V. K., (2020). Microwave-assisted extraction (MAE) used as a tool for rapid extraction of marmelosin from *Aegle marmelos* and evaluations of total phenolic and flavonoids content, antioxidant, and anti-inflammatory activity. *Chemical Data Collections, 30*, 1–13.
Wan, J. J., Qin, Z., Wang, P. Y., Sun, Y., & Liu, X., (2017). Muscle fatigue: General understanding and treatment. *Experimental & Molecular Medicine, 49*(10), 1–11.
Wang, W., & Ballatori, N., (1998). Endogenous glutathione conjugates: Occurrence and biological functions. *Pharmacological Reviews, 50*(3), 335–356.
Webpage – National Library of Medicine NCBI. https://pubchem.ncbi.nlm.nih.gov (accessed on 16 December 2021).
Wilcox, G., (2005). Insulin and insulin resistance. *Clinical Biochemist Reviews, 26*(2), 19–39.
Yau, Y. H., & Potenza, M. N., (2013). Stress and eating behaviors. *Minerva Endocrinologica, 38*(3), 255.

CHAPTER 10

Garlic (*Allium sativum*): A Potential Antidiabetic Agent

PRABHASH KUMAR PANDEY,[1,2] JAYANT DEWANGAN,[2] SHAMBHOO SHARAN TRIPATHI,[1] RANJAN SINGH,[3] FARRUKH JAMAL,[4] and SRIKANTA KUMAR RATH[2]

[1]*Department of Biochemistry, Faculty of Science, University of Allahabad, Prayagraj, Uttar Pradesh, India, E-mails: prabhashpandey@allduniv.ac.in; pandey.prabhash21@gmail.com (P. K. Pandey)*

[2]*Genotoxicity Lab, Division of Toxicology and Experimental Medicine, CSIR-Central Drug Research Institute, Lucknow, Uttar Pradesh, India*

[3]*Department of Microbiology, Dr. Rammanohar Lohia Avadh University, Ayodhya, Uttar Pradesh, India*

[4]*Department of Biochemistry, Dr. Rammanohar Lohia Avadh University, Ayodhya, Uttar Pradesh, India*

ABSTRACT

Diabetes mellitus is one of the world's significant wellness issues, affecting the urban population more than the rural areas. The prevalence of this disease is increasing speedily day by day. Effective regulation of blood glucose levels is the most critical factor in decreasing the dangers of diabetic complexities. Chemically synthesized antidiabetic drugs available in the market have many side effects; therefore, plant-derived substances may provide a better alternative medicine to combat diabetes.

Antidiabetic Plants for Drug Discovery: Pharmacology, Secondary Metabolite Profiling, and Ingredients with Insulin Mimetic Activity. Deepu Pandita, MPhil, Anu Pandita, MSc, & Chander Bhanu, MBBS, PGDDCE (Eds.)
© 2023 Apple Academic Press, Inc. Co-published with CRC Press (Taylor & Francis)

These natural molecules are presumed to be effective, economical as well as with no side effects. Among different accessible plants, garlic (*Allium sativum*), a typical cooking flavor, and a long history as a people cure have antidiabetic potential. According to several reports, garlic's antiglycation, antioxidative, and anti-inflammatory properties have been related to its function in preventing diabetes. Notwithstanding, there is no standard concession to utilizing garlic for antidiabetic purposes, principally due to less logical proof from human examinations and discrepant information from animal studies.

A promising approach to cure this disease by garlic plant molecules focuses throughout this chapter. In this chapter, the authors collected the scientific evidence available throughout the various experimental platforms and literature related to the garlic plant's functional role in improving the blood sugar level of diabetic patients. This chapter focuses on the pharmacology, secondary metabolite profiling, ingredients of garlic plants with insulin-mimetic activity, and its health benefits. Garlic supplements are useful in treating diabetic patients, and this chapter content may disclose a path for the researchers to combat this disease in the future.

10.1 INTRODUCTION

Garlic (*Allium sativum*) has a place with the family Liliaceae utilized as a therapeutic since ancient times and is the oldest cultivated plant. It contains several beneficial compounds such as vitamins, minerals, and others to maintain human health. Because of its medicinal value whole plant is used to treat various ailments of human beings and animals. Researchers focus on the medicinal value of garlic on human health because of its minimal toxicity attribute and a broad spectrum of therapeutic effects.

The prevalence of diabetes mellitus is affecting the worldwide population very rapidly. Several allopathic medicines are available to treat this disease, but traditional medicine from natural resources is in demand because of their minimal side effects. Garlic has been utilized as a society cure since antiquated occasions, and its antidiabetic role has been reported in several studies. Various studies said anti-inflammatory, antioxidative, and antiglycative garlic plant attributes made it a suitable source against diabetes. Clinical studies on a large scale confirm garlic plants' job in the avoidance and treatment of diabetes. The pathophysiology of hyperglycemia is very tricky because of the development of resistance against insulin and the inactivation of pancreatic beta cells; therefore type 2 diabetes is spreading rapidly.

Control of blood glucose level is the main target in the management of diabetes to minimize the micro and macrovascular complications (Stettler et al., 2006; Kirkman et al., 2006). Still, it is not so easy to manage it because of its complicated pathophysiology.

The various bioactive mixtures such as polysaccharides, saponins, phenolic compounds, and organic sulfides found in the garlic have several health benefits (Asdaq and Inamdar, 2011). It is a common spice and used as traditional medicine. In past decades several studies reported the various remarkable biological properties of garlic. Several scientific studies support its role as antioxidant, immunomodulatory, anticancer, anti-inflammatory, antidiabetic, cardioprotective, anti-obesity, and other functions (Al-Qattan et al., 2006).

This chapter covers the pharmacology, health benefits, secondary metabolite profiling, and bioactive compounds of the antidiabetic plant *Allium sativum*, emphasizing its insulin-mimetic/secretagogues activity.

10.2 INGREDIENTS AND SECONDARY METABOLITE PROFILING OF GARLIC PLANT

10.2.1 BIOACTIVE COMPOUNDS OF GARLIC

Polysaccharides, saponins, phenolic compounds, and organosulfur contents are the main bioactive ingredients of garlic. Among these compounds, organosulfur compounds like diallyl trisulfide (DATS), diallyl thiosulfonate (Allicin), diallyl disulfide (DADS), E/Z-ajoene, S-allyl-cysteine (SAC), diallyl disulfide (DADS), and S-allyl-cysteine sulfoxide (alliin) are the main bioactive compounds. A 40 times higher amount of saponin was reported in purple garlic than that in white garlic. Along with this, various saponin compounds are found only in the purple garlic-like sativoside R1, proto-degalactotigonin-rhamnose, degalactotigonin-rhamnose, proto-degalactotigonin, voghieroside D1, and sativoside B1-rhamnose. These saponins show more stability during the cooking process. These organosulfur compounds have higher digestibility in raw garlic than cooked garlic. Besides these organosulfur compounds, garlic contains a higher amount of phenolic compounds than any common vegetable. β-resorcylic acid is garlic's main phenolic compound, followed by rutin, pyrogallol, quercetin, gallic acid, and protocatechuic acid. Garlic also contains polysaccharides in the ratio of 85% fructose, 1% galactose, and 14% glucose.

The flavonoid quercetin is majorly present in the garlic interacts with the vitamin C and E (Al-Jaber et al., 2011). Result changes are seen in transferases and cytochrome P450 isozymes in both *in vitro* and *in vivo* modes. An *in vivo* study uncovers that garlic oil and its some allyl sulfide parts quicken the outflow of those enzymes engaged with the liver's detoxification cascades (Wanwimolruk and Prachayasittikul, 2014).

The most bioactive sulfur-containing substance of garlic is Allicin [S-(2-propenyl)-2-propene-1-sulfinothioate] who is behind the odor and taste of garlic (Slusarenko et al., 2008; Rahman, 2007). Thiosulfinates Alliin (S-allyl-L-cysteine sulfoxide) is the primary forerunner of allicin. The compound allyl mercaptan is liable for the garlic smell. It generates from diallyl disulfide or Allicin interactions with the cysteine residues present in the S-ally-mercapto cysteine (Kaye et al., 2000; Lawson and Gardner, 2005). Allicin is a heat-sensitive substance, and heat effectively damages it during the cooking process; Allicin causes allergic reactions, gastrointestinal tract issues, and intolerance (Rishton, 2008; Batiha et al., 2018; Sobolewska et al., 2015).

Alkaloid allicin is the main beneficial constituent of garlic. Besides this compound, other sulfur-containing substances like S-allyl cysteine (SAC), diallyl trisulfide (DTS), and diallyl disulfide (DADS) also have pharmacological roles (Kumar and Robbins, 2003). In aged garlic, the organosulfur compound SAC presents an outstanding amount (Jin et al., 2013). Ajoene works against the viruses, and DADS and DTS are active against the yeast (Meriga et al., 2012).

10.2.2 SECONDARY METABOLITE PROFILE OF GARLIC

Garlic has numerous health benefits, and its medicinal properties have been widely known for over 5,000 years. Along with water, carbohydrates, proteins, and fibers, it has one of the crucial bioactive elements known as organosulfur compounds. Among various constituents present in the garlic, the ratio of organosulfur compounds is about 2.3%. Alliin, cysteine,

S-allyl cysteine (SAC), S-ally-mercapto cysteine (SAMC), Allicin, Ajoene, dithiins, allyl methyl trisulfide, diallyl sulfides (DAS), diallyl disulfides (DADS), diallyl trisulfides (DATS) are the major sulfur compounds, and their metabolites present in the garlic (Zhang et al., 2020).

Allicin (diallyl thiosulfate) was the first sulfur compound discovered in garlic (Cavallito and Bailey, 1944). It is a volatile and odorous metabolite. Glutamyl L-cysteine is the essential forerunner of allicin, which hydrolyzed to alliin (S-allyl cysteine sulfoxide). In crushed garlic, the alliinase enzyme converts the alliin into alkenyl-alkene thiosulfate allicin. This thiosulfate decomposes and forms sulfide products like ajoenes, DAS, DADS, DATS, and dipropyl disulfide.

Additionally, some other metabolites like SAC and SAMC are also synthesized by catabolism of C-glutamyl L-cysteine. While processing, these compounds also transform to several other sulfur derivatives. The most common example is converting allicin into thiosulfates; thus, it is not present in garlic products (Lanzotti et al., 2014). Preclinical studies proved antibacterial activity of allicin; however, these results cannot relate clinically. Acidic condition in the stomach prevents allicin activation. After the garlic utilization, allicin's serum level could not be identified in human blood tests (Freeman and Kodera, 1995). Besides, few other sulfur derivatives are also present in processed garlic and impact performing several biological functions. Compounds like DATS and SAC are known for antimicrobial properties and pro-apoptotic action. Hence, it is instrumental in fighting several diseases (Liu et al., 2009; Lee et al., 1994).

Allium genus is also a rich source of natural polar compounds like saponins, sapogenins, and phenols. They are more heat-stable as compared to sulfur compounds and do not degrade on cooking. The extraction of these molecules can be done in organic solvent using methanol or acetone, followed by purification with medium pressure liquid chromatography (MPLC) and high-performance liquid chromatography (HPLC). Spectroscopic analyzes like HRFABMS and advanced 2D-NMR are the techniques to confirm their structures (Lanzotti, 2012).

Phytochemical investigation of garlic communicates the presence of various bioactive mixtures. One such metabolite is N-feruloyl amides, which exists in two forms: N-feruloytyrosine and N-feruloyltyramine in the roots of *Allium sativum* L (Fattorusso et al., 1999). N-feruloyltyramine possesses an anti-inflammatory response via blocking arachidonate 5-lipoxygenase pathways (Iwakami et al., 1986). Lanzotti et al. (2012a) showed that both the N-feruloyl amides also have potent anti-fungal activity. A typical Italian variety of garlic, *Allium sativum* L. var. Vogliera was discovered with 10 new furostanol saponins Voghieroside A1, Voghieroside A2, Voghieroside B1, Voghieroside B2, Voghieroside C1, Voghieroside C2, Voghieroside D1, Voghieroside D2, Voghieroside E1, Voghieroside E2. This species has two spirostanol saponins, agigenin 3-O-trisaccharide and gitogenin 3-O-tetrasaccharide, along with eugenol glycosides also. The major phenols and saponins showed an anti-fungal response when screened on two common fungal species, *Botrytis cinerea* and *Trichoderma harzianum* (Lanzotti, 2012; Lanzotti et al., 2012a).

Recently, Farag et al. (2017) studied different phytomolecules present in *A. sativum*, where he used the methanol extract of garlic and purified it by UPLCMS-based approach. The study revealed the garlic contains phenolic acid-like Phthalic acid, Caffeic acid, Ferulic acid, and Caffeic acid dimethyl ether. It has two oxylipins, Jasmonic acid-hydroxy-O-sulfate, and trimethylnonanol sulfate. Methanol extract also showed common fatty acids like Oleic acid, Linoleic acid, Stearic acid, Palmitic acid, Oxo-octadecenoic acid, and 9, 12, 13-trihydroxy octadeca-7-enoic acid. An aromatic compound such as Diethyl phthalate, thiamine derivative Allithiamine, organic acids like citric acid/isocitric acid, and several amino acid conjugates or peptides are the constituent's garlic extract. The metabolites present in garlic show intense antimicrobial action (Farag et al., 2017).

10.3 INSULIN SECRETAGOGUES OR INSULIN-MIMETIC ACTIVITY OF GARLIC

10.3.1 HYPOGLYCEMIC EFFECTS

Diabetes mellitus is a usually happening endocrine, metabolic issue, which prompts high glucose levels. Over 100 million individuals have diabetes around the world. In a long-term effect, it delays wound healing, embryopathy, nephropathy, atherosclerosis, neuropathy, cataract, and retinopathy. Diabetic complications can be embroiled with the accumulation of cutting-edge glycation results, prompting the oxidation of glucose and glycated proteins and the formation of reactive oxygen species. The carbonyl gathering of sugar interacts with an amino group of the protein in hyperglycemic conditions. It produces a labile Schiff base, which transforms into a firm Amadori product, also known as the Maillard reaction. This condition leads to protein glycation. The glycated protein further changes to intermediate 3-deoxyglucosone and finally converts into a complex heterogeneous molecule named advanced glycation end-products (Ahmad and Ahmad, 2006).

Protein conformation is altered by the accumulation of better glycation output, which impairs protein function and enzymatic activity (Vlassara and Palace, 2002). Many cells have receptors for these elevated glycation end-products, which activate the oxidative stress response. Inducing cell inflammation increases vascular permeability and may cause diabetic vascular complications (Yan et al., 1994).

Since ancient times, garlic has been utilized as an antidiabetic or to treat diabetes-related complications (Rahman, 2003). One of the garlic extracts, known as aged garlic extract, is rich in antioxidants and organosulfur compounds like S-allyl cysteine (Imai et al., 1994). Preclinical studies revealed that garlic inhibits advanced glycation end-products. Garlic includes organosulfur compounds that reduce protein glycation and lipid oxidation, such as N-acetylcysteine, S-allyl cysteine, S-ethylcysteine, and diallyl sulfide. Garlic extract also stimulates certain antioxidant enzymes like glutathione peroxidase, superoxide dismutase, and catalase; therefore, it may protect from diabetes-associated implications (Ou et al., 2003; Wei and Lau, 1998).

One of the studies showed 0.25 mg/kg ethyl ether extract of garlic imposed an antihyperglycemic response. It mimics the insulin-like activity, reducing the blood sugar level (Ayodhya et al., 2010).

In an alloxan-induced diabetic rat model, treatment with *Allium sativum* extract was evaluated for hypoglycemic impact. It enhances insulin secretion from partial cells of the pancreas (Chauhan et al., 2010). Similarly, allicin and S-allyl cysteine sulfoxide in garlic stimulates pancreatic beta cells for insulin secretion, enhanced hepatic metabolism, and showed secretagogues activity (Modak et al., 2007). Aqueous-ethanolic essence of garlic was given to the streptozotocin-mediated diabetic male Wistar rat model. Daily treatment of 50 mg/kg garlic extract for four weeks reduced blood glucose significantly (Shariatzadeh et al., 2008). Another study supports that garlic extract orally administered at 100 mg/kg decreased blood glucose level by elevating insulin secretion in the blood plasma (Grover et al., 2002). Aqueous insulin extract also triggers insulin secretion on the isolated pancreas in vitro conditions (Mustafa et al., 2007; Patel et al., 2012).

Alliin, also known as S-allyl cysteine sulfoxide, is an organosulfur bioactive compound in garlic that has been testified to possess a hypoglycemic effect in rats. The alliin content of aqueous garlic leaf extract was determined and regularly administered to alloxan-administered diabetic rats. A five-week treatment with leaf extract reduces blood glucose levels and increases serum lipid profile. The alkaline phosphatase level, aspartate aminotransferase, and alanine aminotransferase enzymes in blood serum are lowered by garlic plant leaf extract. The inhibition was more than the drug glibenclamide. Histological studies revealed that in diabetic rats, pancreatic β-cells destruction is more; however, garlic leaf extract-treated rats show better recovery, and their pancreatic β-cells are less harmed (Nasim et al., 2011).

Allicin (0.25 mg/kg orally), the bioactive ingredient of garlic, shows better hypoglycemic activity in the diabetic rabbits reported in a study performed by Mathew and Augusti (1973). Several organic solvent extracts of garlic (0.25 mg/kg, orally) show in alloxan-treated diabetic rabbits antihyperglycemic outcomes. Garlic induces the pancreas to release insulin from beta cells, which is essential to manage the blood sugar level (Augusti and Sheela, 1996). In alloxan-induced diabetic rats, an antioxidant found in garlic called allyl cysteine sulfoxide displayed activity close to that of the insulin hormone.

After taking Ajoene, a garlic derivative, orally for eight weeks, KK-A(y) mice with a diabetes gene showed promising results in preventing diabetes (Hattori et al., 2005). Tahiliani and Kar (2003) reported that garlic extract at a 500 mg/kg/day dose is beneficial in diabetic conditions. S-allyl cysteine, a bioactive compound found in the aged garlic, has the antioxidant property that can hamper advanced glycation end-products accumulation (Kaleem et al., 2006). Adachi et al. (2006) found that oxovanadium (IV), a derivative of garlic bis (allixinato), was the most effective antidiabetic executrix in type 1 diabetic mice when administered as an oral or intraperitoneal injection.

At a dosage of 0.25 mg/kg, ethyl ether extract of garlic is very effective against diabetes (Ayodhya et al., 2010). This hyperglycemic activity was noticed because of the insulin-like action of garlic. In a study, garlic's insulin secretory function in the pancreas substantially decreases blood glucose status in alloxan-administered diabetic rabbits and rats (Nain et al., 2012). Garlic derivative allicin facilitates hepatic metabolism and increases insulin secretion from the pancreas' parietal cells (Nain et al., 2012). S-allyl cysteine sulfoxide (SACS), the forerunner of garlic oil and allicin, was shown to secrete insulin from the rat pancreas' beta cells in an *in vitro* analysis. This compound's antidiabetic effects are likely attributable to its antioxidant and insulin secretagogue properties (Modak et al., 2007).

Other studies also reported that upon treatment with 100 mg/kg of garlic extract, the plasma's insulin level increases, and plasma glucose level decreases (Grover et al., 2002). A high-fat diet feed rat study revealed that garlic does not exactly work as a hypoglycemic agent but as an insulinotropic (Islam and Choi, 2008). Regular oral administration of 100 mg/kg garlic extracts increased plasma insulin levels and decreased plasma glucose levels (Grover et al., 2002). Aqueous garlic (10% v/v) extracts were found to potentiate glucose-induced insulin secretion in the isolated pancreas (Mustafa et al., 2007). Garlic's two-week effect on rats fed a high-fat diet shows that it is insulinotropic, not hypoglycemic (Islam et al., 2008). Garlic helps prevent diabetes by increasing serum insulin levels, glucose metabolism, and/or pancreatic insulin secretion. This chapter's central motto is to collect the various information available for the garlic plant's hypoglycemic activity through the enhanced secretion of insulin from the pancreas' beta cells or act similarly as insulin action reported various scientific platforms. In this chapter, different *in vitro* and *in vivo* studies performed on the insulin-mimetic or secretagogues activity of garlic plants are mentioned.

10.4 PHARMACOLOGY AND HEALTH BENEFITS OF ALLIUM SATIVUM (ALLIACEAE)

Garlic is an herb, vegetable, and flavoring agent that is used all over the world. The aromatic scent and spicy fragrance properties are attributed to allicin and DADS, which are organosulfur compounds. Allicin, like the sulfuric compounds in garlic, is essential for diabetes management. Allicin act as a hypoglycemic compound. In the year 2011, the scientist Ryan reveals that one-third of diabetes patients who took garlic with their medicines show better improvement against diabetes. A study performed by EI Dimirdash et al. (2011) found that garlic juice and a compound diallyl trisulfide reduce the blood sugar in diabetic rats (Liu et al., 2005). Nephropathy during diabetes causes textural changes in the kidney. The use can cure these textural changes of garlic oil (Koch and Lawson, 1995).

Garlic oil improves the antioxidant system and slows down renal damages. These renal damages caused by diabetes are dependent on the oxidative stress level; upon taking the garlic oil and S Allil Cysteine (SAC), an improvement was noticed in the high blood pressure and renal damages (Guneli et al., 2008).

Garlic contains many bioactive components like allicin and its derivatives. These compounds are beneficial in the treatment of various illnesses of human beings. Garlic is used to treat heart and blood disorders, including artery hardening, high blood pressure, heart attack, and high cholesterol levels (Mikaili et al., 2013). Garlic is also beneficial in treating various types of cancers like bladder cancer, breast cancer, lung cancer, colon cancer, prostate cancer, and stomach cancer. Garlic also improves the immune system as well as used in the prevention of fungal and bacterial infection. Garlic is used to treat benign prostatic hyperplasia, diabetes, traveler's diarrhea, hay fever, osteoarthritis, and cold and flu. The garlic plant effectively treats hemorrhoids, headaches, low, and high blood sugar, snakebites, bronchitis, gout, sinus congestion, malaria, leucoderma, and discoloration of the skin, asthma, and against fighting with stress and fatigue Pendbhaje et al. (2000). Furthermore, Jung et al. (2006) documented garlic's function in liver health, sciatica, tuberculosis, leprosy, fistula, bone fracture, urinary diseases, kidney stone, epilepsy, jaundice, cataract, and night blindness, indigestion, and colic pain.

10.4.1 ROLE OF GARLIC IN INFECTIOUS DISEASES

10.4.1.1 ANTIBACTERIAL ACTIVITY

Staphylococcus aureus, *Salmonella enterica*, *Klebsiella aerogenes*, Mycobacteria, *Enterococcus, faecalis*, and other Gram-positive and Gram-negative bacteria, as well as antibiotic-resistant microorganisms (Kuda et al., 2004; Ross et al., 2001; Cutler and Wilson, 2005; Wallock-Richards et al., 2014). Several studies have indicated that garlic's organic solvent extracts inhibit the growth of many pathogenic bacteria with varying degrees of susceptibility (Mikaili et al., 2013). Still, garlic extracts in chloroform, hexane, and ethyl acetate have no antibacterial impact. Garlic extracts minimize the risk of diarrhea in humans and animals by inhibiting enterotoxigenic and pathogenic bacteria in the human intestine. Garlic extract can also reduce the number of toxins released during bacterial infection (Shokrzadeh and Ebadi, 2006).

Allicin's antimicrobial activity is mainly due to its association with thiol-containing enzymes, including RNA polymerase, alcohol dehydrogenase, and thioredoxin reductase (Gruhlke et al., 2010). Allicin oxidizes cysteine protein or glutathione residues under physiological conditions. Allicin affects cysteine proteases' critical metabolic machinery and destroys all eukaryotic cells due to thiol groups' presence in all essential cells.

10.4.1.2 ANTIPROTOZOAL ACTIVITY

Several scientific studies report the antiprotozoal activity of garlic extracts and their phytochemical constituents. An antihelmintic action against the protozoan *Haemonchus contortus* was founded in an *in vitro* experiment using ethanolic, dichloromethane, and aqueous garlic extracts. The ethanolic section was much effective than others against this protozoan. This study's other extracts were effective against the protozoans *Angiostrongylus cantonensis* and *Trichuris muris* (Zhen et al., 2006). Besides this, the garlic extracts were also used against the *Hymenolepis microstoma, Echinostoma caproni, Fasciola hepatica, Taenia taeniaeformis, H. diminuta* in different *in vitro* and *in vivo* experimental findings Abdel-Ghaffar et al. (2011).

Garlic's active ingredients, thiosulfinates, have antibacterial properties and thus inhibit Blastocystis spp. bacterial outgrowth *in vivo* (Abdel-Hafeez et al., 2015). These phytochemicals inhibit the thiol enzymes contained in several microorganisms. Allicin also prevents parasites from developing DNA and RNA, which slows down the protein synthesis process. A bioactive compound of garlic ajoene inhibits the human glutathione reductase enzyme and trypanothione reductase enzyme of parasite *T. cruzi* (Gallwitz et al., 1999). Hazaa et al. (2016) showed that garlic oil is beneficial against the different parasites like *Plasmodium, Leishmania, Giardia,* and *Trypanosoma.* Garlic phytochemicals like allicin and DATS have been shown to have antiparasitic activity against parasites like *Babesia,*

Trypanosoma brucei, *Giardia lamblia*, *Theleria*, *Plasmodium falciparum*, and *Entamoeba histolytica* (Zhen et al., 2006).

10.4.1.3 ANTIVIRAL ROLES OF GARLIC

Human cytomegalovirus (HCMV), herpes simplex types 1 and 2, influenza B, parainfluenza virus type 3, vaccinia virus, human rhinovirus type 2, and vesicular stomatitis virus, among others, were screened for antiviral activity of garlic extract (Gruhlke et al., 2016). An *in vivo* experiment showed that garlic extract plays a protective role against the influenza virus by producing neutralizing antibodies. This garlic extract activity was due to various bioactive phytoconstituents such as Ajoene, methyl allyl thiosulfate, allyl methyl sulfonate, and allicin (Sawai et al., 2008). Ajoene prevents the leukocyte fusion by checking the adhesion process among the leukocytes, and on the other side, allicin works on microorganisms' thiol enzymes. DATS works on the virus-infected cell and destroys the infected cells by enhancing natural killer cells' activity (Zhen et al., 2006). It is also effective against the replication of HCMV.

10.4.1.4 ANTI-FUNGAL ACTIVITY

Garlic extracts have fungicidal properties. Their efficacy has been tested against the growth and germination of a broad range of fungi like *Cryptococcus, Torulopsis*, Candida *Trichosporon, Rhodotorula Aspergillus* species. A study recently reported the fungicidal property of garlic extracts against the *Rhodotorula mucilaginosa* and *Meyerozyma guilliermondii* species (Parvu et al., 2019). The different solvent extracts of garlic act against many human pathogenic fungi like *Alternaria alternate, Botrytis cinerea, A. flavus, Trichophyton verrucosum, T. rubrum, Penicillium expansum*, ***Microsporum gypseum***, etc. (Fufa, 2019).

Garlic extracts deform and alter the fungal cell wall in an irreversible way, affecting the cell wall's structural integrity and inhibiting the fungus' germination phase. The cytoplasmic material, cell organelles, and the fungus' nucleus are all destroyed due to these changes, resulting in cell death. Allicin and garlic oil enter the membranes of organelles and cells, resulting in cell death (Kuda et al., 2004). Saponins, DADS, and DATS extracted from the garlic work as fungicidal agents against several species like *Ascosphaera apisin, A. niger, Candida albicans, Trichoderma harzianum,* and *Botrytis cinerea* (Zhen et al., 2006).

10.4.2 ANTIDIABETIC ACTIVITY

Different experimental models of rats and mice in which the streptozotocin has induced diabetes and alloxan garlic lower blood glucose level. An Iranian study found that garlic extract has a range of beneficial effects on serum glucose, triglycerides, uric acid, urea, creatinine, alanine transferases, aspartate aminotransferase, and total cholesterol in regular and streptozotocin-induced diabetic rats after a two-week treatment span. Garlic extract in this study increased the insulin level in the diabetic rats (but not in the control animals) and lowered the other parameters mentioned above. In a comparative analysis, the efficacy of the well-known antidiabetic drug glibenclamide and garlic extract was evaluated by Eidi et al. (2006). This group reveals the more potent effect of garlic in comparison to the drug in controlling diabetes.

The ethanolic extract of garlic triggers insulin secretion from the pancreas' parietal cells in alloxan and streptozotocin-induced diabetes in mice and rabbits (Patel et al., 2012). In a clinical study, administering garlic pills at a 900 mg/day dose in diabetes patients showed its good effect against hyperlipidemia and diabetes (Faroughi et al., 2018). Various compounds present in the garlic, such as allicin, cysteine sulfoxide, S-allyl cysteine sulfoxide, and allyl propyl disulfide, work against diabetes via different processes. These compounds promote insulin secretion from the pancreas' beta cells. Allin works similarly to the antidiabetic drug glibenclamide (Zhai et al., 2018). These garlic extracts enhance the liver metabolism and short-acting insulin (Faroughi et al., 2018; Zhai et al., 2018).

10.4.3 ANTIOXIDANT AND ANTI-INFLAMMATORY ROLES OF GARLIC

10.4.3.1 ANTIOXIDATIVE ROLES

Garlic extracts improve the body's antioxidant system, lowering the adverse effects of oxidative stress by reducing free radical output. Garlic extracts protect the liver against the harmful effects of gentamycin and acetaminophen-induced hepatotoxicity. Both these antibiotics used against bacterial infections raise ALT and AST enzymes of the liver function and lower the plasma's albumin levels (Wallock-Richards et al., 2014). Garlic extracts promote glutathione and cellular antioxidant enzymes and modulate reactive oxygen species (Shokrzadeh and Ebadi, 2006). In rats' hepatic tissues, the extracts also increased antioxidant enzymes like superoxide dismutase and decreased glutathione peroxidase activity. Reactive oxygen species are behind the various health issues, and garlic extracts manage these problems by managing these species. Several reports suggest that garlic extracts rich in phenols, flavonoids, and sulfur compounds act as free radical scavengers (Jang et al., 2017). Through the Nrf2 (nuclear factor erythroid-2 related factor)–ARE (antioxidant response element) pathway, AGE activates the expression of additional antioxidant enzymes such as the heme oxygenase-1 (HO-1) subunit and glutamate-cysteine ligase modifier (GCLM). Human endothelial cells are protected from the ARE pathway's oxidative stress (Liu et al., 2018). Allin protects against oxidative stress by providing a broad-spectrum antioxidant activity that inhibits mitogen-activated protein kinase activity and controls ROS production (MAPK). It inhibits the development of NADPH oxidase 1, which affects the generation of ROS through the receptor activator of nuclear factor-kappa B ligand (RANKL)-a mediated fusion of osteoclasts (Chen et al., 2017).

Sulfur compound saponins extracted from garlic extract can scavenge intracellular ROS and mitigate the DNA damages caused by H_2O_2 (Shang et al., 2019). The DAS has potent cytoprotective and antioxidant properties since it inhibits cytochrome P450-2E1 operation. It controls the generation of reactive oxygen species and upregulates mRNA expression for Nef2 and the heme-oxygenase 1 enzyme to aid this attribute.

10.4.3.2 ANTI-INFLAMMATORY

Garlic extracts and phytochemicals linked to garlic have been shown to have anti-inflammatory properties. According to a report, garlic extracts significantly decreased liver inflammation and damage caused by Eimeria papillate infections (Ahmad et al., 2016). Garlic works to minimize inflammation by stopping neutrophilic granulocytes from emigrating into the epithelia. Old black garlic has potent anti-inflammatory and antioxidant properties (Gue et al., 2013). Garlic's anti-inflammatory effects are due to its antioxidative properties. TNF-mediated activation of NF-κB in human umbilical vein endothelial cells is decreased by a chloroform extract of aged black garlic (ABG). In addition to the synthesis of prostaglandin E2 (PGE2), the methanolic extract of aged black garlic inactivates NF-κB, and cyclooxygenase-2 (COX-2) is affected (Jeong et al., 2016). According to (You et al., 2019), aged black garlic suppresses the activation of the toll-like receptor 4 (TLR4) signaling cascade in macrophages directly, improves the cytosolic level of NF-κB and I.B. in LPS activated RAW2647 cells, and reduces the nuclear status of NF-κB, thereby acting as an anti-inflammatory agent. The ABG inhibits the expression of COX-2 and iNOS, and as a result, it can act as an anti-inflammatory agent.

In LPS-activated RAW264.7 cells, ABG inhibits the development of interleukin-6, NO, and TNF-and dermatitis in mice caused by the TPA. Allicin has the ability to improve immune cell function and influence their signaling pathways; thus, allicin shows its defensive activity against pathogens. SDF1α chemokine regulates the actin cytoskeleton's dynamic structure inhibited by the allicin and finally governs the T-cell lymphocytes (Sela et al., 2004). Allicin also inhibits the transendothelial migration of neutrophils. DAS also acts as an anti-inflammatory agent, suppressing the hepatic enzyme CYP-2E1 and inhibiting the expression of inflammatory cytokines such as TNF-α, NF-κB, and IL-1 (Abdel-Daim et al., 2020). Thiacermonone, a sulfur compound present in the garlic, blocks the activity of NF-κB, thus prevents amyloidogenesis and neuroinflammation. This compound's attribute helps treat neurodegenerative disorders concerned with neuroinflammation (Jin et al., 2013).

10.4.4 ANTICANCER ACTIVITY

In a comparative study, raw garlic extract was the most efficient drug against the cancer cells among the 33 extracts of raw vegetables used in the study without causing any harm to the non-cancerous cells (Li et al., 2018). The garlic extracts manage the carcinogens' metabolism, trigger the apoptotic pathways, prevent the invasion, migration, and angiogenesis, and reduce the cell and proliferation process's growth. The garlic extracts control anticancer agents' harmful actions (Shang et al., 2019).

In a study, allicin solution kills the tumorigenic cells after incubation with it (Chhabria et al., 2015). Allicin prevents tumor vessels' formation and expression of surviving genes with apoptosis in cancer cells; thus, allicin suppresses the metastasis process of colorectal cancer. Allicin also promotes the gene silencing process and constrains the cancer cells' cell proliferation process, thereby treating pancreatic cancer (Chhabria et al., 2015).

Allicin may also avoid the proliferation of MGC 803 gastrointestinal cancer cells and cause apoptosis. This can be done by increasing the expression of p38 and cleaving the caspase 3 enzyme (Zhang et al., 2015). Polysulfanes derived from the allicin target the microtubules, leading to the cell cycle's interruption and ultimately towards apoptosis. Tubulin gives the shape to the mitotic spindle, and allicin targets the tubulin stops the cell proliferation process and ultimately inhibits cell division (Rana et al., 2011). The garlic's organosulfur compounds have antitumor properties, and these compounds also inhibit the DNA damage up to 35% in the liver of rats after the intake of N-nitrosodimethylamine. This was achieved by the presence of all in the higher amount (60%) (Iciek et al., 2012). Regular garlic intake decreases the propagation of various kinds of cancers such as colon, lung, breast, stomach, prostate (Fleischauer and Arab, 2001).

Many studies have shown that garlic has antitumor and cytotoxic properties, both *in vitro* and *in vivo*. According to Piscitelli et al. (2002), stable people decreased plasma concentrations by up to 50% in the saquinavir after the garlic administration for three weeks. *Z*-ajoene has antiproliferative activities, and it works against different types of cancers. It is selective against both human breast cancer cells and glioblastoma multiforme cancer stem cells (GBM CSC) (Shang et al., 2019). It induces apoptosis in human leukemic cells by facilitating the activities of peroxides, caspase-8, and caspase-3 (Bayan et al., 2014).

10.4.5 ANTI-ALZHEIMER'S DISEASE ACTIVITY

In older people with cerebrovascular and neurodegenerative disorders, the leading cause of dementia is Alzheimer's disease (A.D.) (Vinaykumar and Robbins, 2003). In the nervous system, the primary enzyme is acetylcholinesterase which converts acetylcholine into acetate and choline (Dall'Acqua et al., 2010). ACh is the crucial enzyme behind the pathophysiology of A.D. Depletion of this essential enzyme in the central nervous system leads to neurodegenerative disorders. An inhibitor of AChE named donepezil is playing a critical use in the prevention and direction of A.D. In *in vitro* experiment, garlic oil hinders the activity of AChE in the synaptosome of the cerebral cortex and shows antioxidative properties (Akinyemi et al., 2018). Besides this, the oil from the garlic bulbs also scavenges diphenyl-1-picrylhydrazyl of (DPPH) free radical. DPPH helps in evaluating any compound's efficiency to donate hydrogen or scavenge the free radicals. This is the criterion by which every food's antioxidant activity is calculated (Singh et al., 2010). The garlic oil's neuroprotective attribute depends on the reduction process of Fe^{3+} to Fe^{2+} (Piscitelli et al., 2002).

Uptake of garlic extracts by the cholesterol-fed rats reduces glycosaminoglycans' concentrations (GAGs) in the aorta and heart significantly. β-N-acetylhexosaminidase arylsulfatase β-glucuronidase and hyaluronidase enzymes are behind the degradation of GAGs. Garlic enhances the production level of these degrading enzymes of GAGs (Mathew and Biju, 2008). Sulfated GAGs bind with the plasma lipoprotein majorly with the LDL, forming lipids' aggregation in result growth of lesions. These sulfated GAGs participate in the neurotoxic activities of several amyloidogenic peptides like A in the pathogenesis of A.D. (Mathew and Biju, 2008). A neuroprotective animal model of AGE performed by Borek, (2006) revealed that the AGE protects the brain against the injury following ischemia, resist against apoptosis and save the neurons, and also inhibits the neuronal death caused by the β-amyloid oxidative stress, ultimately protects against the neurological illness (Jackson et al., 2003). The AGE and SAC present in the garlic preserve the brain's frontal lobe against degeneration and push up the learning retention and memory, thus extending individuals' lifespan (Mbyirukira and Gwebu, 2003). The fundamental cause of Alzheimer's disease (A.D.) onset and progression is the accumulation of β-amyloid (Aβ) in the brain. The imbalance created between the Aβ production and clearance leads toward the synaptic degradation and formation of τ-protein containing the neurofibrillary tangles. Garlic affects the serotonin neurotransmitter's levels, and as a result, garlic intake promotes memory (Haider et al., 2008). In an *in vivo* experiment, garlic uptake inhibits the enzymatic activity of AChE and reduces the oxidative stress level by free radical scavenging, thus improving memory function (Mukherjee and Banerjee, 2013).

The enzyme butyrylcholinesterase breaks down the neurotransmitter choline and AChE; both are responsible for accumulating ACh concentration in the brain inhibited by the allicin compound present in the garlic. Thus allicin delayed dementia and cognitive decline durations (Kumar, 2013). Garlic has anti-inflammatory, antioxidant, and immunomodulatory properties attributed to the allicin, inhibiting the BuChE and AChE enzymes (Sobolewska et al., 2015). In the treatment of A.D., the combination therapy is in use that includes the allicin with the inhibitors of cholinesterase such as the donepezil, galantamine, and rivastigmine (Hogan, 2007) as they improve the cholinergic deficiency that happens in patients with A.D. and having some side effects (Jann et al., 2002). For A.D. treatment, it is very beneficial to search the new derivatives isolated from natural products having minimal side effects and dual function in nature.

10.4.6 ROLE OF GARLIC IN THE MANAGEMENT OF METABOLIC DISEASES

10.4.6.1 IMPACT ON DYSLIPIDEMIA

The primary cause behind cardiovascular diseases and myocardial infarction is dyslipidemia. Dyslipidemia is noticed by the enhanced level of triglycerides, total cholesterol, and LDL, with low HDL levels (Hosseini and Hosseinzadeh, 2015). Garlic preparations suppress cholesterol biosynthesis in the liver and inhibit LDL and HDL oxidation, rendering garlic and its phytochemicals important in preventing hypercholesterolemia. Garlic lowers the cholesterol level by two means either reduces the lipogenic and cholesterogenic effects of 3-hydroxy-3-methyl-glutaryl-CoA reductase, fatty acid synthase, and glucose-6 phosphate dehydrogenase or by pushing the excretion of neutral and acidic steroids in the liver cells (Qidwai and Ashfaq, 2013). The ingestion of ethanolic garlic extract by albino rabbits decreases their body weight and cholesterol levels (Iweala et al., 2005).

Campbell et al. (2001) demonstrate that the AGE will reverse the deformities caused by cholesterol dosing in laboratory rabbits. Sobenin et al. (2010) found that giving garlic at doses of 300 and 60 mg/day for 12 months and 12 weeks decreased increased LDL levels, HDL, T.G., and T.C. The diabetic patients with dyslipidemia who were given garlic tablets at a dosage of 600 mg per day for 12 weeks had a low T.C. and LDL level with a high level of HDL, according to (Ashraf et al., 2005).

10.4.6.2 EFFECT ON OBESITY

Several health issues like hypertension, metabolic syndrome, dyslipidemia, cardiovascular disorders are arisen because of obesity. Garlic extracts help reduce fatty tissues, body weight, and lipid profile in obese mice. The compounds in garlic extracts suppress adipogenesis gene expression while simultaneously increasing the expression of proteins linked to mitochondria's inner membrane (Lee et al., 2011). AMP-activated protein kinase (AMPK) and the thermogenesis mechanism are implicated in garlic extracts' anti-obesity effects. In 3T3-L1 adipocytes, Ajoene activates the apoptosis process, which delays fat accumulation. This mechanism substantially decreases body weight gain in the experimental mice despite not affect food intake. (Han et al., 2011). There are reports which disclose the property of 1, 2-vinyldithiin in lipid accumulation and against the differentiation of human preadipocytes via the expression of PPARγ2, LPL, and C/EBPα (Keophiphath et al., 2009).

Ajoene reduces the body weight gain by lowering the accumulated fat in the 3T3-L1 adipocytes (Han et al., 2011), whereas 1,2-Vinyldithin reduces C/EBPα PPARγ2, LPL expression as well as the effect of PPARγ in the adipocytes of human beings (Keophiphath et al., 2009). Allin promotes short-acting insulin by promoting liver metabolism and reducing insulin release from pancreatic cells (Zhai et al., 2018). The various garlic preparations reduce LDL, T.G., T.C., and enhance the HDL cholesterol level. Thus, these preparations perform hypolipidemic and hypercholesterolemic effects in the users (Qidwai and Ashfaq, 2013). The different garlic extracts prevent platelets' aggregation and bind with the immobilized fibrinogen activated by the ADP. These extracts also inhibit the GPIIb/IIIa receptors and promote cAMP production, thus creating antithrombotic and anti-atherosclerotic factors (Drobiova et al., 2011).

10.4.6.3 ANTIHYPERTENSIVE ACTIVITY

A study performed by Varshney and Budoff (2016) reported that garlic reduces the systolic and diastolic blood pressure efficiently, thus hampers the various risk factors of cardiovascular diseases. Multiple cardiovascular disorders, such as hyperlipidemia, thrombosis, arrhythmia, hypertension, and atherosclerosis, are treated with garlic extracts (Drobiova et al., 2011). These extracts, as well as their bioactive biomolecules, work to reduce blood pressure. Several experiments on humans and other laboratory animals have backed up this argument. Garlic extract increases fibrinolytic function in both healthy and patients with acute myocardial infarction symptoms, according to Sobenin et al. (2008).

The aqueous garlic extract reduced hypertension by lowering the prostaglandin E2 and thromboxane B2 levels (Dubey et al., 2017). After taking a 100 mg/kg dose of garlic extract for five days, the endothelin-1 induced acute hypoxic pulmonary vasoconstriction in the pulmonary arteries of rats wholly cured. This was achieved by reducing angiotensin II production and endothelin I after garlic extract administration (Drobiova et al., 2011). The garlic's active sulfur molecules can manage the endothelium constricting and relaxing factors, thus lowering the blood pressure, and finally, it works as an antihypertensive agent. Garlic extract also works in the vasodilation process by stimulating the production of hydrogen sulfide and nitrogen oxide. These medicinal properties of the garlic plant reduce the users' blood pressure level (Ried and Fakler, 2014). Garlic extract also works against thrombosis and in the aggregation and adhesion process of the human platelets. Garlic extracts hamper both the ADP-activated binding of platelets with the immobilized fibrinogen and platelets' aggregation by increasing the cAMP level and inhibiting the GPIIb/IIIa receptor (Rahman et al., 2016). Garlic extracts also enhance capillary perfusion and blood vessel elasticity. Various garlic extracts have different properties that minimize the risk of multiple disorders such as plasma viscosity, angina dysfunction, and peripheral arterial occlusive diseases (Bayan et al., 2014). By inhibiting the angiotensin-converting enzyme, garlic extracts containing gamma-glutamylcysteine compounds lower blood pressure. A study performed by Dubey et al. (2017) discloses the allicin property in reducing the dexamethasone-induced systolic blood pressure and minimizing the adverse effects of hypertension by improving the body mass food intake in the experimental rats.

10.5 CONCLUSION AND FUTURE DIRECTIONS

We suppose that this information about the garlic mentioned in this chapter will undoubtedly generate interest in readers about the garlic and provide them updated scientific evidence better to utilize garlic in the management of human health. Medicinal plants like *A. sativum* are an excellent source of pharmacologically active molecules; therefore, these plants could provide a potential platform for controlling the pathogens. Besides this, they could also be used as an alternative to synthetic drugs available in the market, which shows side effects and resistance against the pathogenic microorganisms' antibiotics.

STATEMENT OF CONFLICT OF INTEREST

There are no conflicts of interest declared by the authors.

ACKNOWLEDGMENTS

The University Grants Commission, New Delhi (UGC-DSKPDF No.F.4-2/2006 BSR/BL/17-18/0442) and Department of Science and Technology-SERB (PDF/2015/000033) offered financial support to the corresponding author PKP. SST thanks the University Grants Commission (UGC-DSKPDF No. F.4-2/2006 BSR/BL/17-18/381), and JD thanks the University Grants Commission for offering Junior and Senior research fellowships.

KEYWORDS

- ***Allium sativum***
- **antidiabetic**
- **diabetes mellitus**
- **insulin**
- **insulin secretagogues**
- **metabolic syndrome**

REFERENCES

Abdel-Daim, M. M., Abushouk, A. I., Bungău, S. G., Bin-Jumah, M., El-Kott, A. F., Shati, A. A., Aleya, L., & Alkahtani, S., (2020). Protective effects of thymoquinone and diallyl sulphide against malathion-induced toxicity in rats. *Environ. Sci. Pollut. Res., 23*, 10228–102358.

Abdel-Ghaffar, F., Semmler, M., Al-Rasheid, K. A., Strassen, B., Fischer, K., Aksu, G., Klimpel, S., & Mehlhorn, H., (2011). The effects of different plant extracts on intestinal cestodes and on trematodes. *Parasitol. Res., 108*, 979–984.

Abdel-Hafeez, E. H., Ahmad, A. K., Kamal, A. M., Abdellatif, M. Z., & Abdelgelil, N. H., (2015). *In vivo* antiprotozoan effects of garlic (*Allium sativum*) and ginger (*Zingiber officinale*) extracts on experimentally infected mice with *Blastocystis* spp. *Parasitol. Res., 114*, 3439–3444.

Adachi, Y., Yoshida, J., Kodera, Y., Katoh, A., Takada, J., & Sakurai, H., (2006). Bis (allixinato) oxovanadium (IV) complex is a potent antidiabetic agent: Studies on structure-activity relationship for a series of hydroxypyrone-vanadium complexes. *J. Med. Chem., 49*, 3251–3256.

Ahmad, M. S., & Ahmed, N., (2006). Antiglycation properties of aged garlic extract: Possible role in prevention of diabetic complications. *J. Nutr., 136*, 796S–799S.

Ahmad, T. A., El-Sayed, B. A., & El-Sayed, L. H., (2016). Development of immunization trials against Eimeria spp. *Trials Vaccinol., 5*, 38–47.

Akinyemi, A. J., Lekanm, F. A. P., Awonegan, I. O., Anadozie, S., & Oluwasola, T. A., (2018). Antioxidant and anti-acetylcholinesterase activities of essential oils from garlic (*Allium sativum*) Bulbs. *Int. J. Plant Res., 31*.

Al-Jaber, N. A., Awaad, A. S., & Moses, J. E., (2011). Review some antioxidant plants growing in the Arab world. *J. Saudi Chem. Soc., 15*, 293–307.

Al-Qattan, K. K., Thomson, M., Al-Mutawa'a, S., AlHajeri, D., Drobiova, H., & Ali, M., (2006). Nitric oxide mediates the blood-pressure-lowering effect of garlic in the rat two-kidney, one-clip model of hypertension. *J. Nutr., 136*, 774S–776S.

Asdaq, S. M., & Inamdar, M. N., (2011). The potential benefits of a garlic and hydrochlorothiazide combination as antihypertensive and cardioprotective in rats. *J. Nat. Med., 65*, 81–88.

Ashraf, R., Aamir, K., Shaikh, A. R., & Ahmed, T., (2005). Effects of garlic on dyslipidemia in patients with type 2 diabetes mellitus. *J. Ayub. Med. Coll. Abbottabad., 17*, 60–64.

Augusti, K. T., & Sheela, C. G., (1996). Antiperoxide effect of S-allyl cysteine sulfoxide, an insulin secretagogue in diabetic rats. *Experientia, 52*, 115–120.

Ayodhya, S., Kusum, S., & Anjali, S., (2010). Hypoglycaemic activity of different extracts of various herbal plants Singh. *Int. J. Ayurveda Res. Pharm., 1*, 212–224.

Batiha, G. E. S., Beshbishy, A. A., Tayebwa, D. S., Shaheen, M. H., Yokoyama, N., & Igarashi, I., (2018). Inhibitory effects of *Uncaria tomentosa* bark, *Myrtus communis* roots, *Origanum vulgare* leaves and *Cuminum cyminum* seeds extracts against the growth of *Babesia* and *Theileria in vitro*. *Jpn. J. Vet. Parasitol., 17*, 1–13.

Bayan, L., Koulivand, P. H., & Gorji, A., (2014). Garlic: A review of potential therapeutic effects. *Avicenna J. Phytomed., 4*, 1–14.

Borek, C., (2006). Garlic reduces dementia and heart-disease risk. *J. Nutr., 136*, 810S–812S.

Campbell, J. H., Efendy, J. L., Smith, N. J., & Campbell, G. R., (2001). Molecular basis by which garlic suppresses atherosclerosis. *J. Nutr., 131*, 1006S–1009S.

Cavallito, C. J., & Bailey, J. H., (1944). Allicin, the antibacterial principle of *Allium sativum*. Isolation, physical properties and antibacterial action. *J. Am. Chem. Soc., 66*, 1950–1951.

Chauhan, A., Sharma, P. K., Srivastava, P., Kumar, N., & Duehe, R., (2010). Plants having potential antidiabetic activity: A review. *Der. Pharm. Lett, 2*, 369–387.

Chen, Y., Sun, J., Dou, C., Li, N., Kang, F., Wang, Y., Cao, Z., Yang, X., & Dong, S., (2016). Alliin attenuated RANKL-induced osteoclastogenesis by scavenging reactive oxygen species through inhibiting Nox1. *Int. J. Mol. Sci., 17*, 1516.

Chhabria, S. V., Akbarsha, M. A., Li, A. P., Kharkar, P. S., & Desai, K. B., (2015). In situ allicin generation using targeted alliinase delivery for inhibition of MIA PaCa-2 cells via epigenetic changes, oxidative stress, and cyclin-dependent kinase inhibitor (CDKI) expression. *Apoptosis, 20*, 1388–1409.

Cutler, R., & Wilson, P., (2004). Antibacterial activity of a new, stable, aqueous extract of allicin against methicillin-resistant *Staphylococcus aureus*. *Br. J. Biomed. Sci., 61*, 71–74.

Dall'Acqua, S., Maggi, F., Minesso, P., Salvagno, M., Papa, F., Vittori, S., & Innocenti, G., (2010). Identification of non-alkaloid acetylcholinesterase inhibitors from *Ferulago campestris* (Besser) Grecescu (Apiaceae). *Fitoterapia, 81*, 1208–1212.

Drobiova, H., Thomson, M., Al-Qattan, K., Peltonen-Shalaby, R., Al-Amin, Z., & Ali, M., (2011). Garlic increases antioxidant levels in diabetic and hypertensive rats determined by a modified peroxidase method. *Evid. Based Complement. Altern. Med., 2011*, 703049.

Dubey, H., Singh, A., Patole, A. M., & Tenpe, C. R., (2017). Antihypertensive effect of allicin in dexamethasone-induced hypertensive rats. *Integr. Med. Res., 6*, 60–65.

Eidi, A., Eidi, M., & Esmaeili, E., (2006). Antidiabetic effect of garlic (*Allium sativum* L.) in normal and streptozotocin-induced diabetic rats. *Phytomed., 13*, 624–629.

El-Demerdash, F. M., Yousef, M. I., & El-Naga, N. I., (2005). Biochemical study on the hypoglycemic effects of onion and garlic in alloxan-induced diabetic rats. *Food Chem. Toxicol., 43*, 57–63.

Farag, M. A., Ali, S. E., Hodaya, R. H., El-Seedi, H. R., Sultani, H. N., Laub, A., Eissa, T. F., et al., (2017). Phytochemical Profiles and Antimicrobial Activities of *Allium cepa* red cv. and *A. sativum* subjected to different drying methods: A comparative MS-based metabolomics. *Molecules, 22*, 761.

Faroughi, F., Mohammad-Alizadeh, C. S., Javadzadeh, Y., & Mirghafourvand, M., (2018). Effects of garlic pill on blood glucose level in borderline gestational diabetes mellitus: A triple-blind, randomized clinical trial. *Iran. Red. Crescent Med. J., 20*, e60675.

Fattorusso, E., Lanzotti, V., & Taglialatela-Scafati, O., (1999). Antifungal N-feruloyl amides from roots of Allium species. *Plant Biosyst, 133*, 199–203.

Fleischauer, A. T., & Arab, L., (2001). Garlic and cancer: A critical review of the epidemiologic literature. *J. Nutr., 131*, 1032s–1040s.

Freeman, F., & Kodera, Y., (1995). Garlic chemistry: Stability of S-(2)-propenyl-2(propene-1-sulfothioate (Allicin) in blood, solvents, and simulated physiological fluids. *J. Agric. Food Chem, 43*, 2332–2338.

Fufa, B., (2019). Anti-bacterial and anti-fungal properties of garlic extract (*Allium sativum*): A review. *Microbiol. Res. J. Int., 28*, 1–5.

Gallwitz, H., Bonse, S., Martinez-Cruz, A., Schlichting, I., Schumacher, K., & Krauth-Siegel, R. L., (1999). Ajoene is an inhibitor and subversive substrate of human glutathione reductase and *Trypanosoma cruzi* trypanothione reductase: Crystallographic, kinetic, and spectroscopic studies. *J. Med. Chem., 42*, 364–372.

Grover, J. K., Yadav, S., & Vats, V., (2002). Medicinal plants of India with antidiabetic potential. *J. Ethnopharmacol., 81*, 81–100.

Gruhlke, M. C., Nicco, C., Batteux, F., & Slusarenko, A. J., (2016). The effects of allicin, a reactive sulfur species from garlic, on a selection of mammalian cell lines. *Antioxidants, 6*, 1.

Gruhlke, M. C., Nwachwukwu, I., Arbach, M., Anwar, A., Noll, U., & Slusarenko, A. J., (2013). Allicin from garlic, effective in controlling several plant diseases, is a reactive sulfur species (RSS) that pushes cells into apoptosis. In: *Proceedings of the Modern Fungicides and Antifungal Compounds VI: 16th International Reinhardsbrunn Symposium*. Friedrichroda, Germany.

Gu, X., Wu, H., & Fu, P., (2013). Allicin attenuates inflammation and suppresses HLA-B27 protein expression in ankylosing spondylitis mice. *BioMed. Res. Int., 2013*, 171573.

Guneli, E., Tugyan, K., Ozturk, H., Gumustekin, M., Cilaker, S., & Uysale, N., (2008). Effect of melatonin on testicular damage in streptozotocin-induced diabetes rats. *Eur. Surg. Res., 40*, 354–360.

Haider, S., Naz, N., Khaliq, S., Perveen, T., & Haleem, D. J., (2008). Repeated administration of fresh garlic increases memory retention in rats. *J. Med. Food, 11*, 675–679.

Han, C. Y., Ki, S. H., Kim, Y. W., Noh, K., Lee Da, Y., Kang, B., Ryu, J. H., et al., (2011). Ajoene, a stable garlic by-product, inhibits high fat diet-induced hepatic steatosis and oxidative injury through LKB1-dependent AMPK activation. *Antioxid. Redox Signal, 14*, 187–202.

Hattori, A., Yamada, N., Nishikawa, T., Fukuda, H., & Fujino, T., (2005). Antidiabetic effects of ajoene in genetically diabetic KK-A(y) mice. *J. Nutr. Sci. Vitaminol., 51*, 382–384.

Hazaa, I. K. K., Al-Taai, N. A., Khalil, N. K., & Zakri, A. M. M., (2016). Efficacy of garlic and onion oils on murin experimental *Cryptosporidium parvum* infection. *Al-Anbar J. Vet. Sci., 9*, 69–74.

Hogan, D. B., (2007). Progress update: Pharmacological treatment of Alzheimer's disease. *Neuropsychiatr. Dis. Treat., 3*, 569–578.

Iciek, M., Kwiecie´n, I., Chwatko, G., Sokołowska-Jezewicz, M., Kowalczyk-Pachel, D., & Rokita, H., (2012). The effects of garlic-derived sulfur compounds on cell proliferation, caspase 3 activity, thiol levels and anaerobic sulfur metabolism in human hepatoblastoma HepG2 cells. *Cell Biochem. Funct., 30*, 198–204.

Imai, J., Ide, N., Nagae, S., Moriguchi, T., Matsuura, H., & Itakura, Y., (1994). Antioxidant and radical scavenging effects of aged garlic extract and its constituents. *Planta. Med., 60*, 417–420.

Islam, M. S., & Choi, H., (2008). Dietary red chilli (*Capsicum frutescens* L.) is insulinotropic rather than hypoglycemic in type 2 diabetes model of rats. *Phytother. Res., 22*, 1025–1029.

Iwakami, S., Shibuta, M., Tseng, C., Hanaoka, F., & Sankawa, U., (1986). Inhibition of arachidonate 5-lipoxygenase by phenolic compounds. *Chem. Pharm. Bull., 34*, 3960–3963.

Iweala, E. E., Akubugwo, E. I., & Okeke, C. U., (2005). Effect of ethanolic extracts of *Allium sativum* Linn. Liliaceae on serum cholesterol and blood sugar levels of alibino rabbits. *Plant Prod. Res. J., 9*, 14–18.

Jackson, R., McNeil, B., Taylor, C., Holl, G., Ruff, D., & Gwebu, E., (2003). Effect of aged garlic extract on human recombinant caspace-3 activity. *J. Ala. Acad. Sci., 74*, 121, 122.
Jang, H. J., Lee, H. J., Yoon, D. K., Ji, D. S., Kim, J. H., & Lee, C. H., (2017). Antioxidant and antimicrobial activities of fresh garlic and aged garlic by-products extracted with different solvents. *Food Sci. Biotechnol., 27*, 219–225.
Jann, M. W., Shirley, K. L., & Small, G. W., (2002). Clinical pharmacokinetics and pharmacodynamics of cholinesterase inhibitors. *Clin. Pharmacokinet., 41*, 719–739.
Jeong, Y. Y., Ryu, J. H., Shin, J. H., Kang, M. J., Kang, J. R., Han, J., & Kang, D., (2016). Comparison of anti-Oxidant and anti-inflammatory effects between fresh and aged black garlic extracts. *Molecules, 21*, 430.
Jin, P., Kim, J. A., Choi, D. Y., Lee, Y. J., Jung, H. S., & Hong, J. T., (2013). Anti-inflammatory and anti-amyloidogenic effects of a small molecule, 2,4-bis (p-hydroxyphenyl)-2-butenal in Tg2576 Alzheimer's disease mice model. *J. Neuroinflamm., 10*, 767.
Jung, Y. M., Lee, S. H., Lee, D. S., You, M. J., Chung, I. K., & Cheon, W. H., (2000). Fermented garlic protects diabetic, obese mice when fed a high-fat diet by anti-oxidant effects. *Nutr. Res., 5*, 387–396.
Kaleem, M., Asif, M., Ahmed, Q. U., & Bano, B., (2006). Antidiabetic and antioxidant activity of *Annona squamosa* extract in streptozotocin induced diabetic rats. *Singapore Med. J., 47*, 670–675.
Kaye, A. D., De Witt, B. J., Anwar, M., Smith, D. E., Feng, C. J., Kadowitz, P. J., & Nossaman, B. D., (2000). Analysis of responses of garlic derivatives in the pulmonary vascular bed of the rat. *J. Appl. Physiol., 89*, 353–358.
Keophiphath, M., Priem, F., Jacquemond-Collet, I., Clément, K., & Lacasa, D., (2009). 1, 2-vinyldithiin from garlic inhibits differentiation and inflammation of human preadipocytes. *J. Nutr., 139*, 2055–2060.
Kirkman, M. S., McCarren, M., Shah, J., & Duckworth, W., (2006). The association between metabolic control and prevalent macrovascular disease in type 2 diabetes: The VA cooperative study in diabetes. *J. Diabet. Complic., 20*, 75–80.
Koch, H. P., & Lawson, L. D., (1995). *Garlic: The Science and Therapeutic Application of Allium Sativum L. and Related Species* (Vol. 34, p. 108). Baltimore: Williams & Wilkins.
Kuda, T., Iwai, A., & Yano, T., (2004). Effect of red pepper *Capsicum annuum* var. conoides and garlic *Allium sativum* on plasma lipid levels and cecal microflora in mice fed beef tallow. *Food Chem. Toxicol., 42*, 1695–1700.
Kumar, S., (2015). Dual inhibition of acetylcholinesterase and butyrylcholinesterase enzymes by allicin. *Indian J. Pharmacol., 47*, 444–446.
Lanzotti, V., (2012). Bioactive polar natural compounds from garlic and onions. *Phytochem. Rev., 11*, 179–196.
Lanzotti, V., Barile, E., Antignani, V., Bonanomi, G., & Scala, F., (2012). Antifungal saponins from bulbs of garlic, *Allium sativum* L. var. voghiera. *Phytochemistry, 78*, 126–134.
Lanzotti, V., Scala, F., & Bonanomi, G., (2014). Compounds from Allium species with cytotoxic and antimicrobial activity. *Phytochem. Rev., 13*, 769–791.
Lawson, L. D., & Gardner, C. D., (2005). Composition, stability, and bioavailability of garlic products used in a clinical trial. *J. Agric. Food Chem., 53*, 6254–6261.
Lee, E. S., Steiner, M., & Lin, R., (1994). Thioallyl compounds: Potent inhibitors of cell proliferation. *Biochim. Biophys. Acta, 1221*, 73–77.

Lee, M. S., Kim, I. H., Kim, C. T., & Kim, Y., (2011). Reduction of body weight by dietary garlic is associated with an increase in uncoupling protein mRNA expression and activation of AMP-activated protein kinase in diet-induced obese mice. *J. Nutr., 141*, 1947–1953.

Li, Z., Le, W., & Cui, Z., (2018). A novel therapeutic anticancer property of raw garlic extract via injection but not ingestion. *Cell Death Discov., 4*, 108.

Liu, C. T., Hse, H., Lii, C. K., Phi-Sam, C., & Lee-Yan, S., (2005). Effects of garlic oil and diallyl trisulfide on glycemic control in diabetic rats. *Eur. J. Pharmacol., 516*, 165–173.

Liu, J., Guo, W., Yang, M. L., Liu, L. X., Huang, S. X., Tao, L., Zhang, F., & Liu, Y. S., (2018). Investigation of the dynamic changes in the chemical constituents of Chinese "laba" garlic during traditional processing. *RSC Adv., 8*, 41872–41883.

Liu, P., Guo, J., Liu, H., Cheng, Y., Wang, B., Long, C., & Deng, B., (2009). Diallyl trisulfide (DATS) effectively induced apoptosis of postharvest disease *Penicillium expansum* of citrus. *Ann. Microbiol, 59*, 675–679.

Liu, S. G., Ren, P. Y., Wang, G. Y., Yao, S. X., & He, X. J., (2015). Allicin protects spinal cord neurons from Glutamate-induced oxidative stress through regulating the heat shock protein 70/inducible nitric oxide synthase pathway. *Food Funct., 6*, 321–330.

Mathew, B., & Biju, R., (2008). Neuroprotective effects of garlic a review. *Libyan J. Med., 3*, 23–33.

Mathew, P. T., & Augusti, K. T., (1973). Studies on the effect of allicin (diallyl disulphide-oxide) on alloxan diabetes I. hypoglycaemic action and enhancement of serum insulin effect and glycogen Synthesis. *Indian J. Biochem. Biophys., 10*, 209–212.

Mbyirukira, G., & Gwebu, E. T., (2003). Aged garlic extract protects serum-deprived PC12 cells from apoptosis. *J. Ala. Acad. Sci., 74*, 127–128.

Meriga, B., Mopuri, R., & MuraliKrishna, T., (2012). Insecticidal, antimicrobial, and antioxidant activities of bulb extracts of *Allium sativum*. *Asian Pac. J. Trop. Med., 5*, 391–395.

Mikaili, P., Maadirad, S., Moloudizargari, M., Aghajanshakeri, S., & Sarahroodi, S., (2013). Therapeutic uses and pharmacological properties of garlic, shallot, and their biologically active compounds. *Iran. J. Basic Med. Sci., 16*, 1031–1048.

Modak, M., Dixit, P., Londhe, J., Ghaskadbi, S., Paul, A., & Devasagayam, T., (2007). Indian herbs and herbal drugs used for the treatment of diabetes. *J. Clin. Biochem. Nutr, 40*, 163–173.

Mukherjee, D., & Banerjee, S., (2013). Learning and memory promoting effects of crude garlic extract. *Indian J. Exp. Biol., 51*, 1094–1100.

Mustafa, S. S. S., Eid, N. I., Jafri, S. A., El-Latif, H. A. A., & Ahmed, H. M. S., (2007). Insulinotropic effect of aqueous ginger extract and aqueous garlic extract on the isolated perfused pancreas of streptozotocin induced diabetic rats. *Pak. J. Zool., 39*, 279–284.

Nain, P., Saini, V., Sharma, S., & Nain, J., (2012). Antidiabetic and antioxidant potential of *Emblica officinalis* gaertn. leaves extract in streptozotocin-induced type-2 diabetes mellitus (T2DM) rats. *J. Ethnopharmacol., 142*, 65–71.

Nasim, S. A., Dhir, B., Kapoor, R., Fatima, S., & Mahmooduzzafar, A., (2011). Alliin obtained from leaf extract of garlic grown under in situ conditions possesses higher therapeutic potency as analyzed in alloxan-induced diabetic rats. *Pharm. Biol., 49*, 416–421.

Ou, C. C., Tsao, S. M., Lin, M. C., & Yin, M. C., (2003). Protective action on human LDL against oxidation and glycation by four organosulfur compounds derived from garlic. *Lipids, 38*, 219–224.

Pârvu, M., Mot, C. A., Pârvu, A. E., Mircea, C., Stoeber, L., Rosca-Casian, O., & Tigu, A. B., (2019). *Allium sativum* extract chemical composition, antioxidant activity and antifungal effect against *Meyerozyma guilliermondii* and *Rhodotorula mucilaginosa* causing onychomycosis. *Molecules, 24*, 3958.

Patel, D. K., Prasad, S. K., Kumar, R., & Hemalatha, S., (2012). An overview on antidiabetic medicinal plants having insulin mimetic property. *Asian Pac. J. Trop. Biomed., 2*, 320–330.

Pendbhaje, I. S., Amit, P., Shahin, M., Pathan, S., Raotole, A., & Pattewar, S. V., (2000). Ethnopharmacology, pharmacognosy, and phytochemical profile of *Allium sativum* L: A review. *Pharmacology Online, 2*, 845–885.

Piscitelli, S. C., Burstein, A. H., Welden, N., Gallicano, K. D., & Falloon, J., (2002). The effect of garlic supplements on the pharmacokinetics of saquinavir. *Clin. Infect. Dis., 34*, 234–238.

Qidwai, W., & Ashfaq, T., (2013). Role of garlic usage in cardiovascular disease prevention: An evidence-based approach. *Evid. Based Complement. Altern. Med., 2013*, 125649.

Rahman, K., (2003). Garlic and aging: New insights into an old remedy. *Ageing Res. Rev., 2*, 39–56.

Rahman, K., Lowe, G. M., & Smith, S., (2016). Aged garlic extract inhibits human platelet aggregation by altering intracellular signaling and platelet shape change. *J. Nutr., 146*, 410S–415S.

Rahman, M. S., (2007). Allicin and other functional active components in garlic: Health benefits and bioavailability. *Int. J. Food Prop., 10*, 245–268.

Rana, S. V., Pal, R., Vaiphei, K., Sharma, S. K., & Ola, R. P., (2011). Garlic in health and disease. *Nutr. Res. Rev., 24*, 60–71.

Ried, K., & Fakler, P., (2014). Potential of garlic (*Allium sativum*) in lowering high blood pressure: Mechanisms of action and clinical relevance. *Integr. Blood Press. Control, 7*, 71–82.

Rishton, G. M., (2008). Natural products as a robust source of new drugs and drug leads: Past successes and present-day issues. *Am. J. Cardiol., 101*, S43–S49.

Ross, Z. M., O'Gara, E. A., Hill, D. J., Sleightholme, H. V., & Maslin, D. J., (2001). Antimicrobial properties of garlic oil against human enteric bacteria: Evaluation of methodologies and comparisons with garlic oil sulfides and garlic powder. *Appl. Environ. Microbiol., 67*, 475–480.

Sawai, T., Itoh, Y., Ozaki, H., Isoda, N., Okamoto, K., Kashima, Y., Kawaoka, Y., et al., (2008). Induction of cytotoxic T-lymphocyte and antibody responses against highly pathogenic avian influenza virus infection in mice by inoculation of a pathogenic H5N1 influenza virus particles inactivated with formalin. *Immunology, 124*, 155–165.

Sela, U. R., Ganor, S., Hecht, I., Brill, A., Miron, T., Rabinkov, A., Wilchek, M., et al., (2004). Allicin inhibits SDF-1α-induced T cell interactions with fibronectin and endothelial cells by down-regulating cytoskeleton rearrangement, Pyk-2 phosphorylation and VLA-4 expression. *Immunology, 111*, 391–399.

Shang, A., Cao, S. Y., Xu, X. Y., Gan, R. Y., Tang, G. Y., Corke, H., Mavumengwana, V., & Li, H. B., (2019). Bioactive compounds and biological functions of garlic (*Allium sativum* L.). *Foods, 8*, 246.

Shariatzadeh, S. M. A., Mehranjani, S. M., Mahmoodi, M., Abnosi, M. H., Momeni, H. R., Dezfulian, A. R., & Noori, M., (2008). Effects of garlic (*Allium sativum*) on blood sugar and nephropathy in diabetic rats. *J. of Biol. Sci., 8*, 1316–1321.

Shokrzadeh, M., & Ebadi, A. G., (2006). Antibacterial effect of garlic (*Allium sativum* L.) on *Staphylococcus aureus*. *Pak. J. Biol. Sci., 9*, 1577–1579.

Singh, P., Shukla, R., Prakash, B., Kumar, A., Singh, S., Mishra, P. K., & Dubey, N. K., (2010). Chemical profile, antifungal, anti-aflatoxigenic, and antioxidant activity of *Citrus maxima* burm. and *Citrus sinensis* (L.) osbeck essential oils and their cyclic monoterpene, DL-limonene. *Food Chem. Toxicol., 48*, 1734–1740.

Slusarenko, A. J., Patel, A., & Portz, D., (2008). Control of plant diseases by natural products: Allicin from garlic as a case study. *Eur. J. Plant Pathol., 121*, 313.

Sobenin, I. A., Nedosugova, L. V., Filatova, L. V., Balabolkin, M. I., Gorchakova, T. V., & Orekhov, A. N., (2008). Metabolic effects of time-released garlic powder tablets in type 2 diabetes mellitus: The results of double-blinded placebo-controlled study. *Acta Diabetol., 45*, 1–6.

Sobenin, I. A., Pryanishnikov, V. V., Kunnova, L. M., Rabinovich, Y. A., Martirosyan, D. M., & Orekhov, A. N., (2010). The effects of time-released garlic powder tablets on multifunctional cardiovascular risk in patients with coronary artery disease. *Lipids Health Dis., 9*, 119.

Sobolewska, D., Podolak, I., & Makowska, W. J., (2015). *Allium ursinum*: Botanical, phytochemical, and pharmacological overview. *Phytochem. Rev., 14*, 81–97.

Stettler, C., Allemann, S., Jni, P., Cull, C. A., Holman, R. R., Egger, M., Krahenbuhl, S., & Diem, P., (2006). Glycemic control and macrovascular disease in types 1 and 2 diabetes mellitus: Meta-analysis of randomized trials. *Am. Heart J., 152*, 27–38.

Tahiliani, P., & Kar, A., (2003). Mitigation of thyroxine-induced hyperglycaemia by two plants extracts. *Phytotherapy Res., 17*, 294–296.

Varshney, R., & Budoff, M. J., (2016). Garlic and heart disease. *J. Nutr., 146*, 416S–421S.

VinayKumar, D. K., (2003). In: Vinay, K., Ramzi, S. C., & Stanley, L. R., (eds.), *Robbins Basic Pathology*. With Illustrations by James A. Perkins. Saunders: Philadelphia, PA, USA.

Vlassara, H., & Palace, M. R., (2002). Diabetes and advanced glycation end-products. *J. Intern. Med., 251*, 87–101.

Wallock-Richards, D., Doherty, C. J., Doherty, L., Clarke, D. J., Place, M., Govan, J. R., & Campopiano, D. J., (2014). Garlic revisited: Antimicrobial activity of allicin-containing garlic extracts against *Burkholderia cepacia* complex. *PLoS One, 9*, e112726.

Wanwimolruk, S., & Prachayasittikul, V., (2014). Cytochrome P450 enzyme-mediated herbal drug interactions (Part 1). *EXCLI J., 13*, 347–391.

Wei, Z., & Lau, B. H. S., (1998). Garlic inhibits free radical generation and augments antioxidant enzyme activity in vascular endothelial cells. *Nutr. Res., 18*, 61–70.

Yan, S., Schmidt, A. M., Anderson, G. M., Zhang, J., Brett, J., Zou, Y. S., Pinsky, D., & Stern, D., (1994). Enhanced cellular oxidant stress by the interaction of advanced glycation end-products with their receptors/binding proteins. *J. Biol. Chem., 269*, 9889–9897.

You, B. R., Yoo, J. M., Baek, S. Y., & Kim, M. R., (2019). Anti-inflammatory effect of aged black garlic on 12-O-tetradecanoylphorbol-13-acetate-induced dermatitis in mice. *Nutr. Res. Pract., 13*, 189–195.

Zhai, B., Zhang, C., Sheng, Y., Zhao, C., He, X., Xu, W., Huang, K., & Luo, Y., (2018). Hypoglycemic and hypolipidemic effect of S-allyl-cysteine sulfoxide (alliin) in DIO mice. *Sci. Rep., 8*, 3527.

Zhang, X., Zhu, Y., Duan, W., Feng, C., & He, X., (2015). Allicin induces apoptosis of the MGC-803 human gastric carcinoma cell line through the p38 mitogen-activated protein kinase/caspase-3 signaling pathway. *Mol. Med. Rep., 11*, 2755–2760.

Zhang, Y., Liu, X., Ruan, J., Zhuang, X., Zhang, X., & Li, Z., (2020). Phytochemicals of garlic: Promising candidates for cancer therapy. *Biomed. Pharmacotherapy, 123*, 109730.

Zhen, H., Fang, F., Ye, D. Y., Shu, S. N., Zhou, Y. F., Dong, Y. S., Nie, X. C., & Li, G., (2006). Experimental study on the action of allitridin against human cytomegalovirus *in vitro*: Inhibitory effects on immediate-early genes. *Antiviral Res., 72*, 68–74.

CHAPTER 11

Antidiabetic Potential of *Ficus benghalensis*

DIKSHA KAPOOR and TABASSUM KHAN

Department of Pharmaceutical Chemistry and Quality Assurance, SVKM's Dr. Bhanuben Nanavati College of Pharmacy, Mumbai, Maharashtra, India, E-mail: tabassum.khan@bncp.ac.in (T. Khan)

ABSTRACT

Ficus benghalensis, commonly known as East Indian fig tree, Indian banyan tree or vada tree belonging to family Moraceae (Mulberry), is well known in the Indian traditional system for its renowned protective properties. Various parts of *Ficus benghalensis* such as aerial roots, stem bark, fruits, seeds, and leaves possess various pharmacological properties and are used traditionally in various diseases such as diabetes, dysentery, seminal weakness, nervous disorders, and hemorrhages. The Ayurvedic system of medicine describes the medicinal property of the bark of plant in the treatment of diabetes. The current chapter aims to provide information on the scientific and traditional knowledge of *Ficus benghalensis* and explore the pharmacology and phytochemical properties – the bioactive metabolites and the mechanism of action of *Ficus benghalensis* responsible for its antidiabetic effect.

11.1 INTRODUCTION

Plants have been the major source of drugs in traditional system since ancient times. They have been used in traditional system since ages in

Antidiabetic Plants for Drug Discovery: Pharmacology, Secondary Metabolite Profiling, and Ingredients with Insulin Mimetic Activity. Deepu Pandita, MPhil, Anu Pandita, MSc, & Chander Bhanu, MBBS, PGDDCE (Eds.)
© 2023 Apple Academic Press, Inc. Co-published with CRC Press (Taylor & Francis)

Ayurvedic, Unani, Siddha, and Homeopathy systems of medicines. Ficus, the genus belonging to the Mulberry family, is one of the 40 genera of the Urticaceae family. The term *Ficus benghalensis* is given to the plant for its meaning Ficus that is fig and benghalensis which means belonging to Bengal, India. It consists of around 1,000 species and 2,000 varieties with a wide range of shrubs, trees, and climbers (Wagner et al., 2012). *Ficus benghalensis,* one of the largest evergreen trees is widely distributed throughout India in a temperate climate. It is considered as one of a sacred tree to be planted around temples and home. It is well known for its spiritual and mythological contexts (Kilmer et al., 2010). India has a very famous Banyan tree near Narmada known as Kabirvad. The size of the tree is so large that almost 20,000 people can sit inside it. The life span of *Ficus benghalensis* is above 100 years (Figure 11.1(a)).

FIGURE 11.1(a) Banyan tree.

11.1.1 VERNACULAR NAMES

- Botanical name: *Ficus benghalensis Linn*
- English: Banyan Tree
- Hindi: Bad, Bar, Bargad

- Persian: Darakht-e-Reesh
- Sanskrit: Avaroha
- Urdu: Bohar

11.1.2 BOTANICAL CLASSIFICATION (KILMER ET AL., 2010, 2017)

A brief botanical or taxonomic classification of *Ficus benghalensis* is given in Table 11.1.

TABLE 11.1 Taxonomic Classification of *Ficus benghalensis*

Kingdom	Planate-Plants
Phylum	Tracheophyta
Division	Magnoliophyta
Superdivision	Spermatophyta
Family	Moraceae
Genus	Ficus
Order	Rosales
Class	Magnoliopsida
Subclass	Hamamelididac
Species	Benghalensis (L)

11.1.3 MORPHOLOGY

It is a huge tree with a height of about 30 m with branches widely spread having many aerial roots:

1. **Bark:** The bark is thick and smooth in texture. When young, they are green in color and turns grayish-white when mature. When cut, they turn pink. The thickness of the bark is usually 0.8 to 2.5 cm.
2. **Wood:** It is porous and soft with sticky latex.
3. **Leaves:** These are simple and alternate. When matured, they turn leathery and glossy. These are often seen in clusters at the end of branches. These are 10–20 cm long and 5–12.5 cm broad. These are ribbed from base.

4. **Fruits:** These are sessile and axillary. They are in pairs. They enclose male and female flowers. They are small and turn brick red when ripe. They are often enclosed in fleshy receptacles (Figures 11.1(b)–11.1(e)).

FIGURE 11.1(b) Stem bark.

FIGURE 11.1(c) Fruits.

FIGURE 11.1(d) Bark.

FIGURE 11.1(e) Leaves.

5. **Flowers:** These are separate male and female and are very small. Female flowers have short perianth whereas male ones are crowded near mouth of receptacle.

11.2 TRADITIONAL USES

Ficus benghalensis is well known in Indian traditional system for its renowned protective properties. Various plant parts of *Ficus benghalensis* possess various pharmacological activities and are used traditionally in various diseases. A brief summary of the use of its various parts is listed in Table 11.2 (Khaliq et al., 2017).

TABLE 11.2 Traditional Uses of *Ficus benghalensis*

Part Used	Uses	References
Paste of leaves	For ulcers and suppurative wounds	Tripathi et al. (2015)
Aerial roots	Wound healing, antiemetic, topical application on pimples	Tripathi et al. (2015)
Seeds and fruits	Respiratory and skin diseases, rheumatic inflammation	Subodh Kumar et al. (2020)
Bark	Dysentery, diarrhea, and diabetes	Tripathi et al. (2015)
Young stem	As toothbrush (miswak)	Subodh Kumar et al. (2020)

TABLE 11.2 *(Continued)*

Part Used	Uses	References
Latex from aerial parts	Hyperglycemia	Yasmin Khan et al. (2011)
Leaves	Used in decoction as diaphoretic	Mujumdar (2010)
Infusion of small branches	Hemoptysis	Yasmin Khan et al. (2011)
Infusion of bark	Diabetes	Yasmin Khan et al. (2011)
Decoction of root-fibers	Gonorrhea	Mujumdar (2010)
Decoction of leaf buds	Hemorrhages	Mujumdar (2010)
Milky juice	Rheumatism and lumbago	Rajdev et al. (2018)

It is reported to exhibit various pharmacologic activities such as anti-oxidant, anti-inflammatory, antidiabetic, antidiarrheal, etc. An overview of the same is presented here:

1. **Anti-Oxidant Activity:** The anti-oxidant activity of extract of *Ficus benghalensis* was examined using DPPH assay, total phenolic content, hydroxyl radical scavenging activity and reducing capacity. The extract at dose of 250 μg/mL showed 96.7% scavenging of DPPH radical and at 1,000 ppm concentration caused scavenging of hydrogen peroxide by 69.23%. It exhibited greater antioxidant activity when compared to the other compounds (Chaitali et al., 2020).
2. **Anti-Inflammatory Activity:** The ethanol extract of *Ficus benghalensis* at a dose of about 300 mg/kg showed significant anti-inflammatory activity in rats where edema was induced using carrageenan. It exhibited greater anti-inflammatory activity when compared to the standard drug Indomethacin. Thus, ethanol extract is known to have great anti-inflammatory potential (Vv et al., n.d.).
3. **Anti-Diarrheal Activity:** The ethanol extract of roots of *Ficus benghalensis* at a dose of 400 mg/kg showed significant anti-diarrheal activity in rats where diarrhea was induced using castor oil (Kulshreshtha et al., 2011).
4. **Anti-Microbial Activity:** The chloroform extract obtained from fruits of *Ficus benghalensis* showed significant anti-microbial

activity against the strains of *Staphylococcus aureus Micrococcus luleus, Streptococcus faecium, Escherichia coli, Streptococcus faecalis, Pseudomonas aeruginosa* and *Proteus vulgaris* (Vuorelaa et al., 1994).

5. **Anti-Helmenthic Activity:** The aqueous and methanol extract of *Ficus benghalensis* were found to paralyze and kill the earthworms when compared to other anti-helminthic drugs (Chaitali et al., 2020).
6. **Anti-Stress and Anti-Allergic Activity:** The aqueous and ethanol extracts of *Ficus benghalensis* showed significant reduction in eosinophils and leukocytes by milk induced eosinophilia and milk induced leukocytosis in asthma. This shows the anti-stress and anti-allergic potential of bark of *Ficus benghalensis* (Taylor et al., n.d.).
7. **Anti-Atherogenic Activity:** Leucopelargonin derivative, a bioactive compound obtained from the bark of *Ficus benghalensis* at dose of 100 mg/kg increased the atherogenic index, hepatic bile acid and the fecal excretion of bile acids whereas significantly decreased the HMGCoA reductase and lipogenic activity of enzymes in liver, decreased the lipoprotein lipase activity in heart in cholesterol diet-fed rats (Sanjeev et al., 2019).
8. **Cytotoxic Activity:** The fruits of *Ficus benghalensis* are reported to have potent cytotoxic and antitumor activity demonstrated by potato disc bioassay method where tumor was inhibited by 48% and brine shrimp bioassay method with LC50 value of 900 (Vuorelaa et al., 1994).

It is also known for its economic importance and papers and coarse ropes are made from the fibers obtained from the aerial roots and bark of *Ficus benghalensis*. The wood of the plant is used in making furniture and buildings. The prop roots and twigs are used as tooth sticks.

11.3 DIABETES MELLITUS

Diabetes is the third leading cause of death worldwide after cardiovascular disease and cancer. It is a metabolic disease that is characterized by hyperglycemia. Diabetes Mellitus results from either impaired insulin secretion or insulin sensitivity or both. Diabetes mellitus is of two types:

- Type 1 diabetes; and
- Type 2 diabetes.

Type 1 is mainly caused due to degeneration of β-cells of the pancreas producing insulin which causes deficiency of insulin secretion. People with type 2 Diabetes Mellitus have insulin resistance in liver and peripheral tissues (Mahmoodpoor et al., 2018; Inzucchi et al., 2012). There is also a type known as gestational diabetes in which there is an increase in blood glucose levels in early pregnancy.

Diabetic complications are of acute and chronic type and are mainly associated with type 1 and type 2 diabetes mellitus. Acute diabetic complications include hyperglycemia that is high blood glucose, hypoglycemia that is low blood glucose and ketoacidosis. Chronic hyperglycemia leads to chronic complications like nephropathy, neuropathy, retinopathy, peripheral artery disease, coronary artery disease and stroke.

Various factors such as obesity, aging, unhealthy lifestyle, physical inactivity leads to an increase in cases of diabetic patients. Prevalence of this metabolic disorder is rapidly increasing in the world and so is a major concern to health officials. The number of patients of diabetes mellitus and its complications are increasing at a much faster rate and are causing a major economic concern (Dotto and Chacha, 2020).

Treatment of diabetes mellitus is currently considered as a global health concern. Various synthetic drugs like meglitinides, thiazolidinediones, sulfonylureas, incretin mimetics, α-glucosidase inhibitors biguanides, sodium/glucose co-transporter2 (SGLT2) inhibitors, are the current treatment modalities for the management of diabetes mellitus, although effective, their use is associated with many adverse effects (Gacche and Dhole, 2011).

There is a need for novel antidiabetic drugs that are safe and effective with fewer or no side effects. In recent years, more focus has been given to drugs derived from plant materials due to availability, fewer side effects and lower cost. Various medicinal plants have been used traditionally to prevent long-term complications and manage diabetes mellitus. Various bioactive compounds with strong antidiabetic properties have been isolated from medicinal plants. These bioactive compounds belong to various chemical classes such as phenolic acids, flavonoids, triterpenoids, alkaloids, and carbohydrates (Misbah et al., 2013). Traditionally, more than 100 plants are used in the Indian medicinal system for management of diabetes. *Ficus benghalensis* Linn, belonging to the family Moraceae has been extensively investigated for the management of diabetes (Singh et al., 2009).

11.4 PHYTOCHEMISTRY

Phytochemical investigations of *Ficus benghalensis* shows a variety of phytoconstituents ranging from alkaloids, glycoside, carbohydrates, terpenoids, saponins, phenols, sterols, flavonoids, tannins in various parts such as leaf, bark, and root (Arimah et al., 2013). These phytoconstituents are responsible for its diverse pharmacological activity. The phytoconstituents reported in the aqueous extracts of the bark, leaves, and roots are indicated in Table 11.3.

TABLE 11.3 Phytoconstituents Identified in Aqueous Extract of *Ficus benghalensis*

Constituents	Bark	Leaf	Root
Alkaloids	–	+	+
Glycosides	+	+	+
Carbohydrates	+	+	–
Terpenoids	+	+	+
Phenols	+	+	+
Saponins	+	+	+
Xanthoproteic	+	+	–
Tannins	+	+	+
Flavonoids	+	+	+

11.4.1 PHYTOCONSTITUENTS POSSESSING ANTIDIABETIC ACTIVITY

The roots, bark, and leaves of *Ficus benghalensis* contain secondary metabolites responsible for its antidiabetic potential. These phytoconstituents are reviewed in Table 11.4.

The secondary metabolites isolated from leaves of *Ficus benghalensis* are discussed here:

1. **Friedelin:** It is obtained from petroleum ether extract of the leaves of *Ficus benghalensis.* Isolation of Friedelin is mainly done using column chromatography over alumina with benzene and petroleum ether as eluants.

TABLE 11.4 Antidiabetic Phytoconstituents in *Ficus benghalensis*

Part	Class	Phytoconstituents	References
Stem bark	Anthocyanidin	Leucopelargonidin-3-O-α-L-rhamnoside	Kundap et al. (2017)
	Leucoanthocyanidins	Leucocynidin 3-O-β-D-galactosyl cellobioside	Mukherjee et al. (2006)
	Miscellaneous	Meso inositol, 6-heptatriacontene-10-one, β-sitosterol-α-D-glucose, 20-tetratriacontene-2-one	Kundap et al. (2017)
Leaf	Sterols	β-sitosterol	Kundap et al. (2017)
	Flavonols	Quercetin-3-galactoside, rutin	Kundap et al. (2017)
	Flavonoids	Catechin, genistein	Deepika (2016)
	Triterpene	Friedelin	Kundap et al. (2017)
	Phytosterols	Lupeol, α-amyrin acetate, lanostadienylglucosyl cetoleate, bengalensisteroic acid acetate	Naquvi et al. (2015)
Aerial root	Triterpenoids	Bengalensinone, Benganoic acid	DellaGreca et al. (2004)

Structure:

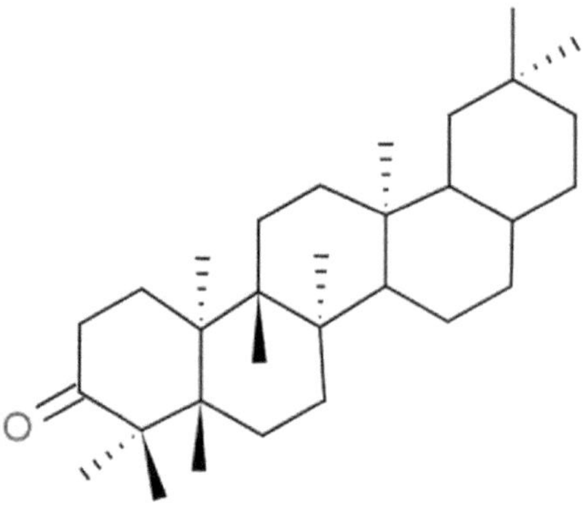

M. P: 235–238C

When treated with Liebermann-Burchardt reagent, it gives pink color.

2. **β-Sitosterol:** It is isolated using column chromatography with benzene as the eluant.

Structure:

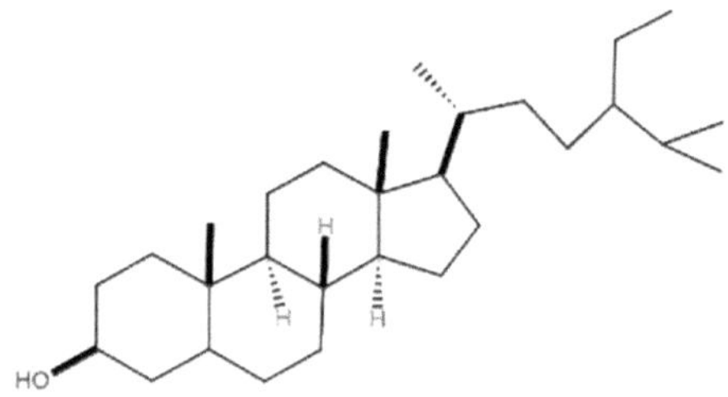

M. P: 135–137C

When treated with Liebermann-Burchardt reagent, it tests positive for sterols.

3. **Quercetin-3-Galactoside:** Obtained from the flavonoid rich fraction. The ethanol extract of the leaves of *Ficus benghalensis* when extracted using ethyl acetate yielded 2 compounds. The major compound obtained was Quercetin-3-galactoside.

Structure:

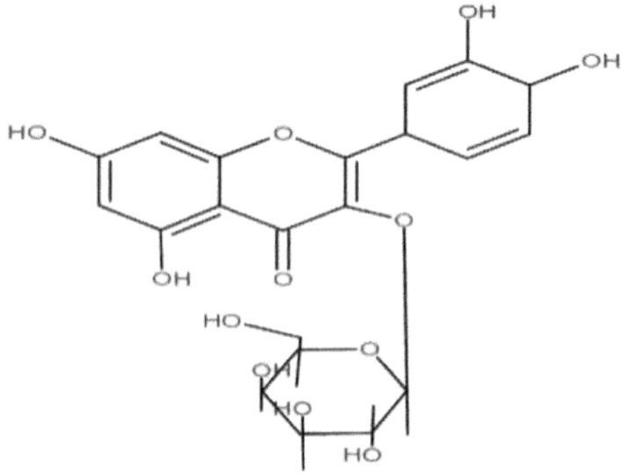

4. **Rutin:** The minor compound obtained from the ethyl acetate fraction of the ethanol extract of the leaves of *Ficus benghalensis* is Rutin.

Structure:

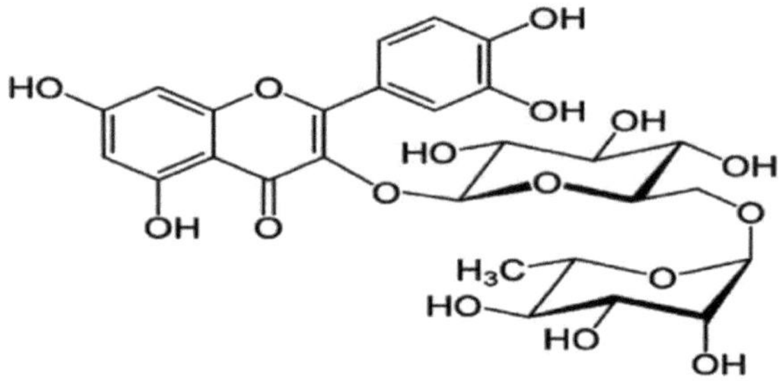

MP: 125°C

5. **Lanostadienylglucosyl Cetoleate**: or lanost-5,24-dien-3b-yl-b-D-glucopyranosid-20-O-yl-docos-1100-enoate is obtained from the methanol extract of the leaves of *Ficus benghalensis* using column chromatography with chloroform as the eluent (Naquvi et al., 2015).

Structure:

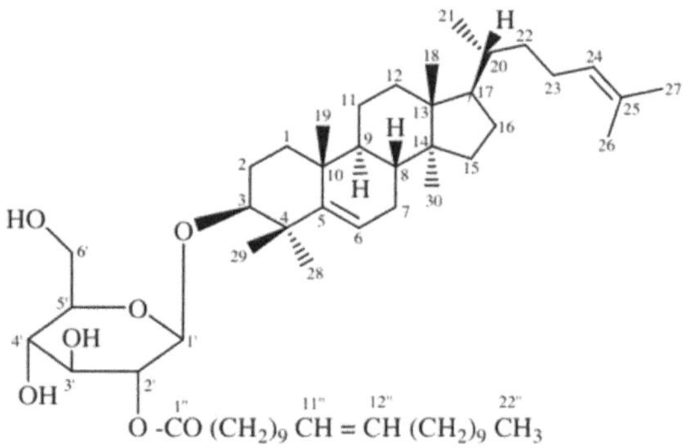

MP: 38–39°C

6. **Bengalensisteroic Acid Acetate:** or 3b-acetoxy-stigmast-22-en-26-oic acid is obtained from methanolic extract of leaves of *Ficus*

benghalensis using column chromatography with chloroform and methanol in the ratio of 97:3 as eluent (Naquvi et al., 2015).

Structure:

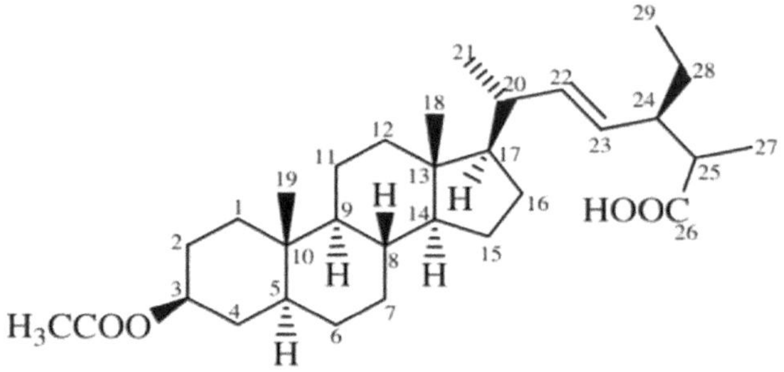

MP: 50–52°C

11.4.1.1 COMPOUNDS ISOLATED FROM STEM BARK OF FICUS BENGHALENSIS

Several anthocyanidin derivatives have been isolated from the ethanol extract of stem bark of *Ficus benghalensis*. They are obtained by column chromatography using silica gel column with mixtures of methanol/chloroform followed by methanol as eluents. The compounds are as follows:

1. 3,'5-Dimethyl ether of leucocynidin 3-O-β-D-galactosyl cellobioside:

Structure:

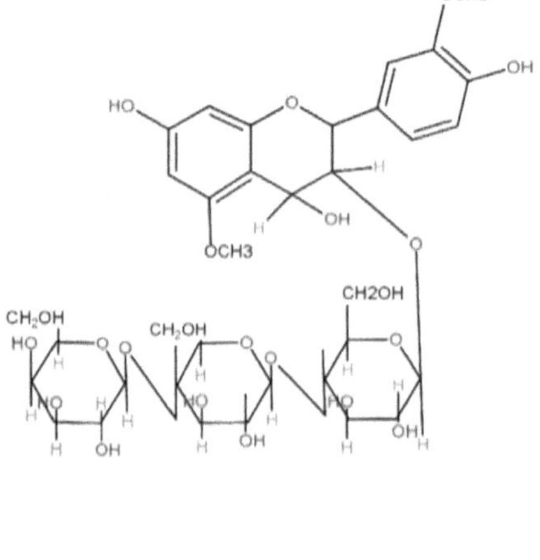

MP: 340°C

2. 5,7-Dimethyl ether of leucopelargonidin-3-O-α-L-rhamnoside:
Structure:

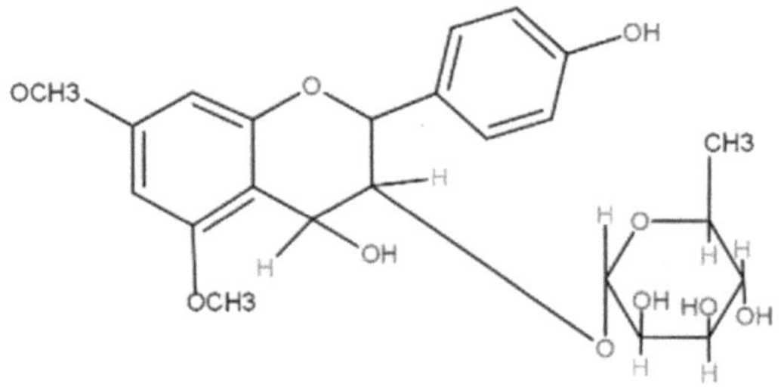

MP: 190°C

3. 3,5,7-Trimethyl ether of leucocynidin:
Structure:

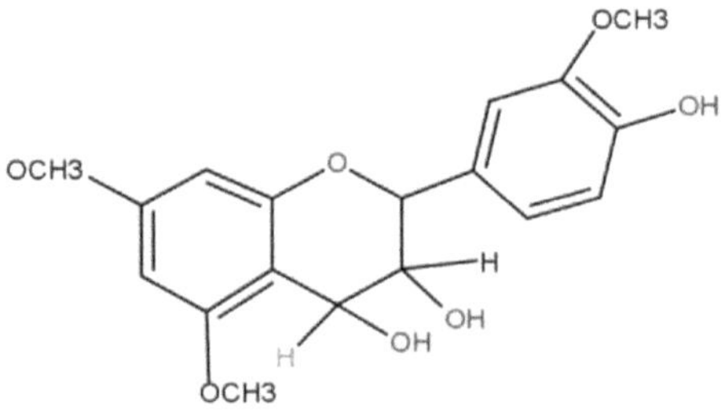

MP: 165°C

4. 3,5,7-Trimethyl ether of delphinidin-3-O-α-L-rhamnoside:
Structure:

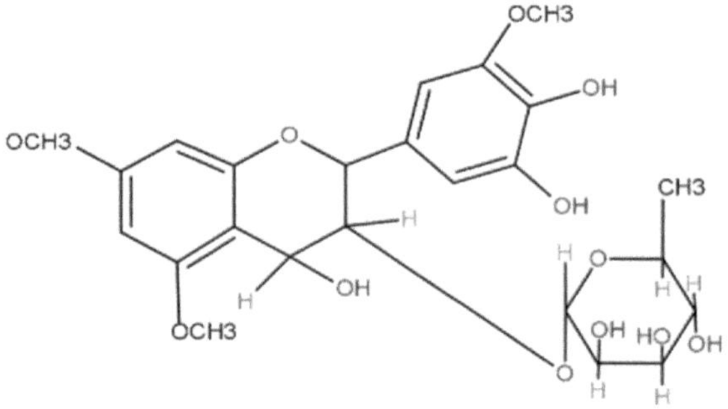

MP: 170–171°C

The other compounds belonging to miscellaneous class are as follows:

1. **Meso Inositol:**

Structure:

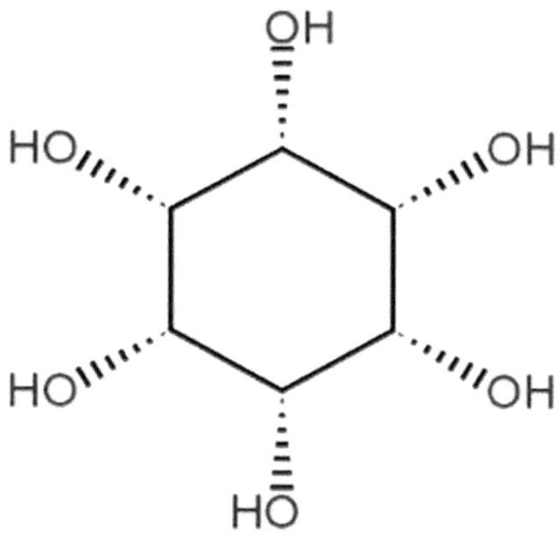

2. 6-Heptatriacontene-10-One:

Structure:

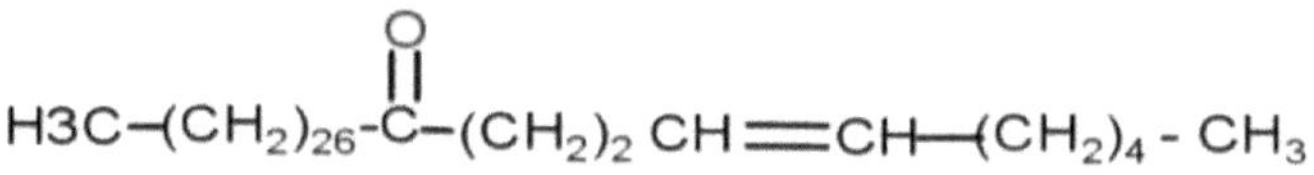

3. B-Sitosterol-α-D-Glucose:

Structure:

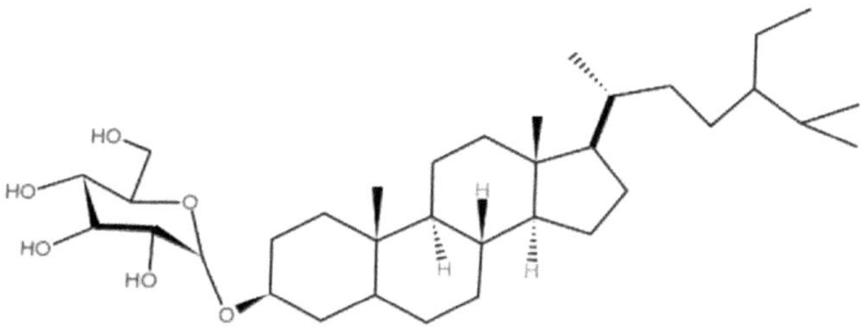

4. 20-Tetratriacontene-2-One:

Structure:

$$H3C-(CH_2)_{12}-CH{=}CH-(CH_2)_{17}-COCH_3$$

5. Pentatriacontan-5-One:

Structure:

$$H3C-(CH_2)_{29}-\overset{O}{\overset{||}{C}}-(CH_2)_3-CH_3$$

11.5 MECHANISM OF ANTIDIABETIC ACTIVITY

The various mechanism through which plant extracts promote anti-diabetic activity includes inhibition of absorption of glucose in the intestine, enhancement of uptake of glucose in skeletal muscle and regulation of glucose homeostasis:

1. **Inhibition of Glucose Uptake:** Overconsumption of dietary carbohydrates like sucrose and starch can cause postprandial glycemia or any other metabolic disturbances. The key enzymes responsible for the conversion of these dietary carbohydrates into glucose are α-amylase and α-glucosidase. Inhibition of glucose absorption can be achieved by inhibiting these carbohydrate hydrolyzing enzymes α-glucosidase and α-amylase. Phytoconstituents such as quercetin, rutin found in the leaf of *F. benghalensis* are known to affect the metabolism, digestion, and the absorption of carbohydrates by inhibition of these enzymes α-glucosidase and α-amylase.

   A study using *in-vitro* models was done to investigate the inhibitory activity of these compounds on carbohydrate hydrolyzing enzymes α-amylase and α-glucosidase.

   The inhibitory effect of *Ficus benghalensis* on α-amylase was studied using in-vitro enzyme starch system. 1% *F. benghalensis* powder was mixed with 25 ml of 4% potato starch and 100 mg of α-amylase was further added to the solution and incubated for 60 minutes at 37°C. 0.1 M NaOH was added after the incubation period to terminate the enzyme activity. The mixture was further centrifuged to determine the glucose content in the supernatant which indicated inhibition of α-amylase by 50% (Ahmed et al., 2011).

2. **Glucose Uptake in Skeletal Muscles:** Insulin resistance in liver, skeletal muscles and adipose tissue plays a very important role in type 2 Diabetes Mellitus. Insulin resistance is mainly caused due to defective glucose transport, fatty acid metabolism, defects in the insulin receptor signal transduction, glycogen synthesis and glucose oxidation. GLUT-4 or Glucose transporter 4 in the skeletal muscle plays a pivotal role in the uptake of glucose by insulin. GLUT-4 is expressed by kidney cells, adipose tissue, and muscle. Enhancement of uptake of glucose in skeletal muscle is achieved by insulin binding to GLUT-4 which then activates PI3K (phosphatidylinositol-3 kinase) and higher the expression of GLUT-4 in skeletal muscle, higher the uptake. Additionally, Peroxisome proliferator activated receptors (PPARs) expressed in liver and skeletal muscles also help in the insulin sensitivity. There are 3 subtypes of PPARs, i.e., PPARα, PPARβ and PPARϒ of which PPARϒ is known to ameliorate insulin resistance. Phytoconstituents such as kaempferol, quercetin, penta-galloyl-glucose enhance the glucose uptake by upregulating expressions of PPARϒ, PI3K and GLUT-4.
3. **Regulation of Glucose Homeostasis:** Regulation of blood glucose levels through balance between gluconeogenesis (glucose synthesis for amino acid metabolites) and glycogenesis (breakdown of glycogen) in the liver plays a major role in management of diabetes mellitus. Increase in Insulin secretion in Pancreas and decrease in hepatic glucose production is mainly caused due to downregulation of glucose 6-phosphate and phosphoenolpyruvate carboxykinase, which are the key enzymes involved in regulation of gluconeogenesis. Thus, keeping a balance between gluconeogenesis and glycogenesis is very important.
4. **Antioxidant Activity:** Diabetic mellitus is also associated with oxidative stress. It is related with changes in levels of lipid peroxidation and antioxidant enzymes. Increase in the levels of

antioxidant enzymes such as glutathione peroxidase, superoxide dismutase and catalase with decrease in levels of lipid peroxidation can lead to decrease in the risk of Diabetes Mellitus. Flavonoids present in the aerial roots of *Ficus benghalensis* are known to cause an increase in the anti-oxidant enzymes and studies have reported its antidiabetic potential through anti-oxidant mechanism (Sharma et al., 2009).

Experimental models with Alloxan or Streptozotocin induced diabetes were used to study the antidiabetic potential of various parts of *Ficus benghalensis*. Rakesh Kumar Singh reported the antidiabetic potential of aerial roots of *Ficus benghalensis*. The aqueous extract of aerial roots of *Ficus benghalensis* at doses of 100, 200, 300 and 400 mg/kg significantly reduced the blood glucose levels after 6 hours of oral administration in streptozotocin-induced rat model. Glipizide was used as a reference to compare the results (Singh et al., 2009). Another study by Dr. R. Shukla et al. reported a significant fall in the blood glucose levels with an increase in glucose tolerance from a partially purified solution of aqueous extract of bark of *Ficus benghalensis* in alloxan-induced diabetic rabbits. The study was conducted for 30 days and the dose of the extract was 150–250 mg/kg (Shukla et al., 1995). Various other studies were performed to check the antidiabetic potential of the plant. α-amyrin acetate isolated from the aerial roots of *Ficus benghalensis* not only lowered the blood glucose levels but also improved glucose tolerance in streptozotocin induced diabetic rats. Various doses of α-amyrin acetate at 15,25, 50 and 100 mg/kg were used for the study. The study was conducted for 10 days and Metformin (50 mg/kg) was used as reference standard (Singh et al., 2009). The ethanol extract of leaves of *Ficus benghalensis* at doses of 200 and 400 mg/kg reduced the blood glucose, triglycerides, and cholesterol levels in Alloxan induced diabetic rats. The study was conducted for 14 days and the reference drug used was Glibenclamide at 5 mg/kg dose. The aqueous extract of the bark of *Ficus benghalensis* at dose of 100,300 and 500 mg/kg decreased the blood glucose levels in streptozotocin induced diabetic rats. 100 mg/kg of tolbutamide was used as the reference drug and the study was conducted for 12 weeks. The extract also decreased serum cholesterol and low-density lipoprotein cholesterol at dose of 50 mg/kg. Blood glucose levels triacylglycerol and glycosylated hemoglobin were also decreased in rabbits where diabetes was induced using Alloxan and the study was conducted for 6 weeks. Table 11.5 presents the antidiabetic potential of various extracts of the plant where diabetes is induced using alloxan and streptozotocin.

TABLE 11.5 Antidiabetic Activity of *Ficus benghalensis*

Part/Extract/ Compound	Dose	Exp. Model	Activity	References
Root	100 mg/kg, 200 mg/kg, 300 mg/kg, and 400 mg kg	Streptozotocin-induced rats	Reduce blood glucose level	Velasco (2002)
Aqueous extract of Bark	150–250 mg/kg	Alloxan-induced rabbits	Decrease blood glucose level and increase glucose tolerance	Singh et al. (2009)
Root/α-amyrin acetate	15, 25, 50, and 100 mg/kg	Streptozotocin-induced rats	Higher HDL-C level and decrease in blood glucose level	Singh et al. (2009)
Ethanolic extract of leaves	200 and 400 mg/kg	Alloxan-induced rats	Decrease triglycerides and blood glucose levels, decrease cholesterol levels.	–
Aerial root/α-amyrin acetate	50 mg/kg	Streptozotocin-induced rats	Decrease blood glucose level	Gayathri et al. (2008)
Bark/aqueous	100, 300, and 500 mg/kg	Streptozotocin-induced rats	Decrease blood glucose level.	Gayathri et al. (2008)
Bark/aqueous	50 mg/kg body weight	Alloxan-induced rabbits	Reduce serum cholesterol, triacylglycerol, LDL cholesterol, blood glucose and glycosylated hemoglobin levels.	Shukla et al. (1995)

TABLE 11.5 *(Continued)*

Part/Extract/ Compound	Dose	Exp. Model	Activity	References
Bark/flavonoids	–	Glucose-induced hypoglycemic rabbits	Control the hypoglycemia and hyperglycemia.	Bengalensls (1963)
Fruit/ethanol	250 mg/kg	Alloxan-induced rats	Decrease blood glucose levels, decrease total cholesterol levels.	Subramanian et al. (2013)
Bark/ leucodelphinidin	250 mg/kg	Alloxan-induced rats	Reduce blood glucose level.	Geetha, Mathew et al. (1994)
Bark and leaves/ aqueous	150 and 300 mg/kg	Alloxan-induced rats	Reduce blood glucose level.	Chikaraddy et al. (2017)
Bark/ethanol	–	Glucose-induced hypoglycemic rabbits.	Control the hypoglycemia	Pari and Latha (2002)
Bark/ hydroalcoholic extract	120 mg/kg	Alloxan-induced rats	Reduce blood glucose and serum cholesterol, triglycerides, LDL, creatinine, urea, alkaline phosphatase, bilirubin, glutamate oxalate transaminase and glutamate pyruvate transaminase levels. Increase HDL and albumin levels.	Saifi et al. (2017)
Bark/ethanol	120 mg/kg	Alloxan-induced rats	Decrease blood glucose level	Sharma et al. (2006)
Bark/ leucopelargonidin-3-0-alpha-L rhamnoside	100 mg/kg	Alloxan-induced dogs	Stimulate secretion of insulin.	Sharma, et al. (2006)

11.6 CONCLUSION

The demand of natural compounds, especially from plant sources to manage diabetes and its associated complications has increased significantly in the past few years. The use of herbal medicines in the treatment and prevention of diabetes mellitus is still continued in modern society. Mitigation of diabetic complications by *Ficus benghalensis* in alloxan and streptozotocin-induced models provides a scientific basis for justification of its traditional use and its utility in complementary and alternative systems of medicine. The observed benefits provide sufficient evidence to warrant further research into understanding the molecular basis of antidiabetic activity of the extracts and isolated secondary metabolites that could be translated to clinical trials and subsequently to patients. The structures of these active compounds can be used as template scaffolds to design new and better antidiabetic drugs using computational drug discovery tools of structure based and ligand-based strategies along with 3D-QSAR. *Ficus benghalensis* can emerge to be a lucrative plant with several medicinal properties along with potent antidiabetic activity if substantiated by intensive research and preclinical testing in suitable animal models along with molecular-level mechanistic studies.

KEYWORDS

- ***Ficus benghalensis***
- **peroxisome proliferator-activated receptors**
- **phosphatidylinositol-3 kinase**
- **phytochemistry**
- **sodium/glucose co-transporter 2**

REFERENCES

Ahmed, F., Chavan, S., Satish, A., & R, P. K., (2011). Inhibitory activities of *Ficus benghalensis* bark against carbohydrate hydrolyzing enzymes: An *in vitro* study. *Pharmacognosy Journal*, *3*(20), 33–37.

Arimah, B. D., Adebayo, M. A., & Ogunlowo, O. P. (2013). Phytochemical analysis and comparison of in-vitro antimicrobial activities of the leaf, stem bark and root bark of Ficus benghalensis. *IOSR Journal of Pharmacy (IOSRPHR)*, *3*(4), 33–38.

Brahmachari, H. D., & Augusti, K. T. (1964). Isolation of orally effective hypoglycemic compounds from Ficus bengalensis Linn. *Indian J Physiol Pharmacol.* (Jan 8), 60–64. PMID: 14248533.

Chikaraddy, A., & Maniyar, Y., (2017). *Evaluation and Comparison of Hypoglycemic Activity of Bark and Leaf of FICUS Bengalensis Linn. in Alloxan Induced Diabetes in Albino Rats, 4*, 138–142.

Deepika, K., (2016). *Studies on Isolation, Purification, and Structure Elucidation of Chemical Constituents from Methanolic Flower Extract of Blepharis Molluginifolia Pers. and Their Biological Activities*, *7*(12), 4893–4904.

DellaGreca, M., Di Marino, C., Zarrelli, A., & D'Abrosca, B., (2004). Isolation and phytotoxicity of apocarotenoids from Chenopodium album. *Journal of Natural Products*, *67*(9), 1492–1495.

Ding, J., Strachan, M. W. J., Reynolds, R. M., Frier, B. M., Deary, I. J., Fowkes, F. G. R., & Price, J. F., (2010). Diabetic retinopathy and cognitive decline in older people with type 2 diabetes: The Edinburgh type 2 diabetes study. *Diabetes*, *59*(11), 2883–2889.

Dotto, J. M., & Chacha, J. S., (2020). The potential of pumpkin seeds as a functional food ingredient: A review: Biofunctional ingredients of pumpkin seeds. *Scientific African*, *10*, e00575.

Forbes, J. M., & Cooper, M. E., (2013). Mechanisms of diabetic complications. *Physiological Reviews*, *93*(1), 137–188.

Gacche, R. N., & Dhole, N. A., (2011). Profile of aldose reductase inhibition, anti-cataract, and free radical scavenging activity of selected medicinal plants: An attempt to standardize the botanicals for amelioration of diabetes complications. *Food and Chemical Toxicology*, *49*(8), 1806–1813.

Gayathri, M., & Kannabiran, K., (2008). Antidiabetic and ameliorative potential of *Ficus bengalensis* bark extract in streptozotocin-induced diabetic rats. *Indian Journal of Clinical Biochemistry*, *23*(4), 394–400.

Geetha, B. S., Mathew, B. C., & Augusti, K. T., (1994). Hypoglycemic effects of leucodelphinidin derivative isolated from *Ficus bengalensis* (Linn.). *Indian Journal of Physiology and Pharmacology*, *38*(3), 220–222.

Inzucchi, S. E., Bergenstal, R. M., Buse, J. B., Diamant, M., Ferrannini, E., Nauck, M., & Matthews, D. R., (2012). Management of hyperglycemia in type 2 diabetes: A patient-centered approach. *Diabetes Care*, *35*(6), 1364–1379.

Khaliq, H. A., (2017). A review of pharmacognostic, physicochemical, phytochemical, and pharmacological studies on *Ficus bengalensis* L. *Journal of Scientific and Innovative Research*, *6*(4), 151–163.

Kilmer, P. D., (2010). Review article. *Journalism: Theory, Practice & Criticism*, *11*(3), 369–373.

Kulshreshtha, M., Goswami, M., Rao, C. V., Ashwlayan, V. D., Yadav, S., & Kulshreshtha, M., (2011). *Anti-Ulcerogenic Potential of Ficus Bengalensis Leaf, Biochemical Parameter & Histopathological Study*, *1*(2), 65–68.

Kumar, S., & Singh, B., (2020). *An Ancient Treasure of Medicinal Properties in Indian Medicine Bargad (Ficus benghalensis Linn.)*, *4*(3), 31–35.

Kundap, U., Jaiswal, Y., Sarawade, R., Williams, L., & Shaikh, M. F., (2017). Effect of Pelargonidin isolated from *Ficus benghalensis* L. on phenotypic changes in zebrafish (Danio rerio) embryos. *Saudi Pharmaceutical Journal, 25*(2), 249–257.
Liu, J., Liu, T., Wang, W., Ma, L., Ma, X., Shi, S., & Wang, M., (2017). Reduced gray matter volume in patients with type 2 diabetes mellitus. *Frontiers in Aging Neuroscience, 9*, 1–10.
Mahmoodpoor, A., Medghalchi, M., Nazemiyeh, H., Asgharian, P., Shadvar, K., & Hamishehkar, H., (2018). Effect of *Cucurbita maxima* on control of blood glucose in diabetic critically Ill patients. *Advanced Pharmaceutical Bulletin, 8*(2), 347–351.
Misbah, H., Aziz, A. A., & Aminudin, N., (2013). Antidiabetic and antioxidant properties of *Ficus deltoidea* fruit extracts and fractions. *BMC Complementary and Alternative Medicine, 13*.
Mujumdar, N. A., (2010). Preface. *Indian Journal of Agricultural Economics, 65*(3).
Mukherjee, P. K., Maiti, K., Mukherjee, K., & Houghton, P. J., (2006). Leads from Indian medicinal plants with hypoglycemic potentials. *Journal of Ethnopharmacology, 106*(1), 1–28.
Naquvi, K. J., Ali, M., & Ahamad, J., (2015). Two new phytosterols from the stem bark of *Ficus bengalensis* L. *Journal of Saudi Chemical Society, 19*(6), 650–654.
Pari, L., & Latha, M., (2002). Antidiabetic activity of *Cassia auriculata* flowers: Effect on lipid peroxidation in streptozotocin diabetes rats. *Pharmaceutical Biology, 40*(7), 512–517.
Priya, S., Thamizhiniyan, V., & Subramanian, S., (2013). Antidiabetic potential of *Ficus bengalensis* fruit extract studied in alloxan-induced experimental diabetes in rats. *Research Journal of Pharmacology and Pharmacodynamics, 5*(2), 110–118.
Rajdev, K., Jain, S., Mahendra., C. H, & Bhattacharaya, S. K. (2018). Antinociceptive effect of ficus bengalensis bark extract in experimental models of pain. *Cureus, 10*(3).
Review, P., & Ficus, O. N., (2020). *World Journal of Pharmaceutical Research, 9*(8), 539–546.
Saifi, A., Chauhan, R., & Dwivedi, J., (2017). Development of a polyherbal formulation FMST and evaluation for antidiabetic activity in alloxan-induced diabetic rats. *Asian Journal of Pharmaceutical Research, 7*(1), 1.
Sharma, R. K., Chatterji, S., Rai, D. K., Mehta, S., & Rai, P. K. (2009). Antioxidant activities and phenolic contents of the aqueous extracts of some Indian medicinal plants. *Journal of Medicinal Plants Research, 3*(11), 944–948.
Sharma, S., Chaturvedi, M., Edwin, E., Shukla, S., & Sagrawat, H. (2006). Evaluation of the phytochemicals and antidiabetic activity of *Ficus bengalensis. International Journal of Diabetes in Developing Countries, 27*(2), 56–59.
Shukla, R., Anand, K., Prabhu, K. M., & Murthy, P. S., (1995). Hypolipidemic effect of water extract of *Ficus bengalensis* in alloxan-induced diabetes mellitus in rabbits. *Indian Journal of Clinical Biochemistry, 10*(2).
Singh, A. B., Yadav, D. K., Maurya, R., & Srivastava, A. K., (2009). Anti-hyperglycaemic activity of α-amyrin acetate in rats and db/db mice. *Natural Product Research, 23*(9), 876–882.
Singh, R. K., Mehta, S., Jaiswal, D., Rai, P. K., & Watal, G., (2009). Antidiabetic effect of *Ficus bengalensis* aerial roots in experimental animals. *Journal of Ethnopharmacology, 123*(1), 110–114.

Taylor, P., Taur, D. J., Nirmal, S. A., & Patil, R. Y., (n.d.). *Natural Product Research: Formerly Natural Product Letters Antistress and Antiallergic Effects of Ficus Bengalensis Bark in Asthma*, 37–41.

Vuorelaa, P., Kivirantab, J., Wahab, S. A., Hiltunena, R., & Vuorelaa, H., (1994). *Bioactivity of Certain Egyptian Ficus Species*, *41*, 71–76.

Vv, P., Rb, P., & Vr, P., (n.d.). *Pharmacognostical Studies and Evaluation of Anti-in ß Ammatory Activity of Ficus Bengalensis Linn.*, 49–53.

Wagner, W. L., Herbst, D. R., Khan, N., & Flynn, T., (2012). *Hawaiian Vascular Plant Updates: A Supplement to the Manual of the Flowering Plants of Hawai'i and Hawai'i's Ferns and Fern Allies*. Flora of the Hawaiian Islands Website. Http://Botany.Si.Edu/Pacificislandbiodiversity/Hawaiianflora/Index.Htm (accessed on 16 December 2021).

Yasmin, K. K., Ajab, K. M., Ahmad, M., Mazari, P., Hussain, I., Ali, B., & Zaima, K. I., (2011). Ethno-medicinal species of genus *Ficus* L. used to treat diabetes in Pakistan. *Journal of Applied Pharmaceutical Science, 1*(6), 209–211.

Index

D

E

O

Q

R

Z